DARWIN AND THE DARWINIAN
REVOLUTION

GERTRUDE HIMMELFARB is a graduate of Brooklyn College and received her Ph.D. from the University of Chicago. She is author of *Lord Acton: A Study in Conscience and Politics*, and editor of Lord Acton's *Essays on Freedom and Power*, *Malthus on Population*, and *Essays on Politics and Culture* by John Stuart Mill. Her most recent book is *Victorian Minds*.

IN THE NORTON LIBRARY

DARWIN AND THE DARWINIAN REVOLUTION

Gertrude Himmelfarb

The Norton Library

W · W · NORTON & COMPANY · INC ·

NEW YORK

CONTENTS

INTRODUCTION

SETTING an example for his future biographers, Darwin once canvassed his scientist friends for information about their lives that would account for their scientific dispositions. Darwin himself has served us well in this matter. As much, perhaps, out of economy as out of consideration for biographers, he has left us a hoard of documents about his life and work. There is the autobiographical memoir begun in 1876 and completed in 1881; the original version, differing in many and important respects from that in his *Life and Letters,* has only recently been published. There are the varying notebooks, diaries, and journals of the voyage of the Beagle. There are notebooks with jottings of his ideas and records of his observations and experiments; notes and transcriptions of his readings; preliminary sketches and original drafts of his works (when the backs of these pages were not thriftily used for other purposes); annotated copies of succeeding editions of these works; and thousands of letters—in original manuscript, typescript copies, and the published edited versions—written by and to him. These illuminate many critical points regarding the origin and development of his theory, his methods of inquiry and character of mind, his intentions and achievements, and the way he stimulated and was stimulated by his contemporaries. The most demanding historian could not have wished for more.

Yet the picture is all the more, not the less, confused by this wealth of information. What was, to begin with, straightforward is revealed to be devious; what was plain and obvious becomes obscure and elusive; while many of the conventional ideas about science in general and Darwinism in particular have to be discarded without any tidy generalizations to replace them. Not one meaning but a diversity of meanings emerges. No one saw this better than Darwin's own contemporaries, and it is in the complexities of the contemporary situation rather than in the specious simplicities that have come down to us that we must seek the truth.

The *Origin* had been out only a few days when the conflicting reports began to appear. Critics quarreled about the accuracy of his facts, the adequacy of his theories, and the logic of his reasoning. Some denied the truth of his hypothesis while asserting its utility for scientific investigation; others denied its utility while conceding its truth; still others denied or asserted both. Religious critics were divided over its religious significance, scientists over its scientific import, and each party against the other over its ultimate philosophical meaning. Even economists, sociologists, and politicians had their say, falling out with each other and thoroughly confusing the issues.

The anomalies of intellectual creation jostle with the anomalies of interpretation. Why was it given to Darwin, less ambitious, less imaginative, and less learned than many of his colleagues, to discover the theory sought after by others so assiduously? How did it come about that one so limited intellectually and insensitive culturally should have devised a theory so massive in structure and sweeping in significance? What were the logic and history of his discovery? Was the new theory inspired by new facts? How did Darwin rise above the antecedents and influences that had shaped him? At what point in this dialectic of discovery did quantity change into quality, the pupil transcend his masters, the past give way to the future? Was Darwin a great revolutionary, and, if so, what was the nature of his revolution?

And what was the later history of the discovery? What

happened when the old heresy became the new orthodoxy? Was Darwinism a legitimate heir of Darwin? By what metamorphosis did a scientific treatise, largely devoted to such abstruse matters as the anatomical variations among different breeds of pigeons, become a metaphysics, politics, and economics? How did it come about that a study of the origin of species could inspire a member of the Austrian Parliament to open a debate on the reconsolidation of the Empire with the words: "The question we have first to consider is whether Charles Darwin is right or no"?

Only the closest reconstruction of the facts can answer these questions. It is necessary to trace the origin and development of Darwin's own views and to correlate them at each stage with the prevailing opinions; to determine the extent to which Darwin was influenced by his predecessors and contemporaries, and the extent to which he and his contemporaries thought he was influenced by them; to show what Darwin intended his theory to mean, what his readers took it to mean, and what in fact it meant and implied. By such slow stages and small observations it may be possible to recapture a sense of how a scientist, with the most innocent of intentions and the best of faith, can give birth to a theory that has an ancestry and a posterity of which he may be ignorant and a life of its own over which he has no control.

This study has occupied me for four years. It has been assisted by grants from the American Philosophical Society and the John Simon Guggenheim Memorial Foundation, for which I am very grateful. I wish also to thank the Cambridge University Library, particularly Mr. H. R. Creswick and Mr. A. T. Tillotson, for permitting me to consult its extensive collection of Darwin manuscripts; the library of the American Philosophical Society for the Darwin-Lyell and Darwin-Romanes correspondence; the Imperial College of Science and Technology, London, for the Huxley papers; Mr. Jocelyn Probie for family letters, including a large number of those of John Tyndall; and the many libraries in England and America where I gathered my materials, particularly the London Library, whose generous facilities and patient service make it the most distinguished of private libraries.

I have also to thank those who were kind enough to read part or the whole of this book in manuscript, and whose comments and criticisms I have taken much to heart: Professor Michael Polanyi of the University of Manchester, whose most recent and brilliant work, *Personal Knowledge*, appeared just too late to be reflected in this book; Dr. Marjorie Grene who has been examining the present status of evolutionary thought in the light of the philosophy of science; Dr. Shirley Letwin of Brandeis University; Mrs. Jane Degras of the Royal Institute of International Affairs; and my husband, Irving Kristol, who has stoically read several versions of this book and listened to still others.

1959

Note to the 1962 edition: For this edition I have made minor changes, particularly in the last chapter.

PRE-HISTORY OF
THE HERO

BOOK I

CHAPTER 1

CHILDHOOD AND YOUTH

THE pre-history of a hero is fraught with danger. On the one hand, the biographer must be careful not to be seduced by his materials into telling all that he happens to know about the youth of his hero. For all too often what he happens to know is not relevant to his purpose, which is to find out how a non-hero was transformed into a particular hero. Nor, on the other hand, must he permit himself to be seduced by history, to let his knowledge of later events influence his judgment of earlier ones, and to try to find intimations of heroism where in fact there were none. In Darwin's childhood and youth, he must frankly confess, there were no intimations of greatness. But there was evidence of the simple, unassuming character which was as much a part of the future hero as his later greatness—which was, in fact, a key to that greatness.

The ancestry of Charles Darwin, as it was known to him, dates back to 1500, when the Darwins (the spelling of the name varied) were yeomen of Lincolnshire. The first evidence of "scientific bias"[1] that he was able to find was in his great-grandfather, Robert Darwin, described by a then celebrated antiquarian as a "Person of Curiosity" and credited with the discovery of "a human Sceleton impressed in Stone . . . the like whereof has not been observed before in this island."[2] The eldest son of this Darwin, another

Robert, shared with his more famous younger brother, Erasmus, a taste for both poetry and botany and was the author of a *Principia Botanica*, "A Concise and Easy Introduction to the Sexual Botany of Linnaeus." According to family tradition, Robert had the book published because he was so proud of the calligraphy in the manuscript that he could not bear to see it wasted. His contemporaries apparently judged it more generously, for it went through three editions, and even Charles praised it for bringing together many interesting facts about biology at a time when the subject was generally neglected.

The most curious and famous member of the family in this generation was Erasmus. In accord with the biographer's private law of heredity, by which a man is thought to be more intimately related to his grandfather than to his father, Charles Darwin has always been taken as the spiritual heir of Erasmus Darwin. Yet, both in personality and in intellectual temper, the two were as far apart as two human beings could be, and this in spite of their affinity of interests. Erasmus died seven years before Charles was born, and Charles seems not to have given much thought to him as a person nor, indeed, to have thought much of him as a scientist. After the *Origin of Species* was published and critics began to comment on the relationship, Charles found himself in the awkward position of having to dissociate himself from Erasmus' scientific theories without seeming to be disrespectful to his own ancestor. As a result, his public testimonials to Erasmus are probably warmer and more generous than his private sentiments would have warranted.

When Charles was a model Victorian—trim, dignified, almost ascetic in feature—Erasmus was more typical of the eighteenth century, having the fat, bulbous, sensual, and dyspeptic appearance so often seen in the portraits of Samuel Johnson and his friends. Erasmus' portrait, ill favored as it is, does not exhaust the physical unsightliness of the man, for what did not appear in it were the deeply pitted pockmarks on his face, a corpulence of body (he was so fat that the dining table had to be cut out to accommodate his paunch) and clumsiness of gait made the worse by lameness later in life, and a pronounced stammer. A biography of Erasmus written by a woman embittered, family tradition had it, by unrequited passion (or at least by frustrated marital designs) provides other unprepossessing physical details, such as the tongue habitually hanging out between his lips; this gratuitous bit of repulsiveness is hotly denied by others, who speak instead of the false teeth which gave a peculiarly unpleasant contour to his jaws.[3]

What seems even more unnatural than such unrelieved ugliness is the fact that it did not, apparently, discourage this biographer, Anna Seward, or other more successful women from courting or allowing themselves to be courted by him. Both of his wives were attractive and intelligent, and, besides the five children of his first marriage (two of whom died in infancy) and the seven of his second, he publicly acknowledged two illegitimate daughters. Discreetly relegating this last bit of information to a footnote in his biographical sketch of his grandfather, Charles observed that, strange as it might seem to his readers, Erasmus suffered no social or professional ostracism as a result of these indiscretions. Nor was this the only occasion on which Charles, with a jocularity he did not really feel, had to invoke the "good old days"[4] to explain the manners and morals of his grandfather. Erasmus was notorious as a voracious eater in a century famous for its gourmands; when Miss Seward called him a glutton, Charles weakly protested that it was mainly vegetables with which he gorged himself. And he indulged his tastes unashamedly. As a physician he was doomed to spend long hours on the road, and he had his carriage fitted with a food hamper on one side and writing facilities on the other, in case either hunger or inspiration should seize him. He did not, however, drink much as an adult, because, it has been suggested, he discovered in his youth that he could not cultivate both women and wine, and he preferred women. He later developed the theory that if wine had to be drunk, it should be homegrown wine; but it is not known whether he thought English wines to be healthier than foreign ones or whether it was simply that they made abstention more attractive.[5]

It is strange to find that the fatigue which was so prominent a feature of Charles Darwin's illness should also have been a complaint of the energetic Erasmus. Whatever Erasmus meant by fatigue, however, it could not have been what Charles experienced. For Erasmus managed to build up an extremely successful practice in the Midlands, first in Lichfield and then in Derby, which brought him what was in those days the large sum of over a thousand pounds a year. Besides the strenuous professional duties suggested by such an income, he organized the local philosophical society and engaged in controversy on such subjects as the immorality of slavery and the virtues of deism. He also elaborated a scheme for female boarding schools and devised a great number of mechanical inventions ranging from the usual flying machine to a horizontal windmill which was actually put to use. And he cultivated the friendship of his prosperous and enlightened neighbors—the manufacturer Matthew Boulton and his partner James Watt, the potter Josiah Wedgwood, the physician and philosopher William Small, and the chemists James Keir and Joseph Priestley; he also corresponded with several other famous but less accessible contemporaries, including Rousseau and Benjamin Franklin. He had met Samuel Johnson, Lichfield's more famous son; but, whether because of temperamental similarity or ideological differences, they disliked each other on sight, and Johnson mortified the local patriots, including Erasmus Darwin, by complaining of the intellectual barrenness of Lichfield.

In addition, he managed to write poetry and scientific treatises by the ream. *Botanic Garden,* including the famous "Loves of the Plants," was published in two parts in 1788 and 1790, and went through several editions. On the strength of the sales in 1788, he received a pre-publication payment of one thousand guineas for the second part, which even today for poetry—and such poetry—would be extraordinary. *Zoonomia,* his work on medicine, was translated soon after its publication in 1794 into German,

French, and Italian, and had the distinction of being placed on the Catholic Index.

In *Botanic Garden* Erasmus gave it as his purpose "to inlist Imagination under the banner of Science; and to lead her votaries from the looser analogies, which dress out the imagery of poetry, to the stricter ones, which form the ratiocination of philosophy."[6] Under the all-embracing aegis of science and philosophy, he was able to discourse on such contemporary issues as slavery and prison reform, as well as botany, zoology, astronomy, and chemistry. Today this volume and his posthumous book of verse, *The Temple of Nature,* are read, if at all, for their scientific theory, as a portent and possibly a source of his grandson's work. At the time, they were read principally as poetic entertainment. The *Edinburgh Review* expressed the consensus when it stated, just after his death: "If his fame be destined in anything to outlive the fluctuating fashion of the day, it is on his merit as a poet that it is likely to rest; and his reveries in science have probably no other chance of being saved from oblivion, but by having been 'married to immortal verse.' "[7] Horace Walpole said of some lines in *Botanic Garden* that they were "the most sublime passages in any author, or in any of the few languages with which I am acquainted." But even his merit as a poet was not conceded by all his contemporaries. And the verdict of posterity has been with the young men in the decade after his death who, like Byron, thought him "a mighty master of unmeaning rhyme."[8]

Whether Erasmus' rhymes were as unmeaning as Byron thought, and what their real meaning for Charles Darwin and for the history of science might be, belong to the discussion rather of Charles' scientific predecessors than of his genealogical ancestors. It is only necessary here to point to the strange coincidence that one of these predecessors should have been an ancestor whom Charles never saw, whose personality was alien to him, and whose mind worked in ways quite unknown to him, but who is important for having bequeathed to his family a tradition of science. Erasmus' eldest son, Charles, inherited, along with his father's stammer, his talent for medicine and his predilection for writing verse. When he died, at the age of nineteen, of blood poisoning contracted while performing an operation,[9] he had already shown great medical promise. His second son and namesake was less promising in his youth and a tragic failure later in life. A desultory collector of coins and statistics, and an unsuccessful lawyer who would have liked, it is thought, to enter the Church but feared to provoke his father's sarcasm, he suffered from the melancholia that afflicted so many of the Darwins; his father complained that he wanted to "sleep away the remainder of his life,"[10] and just one month later his body was found at the bottom of the pool in his garden. The only child of this marriage to survive into maturity was the youngest, Robert (Charles' father), who was as energetic as his other brother was listless, and who carried on the family profession of medicine. Of the children of Erasmus' second marriage, one son was a physician who expressed his taste for natural history by letting half-wild pigs run about in his woods and tame snakes in his house; and one of his daughters became the mother of the famous geneticist, Francis Galton, a young contemporary and admirer of Charles Darwin.

One of the counts in Miss Seward's indictment of Erasmus Darwin was an excessive devotion to money, a charge that Charles dismissed as sheer slander. Yet if he was not miserly, he was conspicuously ungenerous in his treatment of his youngest son. For when Robert was not yet twenty-one, the affluent Erasmus brought him to Shrewsbury, gave him twenty pounds, and parted from him with the words, "Let me know when you want more and I will send it you."[11] Twenty pounds was not much with which to start upon a career, but this, plus a similar gift from an uncle, was all Robert was given during his father's lifetime. After Erasmus' death he received a bequest of twenty-five thousand pounds, but by that time he had already made his way. He had inherited from his father both the will and the ability to make money, and a remarkable faculty for winning the confidence of patients. During his first year of medical practice his earnings were already enough, as he later boasted, to permit him to keep two horses and a servant, and they quickly rose as he became Shrewsbury's leading physician. Before he retired his income was reputed to be the largest of any provincial doctor.

If Robert was the equal of Erasmus as a businessman, he was decidedly his inferior as a scientist. He was a successful doctor for the reason that most general practitioners are: because he had a great respect for common sense, a shrewd insight into people's minds, and a quick sympathy. Charles remembered him as a father confessor to scores of women, and Robert sometimes spoke as if his medical specialty was miserable wives. Apart from a low cunning in administering psychological nostrums, he had an intuitive skill in diagnosing and treating the more common medical cases. Yet he was notably lacking in scientific talent and imagination. A rival physician used to say of him that although he had an uncanny knack of predicting the course of an illness, his mind was wholly unscientific—a judgment in which he would have fully concurred. He once confessed to Charles that he had hated his profession so much at first that, if he had been assured of the smallest pittance or if his father had given him any choice, nothing would have induced him to follow it. And after sixty years of medical practice he was still sickened by the thought of an operation and by the sight of blood—a prejudice that led him to denounce bleeding as a therapeutic technique.

Lacking the temperament and talent for science, he also mocked those who cultivated it. His one contribution to the scientific journals was a paper in the *Philosophical Transactions* on ocular spectra, on the basis of which he was elected a Fellow of the Royal Society when he was only twenty-two; but Charles was given to understand that he had been largely assisted in writing it by Erasmus.[12] "My father's mind," Charles concluded, "was not scientific, and he did not try to generalise his knowledge under general laws; yet he formed a theory for almost everything which occurred. I do not think I gained much from him intellectually."[13]

Having dismissed his father as a scientific model, Charles tried to reinstate him as a moral one: "His example," he said, "ought to have been of much moral service to all his children."[14] This may sound like the conventionally pious tribute paid by a dutiful son to the public memory of his father, but it was apparently intended in earnest. Charles was, indeed, far kinder to his father—and not only in public —than a later, less reverent generation might expect. Robert used to complain that his own father was harsh; if Charles did not make the same complaint of Robert, it was not because Robert was less guilty but because Charles was more forgiving.

Emma Wedgwood, Charles' cousin and later his wife, frequently visited the Darwin home as a child and has left a picture of the tyrannical and unsympathetic Doctor: "Everything in the household had to run in the master's grooves, so that the inmates had not the sense of being free to do just what they liked." Conversation, when he was present, could be neither general nor tête-à-tête but had to be directed to the censorious—and, to make it worse, hard-of-hearing—Doctor, so that no one ever felt at ease; she herself, she confessed privately, was always glad when he went off on a journey and sorry when he returned. With boys particularly he had little patience: "He was a fidgety man and the noise and untidiness of a boy were unpleasant to him." As for Charles, she was convinced that "Doctor Darwin did not like him or understand or sympathize with him as a boy."[15] Charles himself must have had some sense of this, for he once said: "I think my father was a little unjust to me when I was young, but afterwards, I am thankful to think, I became a prime favourite with him."[16]

In a notoriously patriarchal society Dr. Darwin was even more of a patriarch than most fathers. His authority, however, was so naturally exercised and so naturally accepted that it took an outsider to realize just how imperious it was. (And Emma herself, it may be remarked, had no slack ideas of parental authority, her own father being as awesome a character as the Doctor.) When Catherine, Charles' younger sister, then twenty years old, wrote to Emma's sister one cold December day, "I am sure you will feel the full delight for me of what Papa has very good-naturedly given me leave to have; a fire for the morning," she intended no sarcasm and, although she went on to explain that the scheme was less extravagant than it sounded since she was giving up the fire in the dining room in compensation, she felt bound "to make all apologies for such a piece of indulgence."[17] All the Darwin children were to make such apologies as a matter of course. The note of apologetic gratitude punctuated all of Charles' correspondence with his family, from his schooldays when he and his brother had to justify paying one pound and six shillings for lodgings—for which they had the use of two bedrooms and a sitting room—until as a married man and father he had to explain the expenditure of seventy-two pounds for a fence around his house. Nor, as will appear, was it only in money matters that Robert was more than commonly exigent—and Charles more than commonly docile. From the same pinnacle of wisdom from which the Doctor superintended the children's conversation and allowances, he supervised their entire lives, and Charles, for one, never doubted that what he thought or did was "absolutely true, right, and wise." Charles' own children were to recognize in the familiar words, "My father thought or did so-and-so," the accents of finality, the appeal to a superior authority.[18]

The figure of his father, the largest man he had ever seen, Charles recalled—which may be taken as a literal as well as psychoanalytic truth, since Robert was six feet two inches tall and weighed twenty-four stone (336 pounds) at the last recording, after which he continued to gain but gave up weighing himself—loomed large until his death in 1848, when Robert was eighty-two and Charles thirty-nine. That his father had a great deal to do with Charles' modesty, his timidity, and also his invalidism (although there were physical precedents enough for that in the family) is probable. That he served as a moral example to Charles, however, is less obvious. For the characters of father and son were quite different—Robert being assertive, domineering, and irascible in temper, in contrast to Charles' extreme diffidence. Robert also had a cunning that was lacking in the guileless Charles. This expressed itself in such matters as the elaborate stratagems by which the

Doctor sought to cure his patients' emotional and marital ailments. Typical of these was the ruse of telling contradictory stories to husband and wife so that each would find it advantageous to effect a reconciliation—an ingenuity and resourcefulness admired by Charles but which he never sought to imitate.

A more serious divergence of character was revealed in the Doctor's deliberate concealment of his religious beliefs even from those most intimate with him. Robert Darwin was less orthodox in his faith than Erasmus—and also less frank in revealing it, probably because unorthodoxy had become a more serious social offense as a result of the Revolutionary Wars. Having joined the Freemasons as a young man, he later became so secretive about his disbelief that he had his children brought up in thoroughly orthodox fashion, even going so far as to plan a clerical career for Charles. In his characteristically frank and uncensorious fashion Charles spoke of his father's irreligion and of his advice, on the eve of Charles' marriage, to conceal from his future wife whatever religious doubts he might have. It had been the Doctor's experience that a wife, confronted by her husband's sickness and fearing for his life and eternal salvation, would suffer and make him suffer for his disbelief. Seldom, he had discovered, did a husband convert his wife to skepticism; he himself had known only three women who were genuinely enlightened, one of whom was his sister-in-law, Kitty Wedgwood, for whose atheism he seemed to have no evidence save the conviction that "so clear-sighted a woman could not be a believer."[19]

The difference between Robert and his son can be seen in Charles' failure to take his father's advice. He lacked that "sense of expediency," as his son and biographer Francis Darwin delicately put it,[20] which Robert Darwin had in such good measure; and about both his own beliefs and his father's he spoke with great candor. If the fact of his father's irreligion has not been generally known, it was not because Charles sought to conceal it but because, in this matter as in others, his family, after his death, exercised a discretion that Charles himself never thought necessary, eliminating from the published version of his autobiography those passages that dealt with Robert's as well as Charles' disbelief.

Although Robert's mode of expressing, or rather suppressing, his disbelief did not commend itself to his son, the knowledge of that disbelief may have been of some influence. Not only did it make disbelief, when it came, appear to be a natural, acceptable mode of thought, so that loss of faith never presented itself to him as a moral crisis or rebellion; more than that, it seemed to enjoin disbelief precisely as a filial duty. One of the passages which was deleted from the autobiography explained why Charles not only could not believe in Christianity but would not wish to believe in it. Citing the "damnable doctrine" that would condemn all disbelievers to everlasting punishment, he protested that "this would include my Father, Brother, and almost all my best friends"[21]—which made it an unthinkable, to say nothing of thoroughly immoral, idea. There may be more sophisticated reasons for disbelief, but there could hardly have been a more persuasive emotional one.

Charles was predominantly a Darwin rather than a Wedgwood. If he had reason to complain of an intellectual distance from Robert or Erasmus, his mother's family must have seemed even more alien to him. Yet in the history of England the Wedgwoods were the more prominent. If there was such a thing as an industrial revolution, the Wedgwoods helped to make it, and if there was a capitalist spirit, the Wedgwoods epitomized it.

Again, it was the grandfather who was the moving spirit in the family. While Erasmus Darwin was establishing himself in Lichfield as one of the local eccentrics and substantial citizens, thirty miles away in Burslem, Josiah Wedgwood, almost the same age, was mastering the craft and the business of pottery. It was as a modest but rising manufacturer that, sometime around 1759, he first consulted Dr. Darwin professionally and soon became one of his closest friends.

The Wedgwoods had been potters in Staffordshire since

early in the 1600's, and for two centuries half the names in
the parish register of Burslem were Wedgwoods. Burslem
was a primitive and squalid town of mud hovels, each
with its refuse heap before the door, mounds of ashes and
broken pots, and hollows filled with stagnant water from
which the potters had dug their clay. Samuel Smiles, the
Victorian Horatio Alger who specialized in the rags-to-
riches true-life stories of English industrialists (one of
whom was Josiah Wedgwood), described the "saturnalia
of drunkenness" that accompanied every wake in the town,
the "brutal and vicious" morals and manners, the "rude,
barbarous and uncivilised"[22] way of life which prevailed
until Josiah Wedgwood, in the 1760's and '70's, transformed the pottery industry and, with it, the town.

Even before the advent of Josiah, however, his immediate family had succeeded in raising its head above the
morass that enveloped Burslem—Josiah's grandfather by
combining farming with pottery, and Josiah's father
(known locally as "Dr." Thomas Wedgwood) by keeping
an inn. When Josiah, the youngest of thirteen children,
was nine, his father died, leaving the family business to his
eldest son. Josiah immediately left school to work in the
pottery, which was one room in the family home. Two
years later he came down with a particularly virulent case
of smallpox which left him permanently enfeebled and with
a stiffness and pain in one knee. Nevertheless, he finished
his apprenticeship and went on to work as journeyman,
leaving home when he came of age only because his
brother, finding his ideas too unorthodox, refused to admit
him as a partner.

Josiah Wedgwood was the ideal subject for Samuel
Smiles. He had everything the earnest moralist could ask
for: humble origins and physical deficiencies (his leg was
finally amputated, he had spells of blindness, liver attacks,
and a variety of other ailments), transcended by nothing
more than an act of will. His were the simple Puritan virtues of industry and ambition. His industry, to be sure, was
assisted by talent, but it was the familiar kind of talent
available to the slightly superior person, not the exotic
talent of the genius, and so was quite at home in the

democratic universe of Smiles, where although not every man may be elected, every man is potentially of the elect.

His ambition, too, was of the agreeable sort where self-interest was identical with social good. Into an industry having secretiveness as almost the only trade skill, he introduced the principle that both technical and aesthetic innovations should be as widely disseminated as possible, in the belief that what benefited the industry as a whole would necessarily benefit the individual manufacturer. And this was, in fact, what happened. No monopoly of the old industry, no corner on the butter-pot market could have been so profitable, even to Wedgwood alone, as his creation of a new industry. It was this new industry of Etruscan vases, elaborate dinner services, medallions, statuettes, buttons and knickknacks, together with the much improved staples of domestic and industrial use, that made Wedgwood's fortune—as well as that of his competitors. Because the national and international markets which he hoped to exploit were hampered by inadequate transportation, he helped plan and develop a new network of roads and canals, thereby bringing education and sanitation as well as pots and plates to all parts of the country. And because inefficient and unreliable labor was one of his great difficulties, he assumed the paternalistic functions of building houses and schools for his workers, encouraging saving and trying to discourage drinking. In 1760, when John Wesley had come to preach in Burslem, he had been appalled by the ignorance and rudeness of his audience, one of whom pelted him with a clod of mud. Twenty years later he revisited the place and was enraptured by its progress: "The wilderness is literally become a fruitful field. Houses, villages, towns, have sprung up, and the country is not more improved than the people."[23] Wedgwood's personal fortune had improved to the same degree. Besides his exceedingly prosperous business (which was later rather negligently looked after by his sons), he was able to leave twenty-five thousand pounds to each of his three daughters and more to each of his three sons.

Wedgwood may have been a congenial subject for Smiles, but he was also—and this may redeem him for the cynics—a good friend of Erasmus Darwin. Although Erasmus was apt to mock Wedgwood's Unitarianism, deriding it as a "feather-bed to catch a falling Christian,"[24] he was attracted by his intellectual curiosity and even more by his vigorous radicalism. Both men shared a hatred of slavery and a fondness for revolution (American and French), the Wedgwood family going so far in its revolutionary sympathies as to be suspected of Bonapartism. Josiah Wedgwood had married a distant cousin, another Wedgwood, and their six children grew up in great intimacy with the Darwin children. When Josiah died in 1795, Erasmus attended him at his deathbed. Before his own death seven years later, Erasmus had the satisfaction of witnessing the marriage between his son Robert and Susannah Wedgwood, Josiah's oldest daughter, as well as attending at the birth of their first two children.

Of Susannah Darwin little is known save the chronology of childbirth. Married at the age of thirty-one, she had six children in a period of fourteen years. Charles, the next to the youngest, was born on February 12, 1809. When his mother died at the age of fifty-two, he was eight. He later found that he could remember little about her except "her death-bed, her black velvet gown, and her curiously constructed work-table."[25] Later he was disturbed by the thought that her death was less memorable than the burial of a dragoon soldier he happened to witness at about the same time, complete with such stirring details as the horse bearing the boots and carbine of its master and the magnificent sound of firing over the grave. He could only suppose that he had so easily forgotten his mother because she had been ill for some time before her death and because afterward his sisters, out of grief, could never bring themselves to speak of her.

Over a half-century later a former schoolmate recalled that Charles had once brought a flower to school and had boasted that his mother had taught him how to determine its name by looking at the inside of the blossom. The schoolmate, later a well-known botanist, innocently commented that his mother had probably been introducing him to the Linnaean system. Charles himself, strenuously searching his memory for early recollections, particularly as they might illuminate his later career, did not remember this episode. But a more pertinacious psychologist, writing of it a century after the event, has made of it the key to the *Origin of Species* as well as all the rest of Darwin's work. On the assumption that "to know the secret of the name is to know the secret of the child's or flower's origin," he drew the appropriate psychoanalytic conclusion: "Whether or not Darwin's mother actually propounded her enchanting riddle to her boy is not quite so important as the fact that he said she did, showing how keenly his wishes relished the fancy that she had revealed to him the one secret of life that fascinated her—the secret which, if read, would reveal the origin and creation of life and—himself." It was to this secret that the psychologist saw Darwin dedicating his life, peering into the inner recesses of flowers with all the passion of a Peeping Tom, and from essentially the same motives. And the titles of Darwin's works, with their barely concealed sexual overtones—"The Effects of Cross and Self-Fertilization," "On the Various Contrivances by Which Orchids Are Fertilized," "On the Origin of Species by Means of Natural Selection"—were taken to be the answers he gave to "his mother's sacred riddle."[26] It is unfortunate that the ingenuity of this theory should be diminished by the knowledge that it would apply equally well to every botanist and naturalist from Linnaeus down who used this standard method for the classification of flowers.

It is ordinarily difficult enough to establish a relationship between a man and his ideas. To try to establish one between Darwin's mother and his ideas is hopeless. Susannah Darwin was reputedly intelligent, handsome, and good-natured. These conventional tags tell us little enough about her and less about her influence; but they are all we have.

As Freud felt obliged to psychoanalyze himself, so Darwin conscientiously sought in himself the hereditary influences he found so readily in plants and animals.[27] But he succeeded only in ferreting out memories rather than influences. And even the memories were sparse and undramatic, those of the "very ordinary boy"[28] that his father and elders thought him and that he suspected himself to have been. There is no revelation to be found in such commonplace recollections as being startled by a cow running past his window, being given a fig by a shopkeeper for the privilege of kissing the maidservant, receiving plums from the local hermit, or being shut up in a room as punishment and trying to break the window (or perhaps this happened to his younger sister, Catherine; he was not sure).

He was one of six children, and his sisters and brothers played a large part in his memories. Marianne, the eldest, rarely figures in them, partly because she shared the management of the household with Caroline, who was a more domineering character, and partly because she left to get married when Charles was fifteen. It was Caroline who was the real mother-substitute, a figure larger than life and more awesome than the original. Even before the mother's death, while she was an invalid, the education of the youngest children had been entrusted to Caroline, an arrangement that did not do justice to Charles who suffered by comparison with Catherine, a year his junior and, like most girls, an apter, more docile pupil. To Caroline, as to Dr. Darwin, boys were naturally uncongenial. Charles later recalled his sister's well-intentioned but ineffectual and officious attentions: "Caroline was extremely kind, clever and zealous; but she was too zealous in trying to improve me; for I clearly remember after this long interval of years, saying to myself when about to enter a room where she was—'what will she blame me for now'; and I made myself dogged so as not to care what she might say."[29] Nevertheless, Charles remained on good terms with her, although his letters more often betrayed the deference commonly shown to an elder aunt than the normal familiarity of brother and sister. It may be significant that Caroline's marriage to her cousin, Josiah Wedgwood, should have been followed, a year and a half later, by Charles' marriage to Josiah's youngest sister.

Susan, three years younger than Caroline, was the

beauty of the family and the Doctor's favorite. High-spirited and frankly flirtatious (at least by the austere Darwin standards), she was, surprisingly, the only one of the sisters to remain unmarried. With the passing of years, her spirits were subdued, and she became, like the others, prim and censorious; in her case it was Charles' rollicking children who were to prove uncongenial.

Catherine was the youngest and the most querulous of them all, her strong character being able to express itself neither in Susan's vitality nor in Caroline's natural authority. Her father used to praise her "great soul," but when she was finally married, at the age of fifty-three, to an old intimate of the family (whose wife, a Wedgwood, had died the preceding year), the family could not conceal its faintly mocking disapproval and its predictions of disaster. And when she died only three years later, one of her outspoken aunts deplored the fact that she had "failed to work out her capabilities either for her own happiness or that of others . . ."[30]

In this chronicle of domination and frustration, the most interesting figure was Erasmus, Charles' only brother and five years his senior. Unlike his sisters, Erasmus never inflicted his unhappiness on others, preferring to cultivate his melancholia and ailments (both considerable) in private. In public he was the sympathetic brother, the kindly and often gay uncle, the genial man-about-town who had neither wife nor profession to interfere with the leisurely regimen of reading, talking, and visiting. It was perhaps his somewhat sardonic sense of humor that endeared him to Thomas Carlyle, who said of him: "He had something of original and sarcastically ingenious in him, one of the sincerest, naturally truest, and most modest of men," who would have been the intellectual superior of Charles, had not ill-health "doomed him to silence and patient idleness."[31] Charles, who may also have been offended by Carlyle's reference to "this honest Darwin" in distinguishing the elder brother from the younger, also resented the implication that Erasmus was little more than a graceful idler. The description, he protested, had "little truth and no merit."[32] Yet he himself could hardly speak of him without adding, "Poor old Ras."[33]

If this catalogue of the Darwin household makes life at the Mount (the Darwin home) seem oppressively sober and austere, it was not so in the memory of Charles, partly because of the natural resilience of children for whom life is rarely as bad as adults imagine it to be, and partly because life was not, indeed, as bad as these portraits, taken in maturity, would make it seem. Besides, there was the gayer atmosphere of Maer, the home of his Wedgwood cousins, a long day's ride of twenty miles—too long for short visits, but just right for the holidays. Josiah Wedgwood himself, son of the founder of the firm, was by no means gay; silent, reserved, and uncomfortably upright, he is best described in one of Sydney Smith's witticisms: "Wedgwood's an excellent man—it is a pity he hates his friends."[34] Yet Charles, who agreed that he was rather "awful" (in the sense of awesome),[35] surprised everyone by getting along well with him. Nor were the Wedgwood children as much in awe of him as outsiders might have thought. His wife was, to be sure, but then she thought all men were dangerous creatures who had to be humored; Emma Wedgwood, the youngest of the children, resented her mother's habit of acting as if he was to be feared. In spite of him, life at Maer with the eight Wedgwood cousins was more cheerful and also more cultivated than at the Mount. Conversation was freer and wittier, books were more plentiful and more read; there were riding, visiting and party-going, boating in the summer and skating in the winter, and for the boys, particularly Charles, the ineffable glories of shooting and hunting.

Between the Mount and Maer, school impressions were apt to be few. Until he was eight, Charles was taught at home, so that his memories were of Caroline (his teacher) and of Catherine (his rival schoolmate). When he was sent to school for the first time, shortly before his mother's death, it was to a day school at Shrewsbury kept by the Unitarian minister, a Mr. Case. Even then, what he remembered best were such extracurricular activities as going on walks through the fields, accompanying his father on carriage rides, and telling a variety of childish fibs, one of which was an elaborate fiction designed to show how devoted to the truth he was. These fibs he still recalled with shame as an adult. Once, after having looted the orchard, he announced the discovery of a hoard of stolen fruit; fifty years later this was solemnly reported, with all the cadences of a Biblical text: "I once gathered much valuable fruit from my father's trees."[36] Perhaps more than the violation of the truth, it was the violation of property, and parental property at that—the psychological overtones are obvious—that so impressed itself upon his memory and conscience.

Other episodes revealed, he thought, an early interest in natural history. He had boasted to a schoolfellow that he could vary the color of flowers by watering them with colored fluids; but it is doubtful whether this incident merits the portentous comment that even at that age he was interested in the "variability of plants."[37] He also found a taste for natural history demonstrated in his tales of unlikely birds he claimed to have seen and even in his zeal for collecting those traditional schoolboy treasures—shells, seals, stamps, coins, minerals, and pebbles. "The passion for collecting which leads a man to be a systematic naturalist, a virtuoso or a miser, was very strong in me, and was clearly innate, as none of my sisters or brother ever had this taste."[38] In fact, as he elsewhere recalled, his brother collected plants, his father was an enthusiastic gardener who also delighted his son by taking him on his rounds and identifying the wild game, and it was a friend who introduced him to pebbles and minerals—and none of these ever aspired to the title of "naturalist." In a household and community where gardening, fishing, and hunting were normal activities, it would be hard to find a boy without some evidence of a taste for natural history. Even the pigeons, which were to feature so prominently in the *Origin,* appear here as an accepted feature of the environment, the Mount being famous for them. "I was born a naturalist," Darwin asserted.[39] And so, perhaps, he was. But one could not deduce this from the schoolboy's collections of stamps, seals, or even sea shells. It is salutary to remember that shell collectors rarely grow up to be conchologists.

Whether Darwin's taste for natural history was innate or acquired, it is clear that his taste for the more conventional school subjects was simply non-existent. Yet he must have learned something by this time, for the following year, when he was nine, he was withdrawn from Mr. Case's school and was entered as a boarder in the famous Shrewsbury Grammar School, then presided over by Dr. Butler (grandfather of the Samuel Butler who was later to quarrel with Darwin). The ordinances for the administration of the school specified that no scholar would be admitted "before he can write his name with his own hand, and before he can read English perfectly, and have his accidence without book, and can give any case of any number of a noun substantive or adjective, and any person of any number of a verb active or passive, and can make Latin by any of the concords, the Latin words being first given him"[40]—a high standard even for that day. Yet the Shrewsbury School was probably chosen as much for its convenient location as for its scholastic reputation. Only one mile from his home, it permitted Charles to combine the advantages of the boarder and the day boy, since he could lead the "life of a true schoolboy," as he put it, without having to give up his "home affections and interests."[41] The practice of running back and forth between home and school in his few free hours made him an excellent runner, although at the time he attributed his success to his prayers petitioning God to get him back in time.

Darwin entered the school in 1818, shortly before an epidemic of riots broke out in most of the leading public schools of England. By Christmas, the Master of St. John's College, Cambridge, was congratulating Dr. Butler on his firmness in resisting "the turbulence and self-will" of the boys. "Children nowadays," he deplored, "very early imbibe most pernicious notions, if not from their parents and relations, at least from the spirit of the times"[42]—the "spirit of the times" referring to the post-war depression and discontent. More specifically, the riots at Shrewsbury were caused by bad or inadequate food, and if the Darwin boys—both Charles and Erasmus were at the school—were not involved, it may have been because they were among those who enjoyed the "pernicious indulgence," as Dr. Butler thought it,[43] of hampers of game and poultry from home. Yet they too were smitten by the spirit of rebellion. Erasmus, older and bolder than his brother, complained to his father that Charles' bed was "damp as muck," and since Charles had only recently recovered from scarlet fever, Dr. Darwin passed on this complaint to Dr. Butler, with the request for an additional blanket for the boy. The reply was in the best tradition of headmastership. Dr. Butler first denied the charge, claiming that the boys must have got some new whim into their heads, since to his own knowledge the bed linen had been warming before the kitchen fire for two or even three days. And although he admitted that the nights had been unusually cold (and may have seemed even colder to the boys, who had just returned to school after enjoying "domestic indulgences"), he protested that he could not be put in the position of gratifying a boy's fancy or seeming to favor one over another. However, he added conciliatorily, if Dr. Darwin still thought it advisable that Charles have another blanket, he would undertake to provide one for every boy in the school.[44] The rest of this correspondence between Shrewsbury's two leading citizens is missing, so it is not known whether Charles—and the rest of the school—received their blankets or whether they continued to suffer their chilblains in silence. Yet it was not the damp and cold of a Midlands winter in an ill-heated schoolhouse that Charles and his brother later remembered, but the primitive sanitary and living arrangements. There were no baths, only footpans, in which to wash; and twenty to thirty boys had to share a dormitory boasting a single window at one end. Sixty years later they could recall the noxious smell which assaulted them every morning.[45]

Apart from having their appetites curbed and their sensibilities disciplined, the boys were taught Latin and Greek almost exclusively. "Dr. Butler," one of them later remarked, "certainly did succeed in making us believe that Latin and Greek were the one thing worth living for."[46] This was the weekly course of instructions for the fifth and sixth forms under Dr. Butler:

MONDAY:
1. Chapel. History, Grecian, Roman, English. Repeat Greek Grammar.
2. Dalzel's *Analecta Majora*, sixth and upper fifth only. The parts read in this class are Thucydides, Plato, Greek Orators, Aristotle, Longinus. Lecture on Greek Grammar.
3. Cicero's *Orations*.
4. Virgil. Shell attend. Chapel.

TUESDAY:
1. Chapel. Repeat Virgil. Show up Latin theme.
2. Dalzel's *Analecta Majora*. Parts read are the Greek plays, Pindar, Theocritus, Callimachus. Subject for Latin verses given. Remainder of Latin themes shown up. Half-holiday. Masters of accomplishments attend.

WEDNESDAY:
1. Chapel. Tacitus, Demosthenes, Greek play or Plautus, for sixth and upper fifth, Pitman's *Excerpta*, lower fifth; and repeat Dalzel of Tuesday.
2. Greek play. Examination of a class of the lower boys.
3. Horace, *Odes*.
4. *Scriptores Romani*. Chapel.

THURSDAY:
1. Chapel. Repeat Horace. Show up Latin verses.
2. Homer. Shell attend. Lecture in Algebra to sixth and upper fifth. Remaining verse exercises shown up. Half-holiday as Tuesday.

FRIDAY: 1. Chapel. Repeat Homer. Show up
               lyrics.
            2. Juvenal or Horace, the *Satires* and
               Epistles. Shell attend. Show up the
               remainder of the lyric exercises.
            3. Tacitus, Demosthenes, Greek play or
               Plautus, to sixth and upper fifth only.
               Lower fifth Pitman.
            4. Virgil. Shell attend. Chapel.

SATURDAY: 1. Chapel. Repeat Juvenal or Horace.
               Lecture in Euclid to sixth and upper
               fifth.
            2. Open lesson, generally English trans-
               lated into Greek or Latin prose, or
               lesson in Greek play. Preposters of
               the week show up Greek verses.

SUNDAY: Church in the morning. Chapel in the
            evening. Upper boys examined in
            Watts' *Scripture History* or Tomline's
            Theology. Lower boys examined in
            Catechism.[47]

Of this education Darwin later said: "Nothing could
have been worse for the development of my mind than
Dr. Butler's school, as it was strictly classical, nothing else
being taught, except a little ancient geography and
history. The school as a means of education to me was
simply a blank."[48] He never had any aptitude for languages, not even for a modern language such as German
which he was later often obliged to use in his research,
so that a curriculum consisting exclusively of the study
of unfamiliar and unusable languages was obviously not
congenial. Nor did he ever develop a facility with words,
his one desire later in life when he had to write being to
do so as lucidly and quickly as possible, and the constant
versifying in school was a bore and a burden. Fortunately,
he was a good-natured, likeable boy with many friends,
who helped him get together a patchwork collection of
verses which ingenuity could then tailor to almost any
assigned subject.

Yet he was not idle. Except for the hopeless poetry exercises, he worked conscientiously without cribs, committing to memory his 40 or 50 lines of Virgil or Homer. Luckily he had no trouble memorizing and was able to learn the daily stint while at morning chapel—and to unlearn it as expeditiously once it had been delivered. Applied to other subjects, this memory was to serve him well; if psychologists are correct, however, in supposing that learning techniques cannot be transferred from one subject to another, it is unlikely that the recital by rote of passages from the *Aeneid* helped Darwin later to remember the anatomical peculiarities of pigeons.

It is interesting that although the mature Darwin, looking back upon his childhood, subscribed all too willingly to his elders' moral judgments, he was less docile in accepting their judgments of his intellectual worth, or worthlessness. He seems to have agreed that he was an "ordinary" boy, but he balked at the idea that he was "rather below the common standard in intellect," as he suspected his masters and father thought him.[49] He was never intimidated by the fashionable opinion that classical studies were a discipline for the character as well as the mind, the initiation rites through which every English boy had to pass to be admitted to the condition of gentlemanhood. He was convinced that these studies were bad for him, concealing his true worth and suppressing his true interests.

His interests may not have been particularly intense, but at least they were more promising than his school studies. He read on his own and enjoyed, as he could never enjoy Virgil, such books as *Wonders of the World,* the historical plays of Shakespeare, and the recently published poems of Byron and Scott. Years later he distinctly remembered the satisfaction of learning Euclid with a private tutor and having explained to him by an uncle the principle of the vernier barometer. His greatest passions, however, besides shooting, were the collecting of minerals —carried on, as he later remembered, with more enthusiasm than science, for he cared only about the acquisition of differently named minerals and not their classification; the collecting of insects—but only dead ones, his sister

having persuaded him it was immoral to kill merely for
the sake of a collection; and the observing of birds—about
which he reminisced: "In my simplicity I remember wondering why every gentleman did not become an ornithologist."[50] Finally, toward the end of his school days,
he took up the "best part" of his education.[51] In the
garden tool shed, converted into an amateur laboratory,
he assisted his brother in performing chemical experiments. He read the currently authoritative *Chemical Catechism* and helped make up the compounds which earned
him the school nickname of "Gas." His enterprise was not
appreciated by Dr. Butler, who, in punishment for a bad
copy of verse, hauled him up before the school one
day and announced: "This stupid fellow will attend to his
gases and his rubbish but will not work at anything really
useful."[52] He also charged him with being a *poco curante,*
which Charles, not knowing its meaning, took to be a
fearful reproach. But the words that rankled so that Darwin
never forgot them were those of his father: "You care
for nothing but shooting, dogs and rat-catching, and
you will be a disgrace to yourself and all your family."[53]

In the hope of salvaging something from the wasted
years, Darwin was removed from the Shrewsbury school
in 1825, a year before the regular time, when he was
sixteen. And since the only aptitude, or at least interest,
Charles showed was for science, it was decided that he
should join Erasmus (who had already completed his
three years at Cambridge) at Edinburgh to study medicine. Considering how obvious a choice of profession this
was for a Darwin, it is curious how little taste the three
generations of Darwins following the elder Erasmus had
for it. The successful Robert hated it, and neither of his
two sons (nor his grandson Francis) made the least effort
to practice it. Yet for a time it seemed that Charles might
be redeemable; the summer before he left for Edinburgh,
he acted as his father's assistant, attending some dozen of
the poorer patients, consulting with Robert about their
symptoms and treatment, and even preparing their medi-
cines himself. His father, whom Charles held to be the

best judge of character he ever knew, said he would
make a successful physician—"meaning by this," Charles
added with unintentional irony, "one who would get many
patients."[54] What he saw in Charles that would inspire
a patient with confidence—for it was this that Robert regarded as the essential ingredient of a successful doctor—
Charles could not imagine. It may well have been a hope
born of desperation, for if Charles could not succeed as
a doctor (he himself succeeded without any taste for it),
what could the aimless boy do?

Charles remained in Edinburgh for two years before
conceding publicly that this experiment in education was
also a failure. Yet the failure was not so much his, Charles
later implied, as the university's. The instruction was
largely by lectures which were intolerably dull. Dr. Duncan's series on *Materia Medica*, requiring attendance in
a cold lecture hall at eight o'clock on a winter morning,
was "something fearful to remember"[55]—"a whole, cold,
breakfastless hour on the properties of rhubarb!"[56] And
the course on anatomy, scheduled for a more civilized
hour, was unfortunately given by a professor who reproduced, word for word, the exact lectures delivered by his
grandfather in the days when Erasmus Darwin was a
student, including such enticing reminiscences as, "When
I was in Leyden in 1719 . . ."[57] Charles was distressed
by the unhappy cases he saw in the clinical wards, and
horrified by the two bungling operations he witnessed, one
on a child; this was long before the days of chloroform,
and nothing could erase from his mind their haunting
memory or induce him to revisit the operating theater.
The little he saw of dissections also disgusted him, and
he later bitterly regretted that no effort was made to encourage this work. The study of anatomy and the practice
of dissection were the only practical scientific work that
would have been helpful to him, and T. H. Huxley agreed
with Darwin that it was an "irremediable evil" that he
had not had more of them.[58]

What finally persuaded Darwin not to try to overcome
his aversion to medicine was the discovery that his father
intended to leave him enough property to subsist on com-

fortably. It is ironic that the "small pittance" that would
have sufficed to make Robert abandon medicine was to
be Charles', who promptly proceeded to do what his
father would have liked to have done. The further irony
—that Robert never intended this consequence and certainly did not approve of it—completes a not uncommon
pattern of father-son relationship.

Charles remained at the university, however, for a second year, probably without communicating his new resolve to his father. It was then, after Erasmus had completed his course and left, and Charles' studies in medicine
had become frankly perfunctory, that Edinburgh took on
a new interest for him. Left on his own, Charles cultivated new hobbies and new friends. He met several young
geologists, zoologists, and botanists, one of whom treated
him to an animated discourse on the theories of Lamarck.
He later wondered whether this incident could have been
the spark smoldering for years in the recesses of his mind
which finally flared up to produce the Origin of Species;
but although he could remember being astonished by his
friend's passion, he could not remember being impressed by
the theory. What he does not mention is that about this
time, perhaps as a result of this incident, he read Lamarck's
work on the invertebrates and took notes on it.[59] He also
read his grandfather's Zoonomia, a work which he found
vaguely admirable as having made the name of Darwin
illustrious, but which did not engage his attention.

His interest in natural history came in the wake of his
enthusiasm for sports. Even more than most schoolboys,
he was an ardent sportsman. His summer vacations were
given to walking or riding tours, and the autumn holidays
to shooting. His great passion was shooting, and suspecting
something shameful in his zeal (as his father had suggested), he tried to convince himself that shooting was
"almost an intellectual employment"[60] requiring skill and
judgment—a rationalization that persuaded no one. He
had been taught to stuff birds by an old Negro servant of
Dr. Duncan, but he did not pretend that this was of any
particular value; it had only the "recommendation of
cheapness," he apologized to his sister.[61]

Fishing, on the other hand, had a more convincing claim to respectability. When during term time he went out trawling for oysters with the Newhaven fishermen or grubbing around in pools with a zoologist friend, he could persuade himself that he was more interested in discovering some rare variety of sea life than in the sport itself. Even the dissection of his find was not disagreeable; he was only annoyed that his microscope was no better than his skill, and he took detailed notes of his specimens. The stamp of respectability was provided by his membership in the student Plinian Society, where he read two papers correcting some misidentifications. As a matter of form, he sometimes attended meetings of the Royal Medical Society, which bored him, and he was once taken to a session of the Royal Society of Edinburgh, where he was properly awed by the presence of Sir Walter Scott in the chair.

During this second year he also attended some university lectures on geology, which were so dull and incompetent that, as he later recalled, they "completely sickened me of that method of learning"[62] and made him resolve never to read a book on the subject. Yet the professor, Robert Jameson, was a distinguished man, the leader of that school of geology that held the decisive factor in the formation of rocks to be the action of water. Fifty years later Darwin could still reproduce the opening words of his lecture: "Gentlemen, the apex of a mountain is the top and the base of a mountain is the bottom."[63] And he recalled with contempt the sight of the professor standing in a field of volcanic rocks and, against every evidence of his senses, advancing his favorite theory. The state of geological studies at Edinburgh was all the more regrettable, he later felt, because even then he had been ripe for "a philosophical treatment of the subject."[64] He remembered how impressed he had been, two or three years earlier, when an old resident of Shrewsbury had pointed out to him one of the famous local mysteries—a large, peculiarly shaped boulder called the "bell stone"—assuring him that no one would ever be able to explain how a rock of a type found no nearer than Scotland had come to Shrewsbury. When Darwin learned in Edinburgh of the action of icebergs in transporting boulders, he thought back with satisfaction to the old man's prophecy and "gloried in the progress of Geology."[65]

In 1827, when he left Edinburgh, he was eighteen years old and as far from a career as ever. Although he was interested in nature and assorted varieties of wild life, it was not at all clear, at least to his father, that this interest was anything more than the idle "shooting, dogs and rat-catching" of his school days, or, to himself, that it was more than a very ordinary hobby. In his autobiography he recalled only one compliment paid to him at this time: Sir James Mackintosh, the Whig politician and historian and one of the leading lights of the Wedgwood circle, once said of him, "There is something in that young man that interests me."[66] Darwin glowed with pride when he was told this, but reflected afterward that it was probably prompted by his having listened with great attention to the brilliant conversationalist, "for I was as ignorant as a pig about his subjects of history, politics, and moral philosophy."[67] It is pathetic to think of him cherishing the memory of this one compliment, and writing, fifty years later, that a bit of praise is a good thing for a young man in helping him keep to the right course.

If Darwin managed to keep on some kind of course, it was without any such encouragement. Neither his family nor his masters saw anything in the least praiseworthy in him. He had already failed—in all but the formal sense —in two experiments in education, having shown little ambition and less perseverance in both. It was with his customary self-depreciation that he recalled his father's state of mind when he was told that Charles had decided not to practice medicine: "He was very properly vehement against my turning into an idle sporting man, which then seemed my probable destination."[68]

CHAPTER 2

CAMBRIDGE

THE final recourse of Victorian society for the maintenance of misfits and dullards was the church. Young men with no other discernible calling were graced with the highest calling of all. That the church was, at the same time, the refuge of the talented and brilliant did not in any way hinder it from performing the humble but useful service of relieving despairing fathers of surplus sons. So it is not as absurd as might at first appear that Dr. Darwin should have conceived the plan of making Charles a clergyman. Nor did he feel any religious compunctions on this score. He respected neither the clergy nor his son enough to credit them with any profound religious convictions; and he was sufficiently aristocratic in his own sentiments to feel that, while he personally was superior to the irrational dogmas of religion, the masses of men required the stabilizing influence of a church—and in any event that his son required the stabilizing influence of a career.

Nor did Charles think the plan absurd, as was demonstrated by the equanimity with which he received it. He may not have been a particularly creditable son, but he was a dutiful one, and he was prepared to believe that his father knew what was best for him. He even found the idea rather attractive, on condition only that he would be a *country* clergyman.[1] When he asked for time to consider the matter, it was because it occurred to him that he might have scruples about accepting all the dogmas of the Church of England—not that he was troubled by a lack of faith as such (he was probably not then aware of his father's skepticism, and years later he was to be shocked when his shipmate confessed that he did not believe the Biblical account of the flood), but rather because of the Wedgwood tradition of Unitarianism. He set himself, therefore, to read "Pearson on the Creeds" and a few other standard works in theology, from which he emerged with the happy conviction that he did not "in the least doubt the strict and literal truth of every word in the Bible" and that he fully accepted the Anglican creed.[2]

Looking back upon this episode after half a century, Darwin reflected how illogical it had been of him to say "that I believed in what I could not understand and what is in fact unintelligible." It would have been more sensible to say merely that he had no wish to dispute any dogma. But at least, he consoled himself, "I never was such a fool as to feel and say 'Credo quia incredibile.'"[3] This, however, was the reflection of hindsight. At the time that he read Pearson and was confirmed in his orthodoxy, he was not sophisticated enough—or skeptical enough—to distinguish between the act of subscribing to dogmas and the refusal to challenge them, between the faith that embraces belief and that which suspends disbelief. If he had any doubts at all, it was not on the matter of faith but on the matter of a calling. For one short period he was not certain that he could give an honest, affirmative answer to the question in the ordination ritual: "Do you trust that you are inwardly moved by the Holy Spirit?"[4] But this was a transient doubt, induced, perhaps, by a sudden and unaccustomed probing into an area which he was generally content to leave to the wisdom of his elders and the discretion of his unconscious.

Nor was it as ludicrous then as it seemed later that he who was to be so strenuously attacked by the clergy should once have seriously undertaken to become a clergyman himself. At the time the proposal was both sensible and honorable. Indeed, if the phrenologists are to be trusted, it would have been an eminently suitable profession—as a German society of psychologists discovered many years later when, at a public meeting, the shape of Darwin's head was discussed and it was discovered that he had "the bump of reverence developed enough for ten priests."[5]

The only ludicrous aspect of this proposal was not that one who was indifferent to religion should decide to become a clergyman, but rather that one who was indifferent almost to the point of incompetence to conventional academic studies should engage to enter the most conventional and academic of disciplines. For to become a clergyman meant that Charles would have to take a degree requiring a three-year university course in subjects which he found even less congenial than medicine. Before he could so much as enter the university, he had to have special tutoring in the classics, having managed to forget, in his two years at Edinburgh, almost everything he had ever been taught, including part of the Greek alphabet. Yet he was undaunted. In several months he recovered his school knowledge of Greek and Latin and was ready to take up residence at Christ's College, Cambridge, in the Lent term of 1828.

Charles went to Christ's because his brother had; otherwise there might have been good reason to avoid that particular college. Having acquiesced in the drastic expedient of a university education in order to be delivered from the fate of the idle sporting man, he would have done better to avoid a college that had a reputation for "horsiness," a senior tutor who spent the racing season at Newmarket, and a dean who did not hesitate to abridge the chapel services when they seemed too long—a college that catered to men with much money and little liking for discipline. Yet Christ's, while not the best college for his purpose, was by no means the worst. Cambridge as a whole seems to have been in a bad way at that time, with horsiness the least of the perils lying in wait for the undergraduate. One pamphlet described in lurid detail the "corrupt state" of the university: habitual drunkenness, gambling, and falling into debt; a profligacy so common that one could hardly

find a female servant in a university lodging house who
had managed to preserve her virtue; and a condition of
moral laxity in which the highest aspiration was to be
recognized as an authority on food and drink. Some of
these charges were dismissed as Evangelical rantings, but
there was enough substance in them to be embarrassing.
Just as one of Cambridge's most esteemed professors was
solemnly assuring the public that there was no gambling
at the university, two undergraduates were being sent
down on that charge, one admitting to losses of nearly
eight hundred pounds. A more reliable witness and Darwin's own contemporary at Cambridge, Thomas Macaulay, once exchanged reminiscences with the Master of
Trinity: "You will remember," wrote Macaulay, "two reverend gentlemen who were high in college office when I
was an undergraduate. One of them never opened his
mouth without an oath, and the other had killed his
man."[6]

Darwin's own career in debauchery was considerably
less sensational. Apart from riding and hunting, it seemed
to revolve about a society facetiously called the Gourmet
Club. The club, in parody of another which had as little
right to the title, met weekly to dine on unconventional
fowl and beasts, to play cards, and to drink rather more
than a mature and virtuous man liked to remember. It included, he recalled, some "dissipated low-minded" young
men, in whose company he was ashamed to say that he
enjoyed himself immensely.[7] He was also later disturbed
by the memory of some modest debts, which must have
weighed far more heavily upon his mind than they did
upon his father's bank balance. With these dissipations and
some more innocent employments, such as listening to
music (which he enjoyed, although he had no ear at all)
and admiring the engravings in the Fitzwilliam Gallery, he
looked back upon Cambridge, shamefacedly, with great
pleasure. In excellent health and in high spirits, he enjoyed
life then as he was never to enjoy it again. By his contemporaries he was remembered for his good nature and even
temper; he was never known to have offended anyone or
to have taken offense. "Placid, unpretending, and amiable"

are the words that appear repeatedly in their recollections of him.[8]

Life at Cambridge was so agreeable that even the tedium of the academic studies could not long dispirit Darwin. Studies might be a bore and a nuisance, but they were not generally a serious grievance. There was no university or college entrance examination, the colleges having jealously insisted upon their right to select their own students, so that they might not be obliged to turn away someone with better breeding than wit. And the standards that were lax before matriculation continued to be lax afterward. In classics Darwin did next to nothing, only attending the few compulsory lectures and going through the routine cramming before examinations. Mathematics he applied himself to with more interest, going so far as to engage a private tutor one summer; unfortunately, he bogged down in the early stages of algebra, and, although he continued to enjoy Euclid, one of the regrets of his later years was his failure to master mathematics, "for men thus endowed seem to have an extra sense."[9]

With theology, as it was understood for purposes of the B.A. examination, he had more success. He mastered the two required works, Paley's *Evidences of Christianity* and *Moral and Political Philosophy*, so that he was able to reproduce their arguments perfectly, discovering in them the same satisfaction of orderly, logical reasoning that attracted him in Euclid. He enjoyed the experience so much that he was inspired to take up another of Paley's books, *Natural Theology*, which was not required reading. For the youthful Darwin, as for Cambridge, right thinking on religion began and ended with Paley. When he later wrote to one of his Cambridge professors, "It would, indeed, be a grand step to get a little more divinity in study for men of different minds,"[10] it was the study of Paley he had in mind. And toward the end of his life he reaffirmed his early impression that the analysis of Paley was the only part of the university course which was "of the least use to me in the education of my mind."[11] He did not, to be sure, concern himself with the question of the truth of Paley's premises; metaphysics were as little his forte then as later. What he would have liked in the way of an education was instruction in the methods and logic of thought, and it was this that the study of Paley came closest to providing. For the rest, however, he judged Cambridge to be as much a waste of time as Edinburgh and Shrewsbury.

When Darwin later reflected that his time was sadly wasted at Cambridge—"and worse than wasted"[12]—it was the sportiness and horsiness of it that he had in mind. With as much or more justice, the same judgment might have been applied to his studies there. This was time truly wasted, and worse than wasted. For not only did he not profit from what Cambridge had to offer (Paley excepted), so that he never remotely resembled the model of the cultured Englishman who could quote an aphorism from Homer or compose a Greek elegy as the occasion might demand; but he also failed to engage and develop his own interests. It was not so much the waste of one kind of education that he might have deplored, as the failure to provide another.

These, of course, are retrospective judgments, which few people at the time, let alone so compliant a young man as Darwin, would have thought to express. There was only one course of studies which entitled a man to regard himself as educated, and this was the venerable trinity of classics, mathematics, and theology, of which classics was decidedly the superior member, mathematics and theology being included only by courtesy of Euclid and Paley. To have supposed that science might take a place next to these would have been a gross impertinence. The very word "scientist" had not yet been coined, and "science" it-self still meant philosophy in general or any knowledge not derived from revelation. (As late as 1885, the historian E. A. Freeman could write: "I remember him years ago as a logic and science coach. I don't mean cutting up cats, but what science meant then, Ethics, and Butler, and such like."[13]) Most of the educated men still shared Addison's contempt for those "whimsical philosophers" who set more store on a collection of spiders than on a flock of sheep:

I would not have a scholar unacquainted with these secrets and curiosities of nature, but certainly the mind of man, that is capable of so much higher contemplations, should not be altogether fixed upon such mean and disproportionate objects . . . Studies of this nature should be diversions, relaxations and amusements; not the care, business and concern of life.[14]

As a hobby, to be picked up casually and pursued privately, science was a respectable employment for an otherwise unoccupied gentleman. But as a formal discipline, taught by professionals for the training of professionals, it was almost vulgar. Nonconformist academies were teaching science as early as the eighteenth century, but gentlemen did not attend these academies. Still less did they attend the Mechanics' Institutes for Working Men, established early in the nineteenth century by public-spirited gentlemen in the interests of the lower classes. Nor were they inspired by the Society for the Diffusion of Useful Knowledge, founded two years before Darwin entered Cambridge by that unregenerate plebeian Henry Brougham, which published penny magazines and a penny encyclopedia for the instruction and advancement of the industrious poor.

The universities, not having the incentive of "self-help," came much later to the sciences. There were, to be sure, both in Cambridge and Oxford, chairs of "experimental philosophy," anatomy, chemistry, mineralogy, geology, and botany; but as these subjects were not included in the curriculum and could not be turned to account in the examinations, neither students nor professors were inclined to take them seriously. It had long been customary in both universities to regard these professorships as sinecures. In the 1820's, however, a new group of Cambridge professors experienced a sudden accession of conscience and resolved to take their duties more seriously, even going so far as to acquire a rudimentary knowledge of the subjects to which their endowments committed them, and to announce their intention of delivering several lectures each year. (The older incumbents were not converted: The Lowndean Professor of Astronomy and Geometry, who held the chair from 1795 to 1837, did not deliver a single lecture in the course of his entire career.) In Oxford the situation was worse. Possibly because of the Tractarian movement which diverted attention to theology, but more probably because of the introduction of honors degrees in 1830, which encouraged distinction only in classics and mathematics, attendance at science lectures actually declined; and in 1839 the professors of science sent a joint petition to the heads of the colleges asking to be excused from giving lectures since so few students attended. It was not until the 1850's that the sciences were formally accredited in both universities—in Oxford with an honors school in natural science, and in Cambridge with the natural science tripos.

Until the sciences became respectable enough to merit a degree, they led a somewhat furtive life in the universities, existing in the interstices of the curriculum and on the fringes of such extra-curricular activities as hunting and fishing. To Darwin at Cambridge, science first came in the form of beetle collecting. He was introduced to beetles by his cousin and fellow student at Christ's, W. Darwin Fox, and beetles soon made all his earlier collecting passions appear to be mere schoolboy infatuations. He went on field expeditions during the holidays and plagued his friends with requests for specimens. In his old age he could still remember the exact appearance of trees and banks where he had made notable conquests, and he could distinguish closely allied species that he had not seen since his youth. But beetle hunting, however passionately pursued, was almost more a sport than a science. It was still the mere love of collecting that inspired Darwin. He did not dissect his specimens, and he was even unsystematic in his methods of identifying them, not bothering to compare their external appearance with published descriptions but being content to get them named by whatever haphazard method suggested itself. Collecting was a schoolboy game which, like many amateur games, entailed a considerable expenditure of time and energy and a high degree of professional skill.

Yet there were some professionals at the university, and

it was inevitable that an enthusiast, even an amateur like Darwin, should eventually meet them. His first introduction, which came by way of the same friend who had introduced him to beetles, was to the Professor of Botany, the Rev. John Stevens Henslow. From Darwin's letters at the time, as well as from his later recollections, one has the impression of a venerable sage, bountiful in his wisdom, generous in his sympathies, with an impeccable sense of morality, an unfailing benevolence, and the charming faculty of putting young people at their ease. In fact, however, the sage, when Darwin met him, was all of thirty-two, a sober and diligent young man, with a respectable but conventional knowledge of the natural sciences. Darwin remembered him as being particularly well versed in botany, entomology, chemistry, mineralogy, and geology.

To the reader today, this catalogue of talents is suspicious in the same measure that it is impressive; and the history of Henslow's academic career does nothing to allay these suspicions. As a student at Cambridge preparing to take orders, most of his time was devoted to mathematics and theology, but in the intervals left him from these studies he managed to acquire, as his biographer quaintly put it, "a tolerably correct knowledge of the two cognate sciences of chemistry and mineralogy."[15] It is doubtful that he had more than a casual acquaintance with botany (which by no stretch of the imagination could have passed as a cognate science of chemistry or mineralogy) when in 1818, the same year he received his B.A., he was elected a Fellow of the Linnaean Society. To geology he paid no attention until the Easter vacation of the following year, when he accompanied Professor Sedgwick on a tour of the Isle of Wight. Although this was his first practical acquaintance with geology (and, perhaps, his first theoretical acquaintance), he proved to be "so ready a learner"[16] that by the time the long summer vacation arrived he was ready to reverse roles and conduct a group of students on an exploration of the Isle of Man. With his election as a Fellow of the Geological Society, his professional accreditation was complete; in less than a year he had progressed from novice to expert. In the next year he added to his repertoire zoology, with a particular emphasis on entomology, and two years later he was the leading candidate for the professorship of mineralogy; he was elected after some controversy which had nothing to do with professional qualifications. When the chair of botany became vacant, he annexed that as well; he later publicly admitted that he "knew very little indeed about botany" at this time, adding, however, that he "probably knew as much of the subject as any other resident in Cambridge."[17] For three years he held both chairs simultaneously, resigning as professor of mineralogy only in 1828, the year Darwin arrived at Cambridge. In the ten years that had passed since taking his degree, Henslow had managed to accumulate what today appears as a singular collection of academic trophies. Yet neither Darwin nor any of his contemporaries was under the illusion that Henslow was a genius. Indeed, his career was so far from being singular as to be almost typical.

Similarly autodidactic, and even more eminent than Henslow, was Adam Sedgwick, a Fellow of Trinity. When the Woodwardian Professorship of Geology became vacant in the summer of 1818, Sedgwick was one of the candidates, and although he confessed to complete ignorance of the subject, while his rival was known to have some familiarity with it, he was elected to the chair by a considerable majority. He later tried to brazen it out: "I had but one rival, Gorham of Queens', and he had not the slightest chance against me, for I knew absolutely nothing of geology, whereas he knew a good deal—but it was all wrong!"[18] At the time, neither Sedgwick nor anyone else could have known whether Gorham's geology was wrong or right; more important was the fact that Gorham had few influential friends and came from a small college with an unfortunate reputation for evangelicalism. The campaigning conducted by both sides before the election would have been worthy of a parliamentary contest. It included elaborate canvassing and circularizing of electors, maneuvers for the resignation of a weak third candidate so that Sedgwick's vote would not be split, arrangements for the transportation of fellows who had promised their votes, and finally the solemn pledge of both candidates to deliver a series of lectures if elected.

These proceedings aroused no surprise; what did surprise and please the university was that Sedgwick, when elected, instead of regarding the chair as a sinecure, promptly applied himself to the study of geology—and as promptly mastered it. "Hitherto I have never turned a stone," he was reported to have said during the heat of the campaign; "henceforth I will leave no stone unturned."[19] That very summer he went off alone on a geologizing trip to Derbyshire and Staffordshire. He returned feeling competent enough to announce a course of lectures for the following Easter term and to set about initiating into the mysteries of the trade such neophytes as Henslow. In a matter of months Sedgwick had become Cambridge's leading geologist. Much later, when the progress of geology left him behind, some of his admirers felt called upon to apologize for the tenacity with which he held to outmoded theories and the brashness with which he defended them. What else, they asked, could be expected of a man who was at once a practicing scientist, a university professor with teaching, lecturing, writing, and proctoring duties, and a canon actively engaged in the affairs of the cathedral? They might have added: a politician with a penchant for getting involved in both university and national controversies, and a man with many ailments, not all of them hypochondriacal, and with a constitutional dilatoriness that was the bane of his collaborators. In 1831, however, when Darwin first became intimate with him, it occurred to no one to apologize for the eminent scholar who had just been elected Fellow of the Royal Society and President of the Geological Society.

The third of Cambridge's scientific worthies was William Whewell, who never recovered from the badly inflated ego with which he was afflicted as a child prodigy. Fellow of Trinity, Fellow of the Royal Society, and mathematics lecturer at Cambridge, he confessed that his great interest was ecclesiastical architecture. When it seemed that Henslow might resign the professorship of mineralogy in 1825, Whewell announced himself a candidate for the chair. He then hastened to Germany for a three-month visit to learn something of the subject he was proposing to teach, and, when Henslow did finally resign, Whewell's qualifications earned him the appointment. Within a decade, geology, philology, astronomy, physics, economics, pedagogy, scientific methodology, and moral philosophy had all been appropriated by his marauding intellect, until Sydney Smith was provoked to the famous quip that science was his forte and omniscience his foible.

If Darwin did not officially become a student of the sciences at Cambridge, it was not because he had doubts about the competence of the professors. On the contrary, as university teachers of science went, Henslow, Sedgwick, and Whewell were men of exceptional earnestness and erudition. It had never been pretended that the universities were an appropriate place in which to receive instruction in the sciences. And natural science, in particular, was assumed to be the province of the individual, the research equipment and the abstract concepts being presumed available to anyone with a moderate amount of patience and common sense. (Botany, since it demanded the fewest resources of knowledge and equipment, was especially popular. Thus the Linnean Society was the first of the several scientific bodies to be established.) Indeed, Cambridge at this time proved to be more fortunate in its appointments than the famed Royal Society, which had been founded to promote "by the authority of experiments the sciences of natural things and of useful arts, to the glory of God the Creator, and the advantage of the human race,"[20] but which had degenerated, by the 1830's, to what one member described as "a mere club-like association of highly respectable, well educated and very honorable men, with every kind and no kind of scientific knowledge."[21] Between 1830 and 1835, of the 158 fellows named, only ten contributed to the *Transactions* either before or after their election. Compared with this record, Cambridge, professing no such exalted aims, had done very well.

Darwin fitted very well in this community of amateurs turned professionals. Without being obliged to announce himself formally as a student of botany, he was free to consult Henslow and to join his classes in their botanical excursions, the two becoming such constant companions that he came to be known as "the man who walks with Henslow."[22] It was to Henslow that he rushed with the news of a great botanical discovery he had just made—only to be told, with great delicacy, that the discovery was, in fact, common knowledge. And it was at the weekly soirees at Henslow's home that he was privileged to hear, as he later described it, "the great men of those days, conversing on all sorts of subjects, with the most varied and brilliant powers."[23] With the somewhat ingenuous honesty that characterizes his memoirs, he recalled: "Looking back, I infer that there must have been something in me a little superior to the common run of youths, otherwise the above-mentioned men, so much older than me and higher in academical position, would never have allowed me to associate with them."[24] He could not afterward explain this superiority, nor was he at the time aware of it; and he remembered how preposterous the idea seemed to him when one of his sporting friends jokingly predicted that he would one day be a Fellow of the Royal Society.

Through Henslow, Darwin met the other "great men" of the circle, including Sedgwick. It was Sedgwick who initiated him, during his last year at Cambridge, into the study of geology. In the short time left, Darwin could not learn much from the professor, even if the latter had had much to teach him. But he did remember two episodes that were probably more instructive than any formal tuition. The first occurred in the summer of 1831, the evening before they left for a geological expedition in North Wales. They were spending the night at the Darwin home in Shrewsbury when Charles related how a local workman had recently approached him with a specimen of a tropical shell that had been found in an old gravel pit. Convinced that the shell had really come from the pit, mainly because the laborer had refused to sell it, he was enormously excited at the wonderful circumstance of a tropical remain being found near the surface in the middle of England. Sedgwick, however, refused to be enticed, insisting that the shell must have been thrown into the pit, for if it had genuinely originated there, it would nullify everything that was already known about the deposits of the Midlands. Later Darwin realized that the shell was genuine and that the gravel beds dated back to the glacial period. At the time, however, Sedgwick's error was even more instructive than the truth, for Darwin felt the dramatic force of what he had read but had not properly appreciated: that science consists not in the mere collection of facts but in combining them in generalizations and laws.

The significance of the second episode did not fully emerge until years later. Sedgwick and Darwin had been carefully searching the rocks in one district for fossils. But even as they pored over them, they neglected to see the most glaring evidence confronting them—the scored rocks and perched boulders that plainly told the glacial history of the area. Later Darwin was to recognize this as one of Sedgwick's blind spots; ten years after this expedition, in a paper on the ancient glaciers of Wales, he wrote that a house burnt down by fire could not have told its story so clearly as did the rocks of that valley.[25] Even at the time, however, Darwin felt that all was not well with the science of geology. A Cambridge friend recalled his saying: "It strikes me that all our knowledge about the structure of our earth is very much like what an old hen would know of a hundred acre field, in a corner of which she is scratching."[26] Between the Edinburgh professor who denied the evidence of volcanoes and the Cambridge professor who denied that of glaciers, Darwin's initiation into geology was not entirely fortunate.

If much remained to be done in geology, the prospects for an enterprising young man were even more promising in the other natural sciences. In his last year at Cambridge, Darwin read two books which convinced him of this. One was the recently published *Preliminary Discourse on the Study of Natural Philosophy* by John Herschel (later Sir John Herschel), the Cambridge astronomer and son of a still more famous astronomer. For Herschel, as for his contemporaries, "natural philosophy" was what today would go under the name of "natural science." As the terms have become restricted in meaning in the past hundred years, so has the spirit in which they are conceived. Herschel's notion of science was not at all the neutral, passionless ideal so often professed today. He did, to be sure, go through the then conventional motions of dividing his subject into its several branches and classifying it according to the principles of experience, observation, generalization, analysis, and so on. His main purpose, however, was to establish the utility of the sciences for the well-being and progress of society—and even for the well-being of the scientist himself:

> A mind which has once imbibed a taste for scientific inquiry, and has learnt the habit of applying its principles readily to the cases which occur, has within itself an inexhaustible source of pure and exciting contemplations. . . . Every object which falls in his way elucidates some principle, affords some instruction, and impresses him with a sense of harmony and order. Nor is it a mere passive pleasure which is thus communicated. A thousand questions are continually arising in his mind, a thousand subjects of inquiry presenting themselves, which keep his faculties in constant exercise, and his thoughts perpetually on the wing, so that lassitude is excluded from his life, and that craving after artificial excitement and dissipation of mind, which leads so many into frivolous, unworthy, and destructive pursuits, is altogether eradicated from his bosom.[27]

"Frivolous, unworthy, and destructive pursuits"—Darwin may have heard the echoes of his father: "You care for nothing but shooting, dogs and rat-catching, and you will be a disgrace to yourself and all your family." To the family wastrel, Herschel offered the promise of redemption.

Herschel also held out a promise of scientific achievement and fulfillment. It was reassuring to know that, although physics required a knowledge of mathematics and geometry "altogether unattainable by the generality of mankind,"[28] the sciences of natural history, such as chemistry and geology, required no such esoteric knowledge.

And it was gratifying to know that a century and a half after Newton there was still a boundless, unexplored territory to be conquered by men with imagination: "We remain in the situation in which he figured himself—standing on the shore of a wide ocean, from whose beach we may have culled some of those innumerable beautiful productions it casts up with lavish prodigality, but whose acquisition can be regarded as no diminution of the treasures that remain."[29] Novice and expert alike found this an exhilarating doctrine; in later years Darwin was to reread the book with an admiration enhanced by his respect for Herschel as a person.

The other book Darwin read in his last year at Cambridge was of more immediate and dramatic appeal. This was Alexander von Humboldt's *Personal Narrative of Travels to the Equinoctial Regions of the New Continent,* which seemed to transform science from the petty enterprise of catching, pinning, and tagging insects into a glorious adventure of romantic dawns and magnificent vistas. Darwin was so moved by Humboldt's glowing descriptions of Tenerife, one of the Canary Islands, that he copied them out and read them to Henslow and the rest of the party on one botanical excursion, infecting them with his enthusiasm to the point where some of them earnestly declared their intention to visit Tenerife—although only Darwin went so far as to make inquiries about travel arrangements and even to study Spanish. Many years later Darwin wrote: "My whole course of life is due to having read and re-read as a youth his 'Personal Narrative.' "[30] In the dramatic form of the travelogue Humboldt had tried to do what Darwin himself, in a different way, was later to attempt— to view nature not as a collection of separate sciences and problems but as a single event. In Humboldt's case, this resulted in natural history in its most ambitious and romantic form: "I have conceived the mad notion of representing in a graphic and attractive manner the whole of the physical aspect of the universe in one work, which is to include all that is at present known of celestial and terrestrial phenomena, from the nature of the nebula down to the geography of the mosses clinging to a granite rock."[31]

Humboldt's ambition—to embrace and contain all of nature—was as appealing to the mature Darwin as to the youth. If Darwin was later disillusioned with Humboldt, it was because his scientific achievement was not equal to this ambition. Their first meeting (it had been Humboldt who had asked to see Darwin, then a rising young scientist) was disappointing—perhaps, Darwin sensibly reminded himself, because he had expected too much. Later he came to know him as a vain and pretentious man. Yet when he reread the great inspiration of his youth, shortly before his own death, he felt once again his fascination: Humboldt's geology may have been "funny stuff," and his scientific work more remarkable for its omniscience than its originality, but he was "the greatest scientific traveller who ever lived," the father of that vast progeny of explorers who had done so much for science.[32]

No other book, no dozen books influenced the young Darwin so much as these two of Herschel and Humboldt. They inspired in him, he later recalled, "a burning zeal to add even the most humble contribution to the noble structure of Natural Science."[33]

At the time, Darwin's "humble contribution" to science seemed destined to be beetles. Geology and botany he did not take very seriously. Although he went to some of Henslow's lectures, he did not actually study with him, and Sedgwick's lectures he did not even bother to attend. And even beetle collecting did not imply the systematic study of entomology. It was still not far removed from shooting —but less troublesome to the conscience, for one did not have the anguish of paining and killing (once, after finding a wounded bird, he vowed never to shoot again), or the uneasy feeling that one was being dissolute. Yet the passion and pleasure were the same. And as shooting, however blissful, could not be made a career, neither could beetle collecting. Nor did Darwin and his fellow enthusiasts have any illusions on this score. Among the more avid beetle collectors at Cambridge were a cousin who went on to be ordained, a friend at Trinity who became an archeologist, and another who attained fame as an agriculturalist, businessman, and Member of Parliament. Looking back upon these varied destinations, Darwin inferred that collecting beetles might be an "indication of future success in life."[34] However this might be, certainly one could not have inferred that beetle collectors were apt to develop into entomologists or that Darwin himself would become a professional naturalist.

The most that Darwin hoped for was a country parish where he could lead the life of a country gentleman *cum* clergyman. And this was an entirely reasonable expectation. Today, when all callings have become so highly professionalized that there is hardly an idiosyncrasy of interest that is not featured in the university curriculum or employment registry, and when religion itself is an arduous occupation, the church as a catch-all profession has become obsolete. But in Darwin's time the clergy were blessed with an abundance of leisure. And as they were also blessed with a puritanical conscience that made the waste of time as sinful as the waste of money, they were given to the cultivation of "hobbies"—a frivolous word that does not adequately express the seriousness and strenuousness of many of these leisure-time occupations. Thus it was from the ranks of the clergy that there emerged some of England's leading scientists, mathematicians, economists, and novelists—to say nothing of the scores of less illustrious, though no less industrious, beekeepers, bird watchers, Latin prosodists, and amateur antiquarians.

For scientists in particular—especially those "natural historians" who, as Herschel pointed out, needed no specialized knowledge or equipment—the church was, in more ways than one, a God-given vocation. A good catch of beetles or game could go far to relieve the tedium of catechisms and sermons. And if catechisms and sermons proved too tiresome, someone like Darwin might even look forward to a university fellowship (also requiring ordination). In either case, it was clearly expedient to take orders. And it would not have occurred to Darwin that there was anything unseemly about combining the ministry with science. Later, when the religion-science antithesis had become firmly fixed in men's minds, many a prelate must have reflected upon the irony that the church itself had done so much to foster and nurture its rival. But such reflections were well in the future. In Darwin's youth there was nothing anomalous in the appearance of a scientist in clerical garb, no taint of irreligion in science as Darwin or any of his masters understood it. Indeed, the most deeply religious man Darwin had yet met was Henslow. During one of their many talks on religion, Henslow confessed that he would be grieved if a single word of the Thirty-Nine Articles should be altered—a sentiment that surprised Darwin not because it came strangely from a professor of botany but only because it was so passionately expressed. The warfare between science and religion was at all times more complicated than rationalist historians have made it out; in the 1820's and '30's it had not yet affected the natural sciences, which were still on the best terms with religion, both members of one happy family. Perhaps some of the bitterness of the struggle, when it did come, was the bitterness of fratricidal warfare.

In 1831, when Darwin was graduated, he could contemplate with serenity a future as a country cleric. For technical reasons having to do with residential requirements, he had remained on at Cambridge for two extra terms. He had taken the examinations early in 1831 and, by doing well in Paley and Euclid and not too miserably in classics, managed to place tenth among those who did not stand for honors. This was by no means a distinguished record, but it seemed to satisfy Darwin and his elders, who apparently expected no better. For a prospective country clergyman, at any rate, it was sufficient.

EMERGENCE OF
THE HERO

BOOK II

CHAPTER 3

THE VOYAGE OF THE BEAGLE

THE transition between pre-history and history occurred in the life of Darwin during the voyage of the Beagle. It was then that the shape of the hero began to come into focus and the quality of his mind to reveal itself. The temptation here, and the common way of biographers, is to read the story of the voyage as simply a prelude to the *Origin*, and to cite the documents of the voyage only when they happen to anticipate or coincide with passages of the *Origin*. This would be to read history backward, and we would finish knowing no more than when we started. If all that were important were the ideas of the *Origin*, then it is in the *Origin* itself that they should be sought. The voyage is important not only as it may suggest the *Origin* but also as it may contradict it or even be irrelevant to it. Only in this way can the mind of the creator, as distinct from the ultimate creation, be exposed. And with the mind of the creator, perhaps also the pattern of creation, the pattern of discovery.

In the summer of 1831, when Darwin came down from Cambridge, he was twenty-two and a half—a tall, lean young man who looked even more than his six feet. He was healthy, good-natured, unambitious and unassuming, the prospective heir to a substantial income of whose magnitude he was ignorant. Behind him was the memory of a happy three years at Cambridge, a satisfactory pass degree, and a pleasurable month geologizing in Wales. In the distant future he could see himself comfortably settled in a country parish, but at the moment his dreams were of a trip to Tenerife and, more excitingly, of the opening of the partridge season. On Monday, August 29, he returned home from Wales to prepare himself for the first day of shooting, the sacred September 1, and found waiting for him two letters which started him upon what he afterwards took to be "the most important event of my life."[1]

The two letters were in one enclosure. One was from George Peacock, Professor of Astronomy at Cambridge and an influential member of the University Senate, explaining that he had been asked to recommend a naturalist for a scientific expedition leaving within the month for a survey of the southern coast of Tierra del Fuego, which was to stop at the South Sea Islands and return to England by way of the Indian Archipelago. The expedition was to be entirely for scientific purposes and was to be commanded by Captain FitzRoy, "a public-spirited and zealous officer, of delightful manners, and greatly beloved by all his brother officers."[2] The Admiralty was not prepared to pay a salary to the naturalist, although he would have an official appointment and free accommodations.

The second letter was from Henslow, who said that he had proposed Darwin as "the best qualified person I know of who is likely to undertake such a situation"[3]—a scrupulously exact statement which did not imply that there was no one better qualified. In fact, as Darwin soon discovered, the offer had been made first to Henslow himself, who would have accepted but for having to leave his wife, and then to Leonard Jenyns, Henslow's brother-in-law, a clergyman naturalist who did not think it right to leave his parish.[4] Henslow had recommended Darwin, he frankly explained, not as a "*finished* naturalist, but as amply qualified for collecting, observing and noting anything worthy to be noted in Natural History."[5] Captain FitzRoy, Henslow added, was a young man who wanted a companion more than a mere collector; a gentleman, therefore, was required, and this qualification at any rate Darwin met.

Warning Darwin against excessive modesty, Henslow assured him that "there never was a finer chance for a man of zeal and spirit."[6]

The circumnavigation of the globe, botany, zoology, and geology to his heart's delight, two years of bliss for the naturalist and adventure for the young man—the proposal was irresistible, far better than anything he had fancied while reading Humboldt or dreaming about Tenerife. Of course Darwin would accept. Or so he thought until he consulted his father. But Dr. Darwin had objections which Charles had not thought of and convictions far more decided than those held by his son. His objections, as Charles related them to his uncle (although not to Henslow or Peacock), were that: (1) the trip would be "disreputable" to his character as a clergyman; (2) it was a wild scheme; (3) the place of naturalist must have been offered to many others before Charles; (4) and from its not having been accepted, there must be some serious objection to the vessel or the expedition; (5) Charles would never settle down to a steady life afterward; (6) his accommodation would be most uncomfortable; (7) it would constitute yet another change of profession; and (8) it would be a useless undertaking.

The next day Charles wrote to Henslow and Peacock thanking them for the offer and regretfully declining it. His father had not absolutely refused permission, he explained, but he did not feel free to act against his strong advice. The following morning he left Shrewsbury for Maer to be ready for the first day of shooting.

Charles' ready acquiescence is interesting in revealing what today appears to be not only an exaggerated sense of filial duty but a correspondingly feeble sense of professional dedication. It may be that a generation in which the filial instinct has been completely supplanted by the professional is in no position to judge Darwin's behavior. Or perhaps it is that our hindsight knowledge of what the voyage was to mean in his life has distorted our vision; Darwin may have been acquiescent only because he did not think the matter important. In any event, he displayed none of the anguish which we expect of a young man torn between duty and inclination. Nor was he outraged that his father should have had the low opinion of him expressed in his objections—perhaps because he unconsciously shared it. Dr. Darwin, apparently, was no more impressed by his son's sponsors than by his son's talents, and seemed to suspect that beetle hunting was not far removed from rat-catching. Between the naturalist and the idler, he saw not enough margin for safety.

Perhaps from overconfidence, his father had left a loophole in his decision. He had said that if Charles could find "any man of common sense" who advised him to go, he would give his consent.[7] Neither Henslow nor Peacock apparently qualified as men of common sense, but at Maer Charles found such a man. Having told his uncle, Josiah Wedgwood, about the rejected offer, he found, to his own surprise, that Wedgwood shared his opinion and not his father's. His uncle saw nothing in the offer that was demeaning to Charles and nothing to suggest that the ship would be unseaworthy or that Charles' accommodation would not be adequate. Moreover, so far from it being a wild or useless scheme, he thought it would encourage Charles to develop "habits of application" and an "enlarged curiosity" that would be all to the good. As for his future in the church, the position would not be directly useful, to be sure, but it would reflect honorably upon him; "and the pursuit of Natural History, though certainly not professional, is very suitable to a clergyman."[8] Charles hastened to write to his father of this unimpeachable advocate, enclosing a letter from Wedgwood commenting on Dr. Darwin's objections. He apologized for presuming once more on his father's indulgence, promised to yield happily to his decision, and begged only for an immediate reply so that he might communicate with Henslow if he was granted permission to go.

The letter was dispatched early in the morning. Charles promptly went out shooting, and several hours later he was called for by his uncle who told him that he intended to go personally to Shrewsbury to plead the case. That same day Dr. Darwin gave his consent, and Charles wrote to Peacock saying that he would accept the position if it was still available. Five years later, on his return from the expedition, Darwin recalled those anxious two days when he did not know whether he would go or, indeed, whether he *should* go. He confessed that apart from being more than half persuaded by some of his father's doubts, his heart sank at the thought of leaving England for as long as two years.

There were many more decisions and indecisions before the Beagle finally set sail with Darwin officially installed as naturalist. It soon became evident that Peacock had unwittingly misrepresented the situation, making it sound both more attractive and more certain than in fact it was. For a while FitzRoy was minded to take on an old friend who, fortunately for Darwin, could not get leave from his office. More serious was the fact that Darwin's nose did not initially recommend itself to FitzRoy, who believed that a man's character showed itself in the shape of his features. Nor was he reassured by Darwin's introduction as the "grandson of Dr. Darwin the poet,"[9] and it was only after a common friend, the nephew of Lord Londonderry, had vouched for him, that the captain was satisfied he was truly a gentleman. And on Darwin's part there was cause for hesitation. The sight of the vessel was anything but reassuring. There hardly seemed to be enough room in it for himself and his personal effects, let alone for equipment and specimens. With unintentional sarcasm, he wrote to Henslow that he could not honestly call it one of the very best opportunities for natural history that had ever occurred, the want of room being an insurmountable evil. Nor was the seaworthiness of the ship beyond question. When Darwin first saw her lying in the dockyard, she "looked more like a wreck than a vessel commissioned to go round the world."[10] The Beagle had just returned from a five-year voyage, and Darwin later learned that she was found to be so rotten that she practically had to be rebuilt. Originally she was a three-masted, 235-ton brig carrying six guns and of a class familiarly known to the sailors as "coffins" because they were likely to go down in bad weather. (The bulwarks were high in proportion to the size of the ship, so that a heavy sea breaking over her might be fatal.) The necessary repairs turned out to be far more extensive than had at first been thought, and, by the time she was ready for her second voyage, she had gained seven tons and her upper deck had been raised to make her safer in heavy weather and to give better accommodation below.

There were other impediments as well. The voyage, it was now calculated, would take three rather than two years, and it was not certain that it would circumnavigate the globe. Until Darwin was reassured on this last point, he would not undertake to go. And the cost to him promised to be greater than he had anticipated. One of the arguments he had used to win over his father was that he would have to be very clever to overspend his allowance on board the Beagle as he had done at Cambridge—to which his father had retorted: "But they tell me you are very clever."[11] It now appeared that not only would his appointment carry no salary, but it would entail considerable expenditure for equipment and the cost of the officers' mess. And the companionship of these officers, he soon discovered, would not be the genial friendship he had known at Cambridge. Except for the captain, they had all the bad characteristics and none of the good of the "freshest freshmen."[12] Because of either their sensitivity to gradations of rank or their experience that intimacy breeds antipathy, they were careful to preserve among themselves the most formal relations, and their conversation was duller than Darwin thought possible for men who had seen so much of the world. There was apparently more in such a voyage than Humboldt had dreamt of—or than he had thought fit to reveal.

There was also, Darwin was becoming aware, a real risk to life and health. During his two-month stay at Plymouth, while the ship was being fitted and favorable sailing conditions were awaited, he learned something of the precariousness of life at sea. One captain, who had seen much the same service Darwin was about to embark upon, told him that in his eight years surveying the African coast, he had buried thirty young officers, and not once had a boat sent up the river returned with its full complement of men. Two days after this unnerving conversation, a drunken sailor from the crew of the Beagle slipped overboard, his body never to be recovered. And there were other inconveniences short of fatality to put up with. Twice the ship set sail, only to turn back because of bad winds, and the few days at sea were enough to give Darwin a most unsavory taste of seasickness. He would not soon forget, he wrote, the memory of his torment, made more agonizing by the raucous concert of whistling wind, roaring seas, the hoarse screams of the officers, and the shouts of the men. With utter dismay, he looked forward to a more or less permanent state of misery, aggravated by an inability to work. FitzRoy's assurance, later put in writing in Darwin's contract, that he could leave the ship and board a boat for England any time his seasickness became unbearable, and that he would be left on land for the two months or so of bad weather each year, was little solace. To make matters worse, he suddenly experienced more disturbing promptings of mortality in the form of palpitations and pains about the heart. To a young man with a smattering of medical knowledge, this seemed ominous, and he deliberately refrained from consulting a doctor because he did not want to hear the expected verdict that he was not fit for the voyage. A skeptical reader might suspect that the palpitations and pains were symptoms not of heart disease but of Darwin's fear of a long and lonely journey into the unknown, a fear which he did not attempt to conceal from himself.

It was thus with mixed feelings of elation and apprehension that he approached the actual sailing date. To Henslow he confessed that he thought Jenyns had been wise not to come—"that is judging from my own feelings, for I am sure that if I had left college some few years, or been those years older, I *never* could have endured it."[13] To his sister Susan (in a letter intended also for his father), he put a better face on it: "My spirits about the voyage are like the tide, which runs one way, and that is in favour of it; but it does so by a number of little waves, which may represent all the doubts and hopes that are continually changing in my mind."[14] On the whole he thought that if he managed to persevere in his aims, mastering as many branches of natural history as possible, using his spare time to study French, Spanish, mathematics, and a little classics, and disciplining his mind so as to compensate for the many opportunities he had lost at Cambridge—and if, in addition, he succeeded in keeping his health and returning alive— then the trip would have been worthwhile. But in the meantime he could not remove a book from the shelf in his cabin without wondering whether it was worth the trouble to read it, or take the soap from its container without doubting the need to wash his hands. In one breath he rejoiced in his "extraordinary good fortune in obtaining what in the wildest castles in the air I never had even imagined"; and in the next he was driven to such dubious comforts as: "If it is desirable to see the world, what a rare and excellent opportunity this is. It is necessary to have gone through the preparations for sea to be thoroughly aware what an arduous undertaking it is. It has fully explained to me the reason so few people leave the beaten path of travellers."[15] By the time the ship finally sailed, he had been reduced, partly by a dinner of mutton chops and champagne, to a state of near-insensibility, so that he was obliged to apologize for the "total absence of sentiment" he felt on leaving England.[16]

The Beagle finally sailed on December 27, 1831—a date Darwin vowed to commemorate as the birthday of his "second life."[17] Three days later he reported on his new life: "I often said before starting that I had no doubt I should frequently repent of the whole undertaking. Little did I think with what fervour I should do so. I can scarcely conceive any more miserable state than when such dark and gloomy thoughts are haunting the mind as have today pursued me."[18] It was not the vexations and discomforts of living in confined quarters that threw him into such black despair; with these he could cope, and he could even take a perverse satisfaction in accomplishing so much with so few facilities. Indeed, he discovered that physical concentration stimulated intellectual concentration, so that in later life he was apt to attribute his methodical habits of work, which he took to be an important ingredient in his success, to the necessity of tidiness imposed on him on the Beagle. What made life on board a misery were the acute attacks of seasickness. There was no appeasing the sea or adjusting to its temper, he discovered. When it chose to act up, he could only take to his hammock and hope that time and insensibility would mitigate its horrors. Looking back on his trip, he warned those young men who might be tempted to follow in his path that seasickness was no trifling evil that would pass in a week, he himself having suffered even more at the end of the trip than in the beginning.

There were other grievances of a lesser order which afflicted him. One of these was Captain FitzRoy. Before embarking, Darwin had looked upon the captain as a refuge from the dull and disagreeable company of officers. It was not long, however, before he came to look upon the officers as a haven of sanity and good humor. When Darwin met him, FitzRoy was twenty-six,[19] only four years older than Charles himself, and he had already commanded the Beagle for two years in the course of its previous voyage to South America. A direct descendant of Charles II,[20] grandson of the third Duke of Grafton and of the first Marquis of Londonderry, and nephew of Lord Castlereagh, FitzRoy combined the normal authoritarian attitude of a naval captain with the abnormally authoritarian disposition of an intense, high-minded, and puritanical aristocrat. Oddly enough, those who knew FitzRoy had made a point of praising his good temper, and Darwin had begun by regarding him as his "beau-ideal."[21] His disillusion came before they had set sail. On Christmas Day, FitzRoy (himself a teetotaler) had given the crew a holiday, with access to a large stock of liquor—which had the obvious consequences. The next day he exacted penance in the form of floggings and chainings. At first Darwin thought of the episode as an unfortunate but necessary exercise of naval discipline. A little later, tormented by seasickness and the echo of the cries of angry and humiliated men, he wondered whether there was not something unseemly, or worse, in FitzRoy's fluctuation between indulgence and retribution.

It was not long before Darwin came to see in the captain's exaggerated sense of authority a morbid and violent disposition. Whether there was in fact a hereditary taint of insanity in his family (suggested by the suicide of Castlereagh) is uncertain, but it is certain that FitzRoy himself thought there was and confided his fear to Darwin. At one point during the voyage the fear was very nearly realized. He was inspired to hire some additional small vessels at great expense, an act which the Admiralty had not authorized and whose cost it refused to bear. Partly as a result of this misadventure and of the strain of navigation in difficult and uncharted waters, FitzRoy lapsed into a deep melancholia which he himself identified as a symptom of imminent insanity. He went so far as to turn over his command to his lieutenant and was persuaded to resume it again only when he was assured that the survey would suffer in his absence. This was his most severe attack during the voyage, but there were other fits of moroseness, alternating with passionate outbursts of temper, that were a constant trial to those who had to live in close quarters with him. He was the only man Darwin could fancy in the role of a Napoleon or a Nelson. His vanity and petulance were as great as his candor and sincerity, and Darwin did not know which was more uncomfortable: his speech—the effect of his words on the crew was like the blows of a waggoner on his dray horses—or his silence, which Darwin charitably attributed to "excessive thinking."[22] Later he began to suspect that something more serious was wrong, that "some part of the organization of his brain wants mending."[23]

With his exalted heritage and proud temper, it was little wonder that FitzRoy was an ardent Tory, as ardent a Tory as Darwin, in his quieter way, was a Whig. This political disagreement, added to the normal difficulties of living on good terms with the captain of a man-of-war who regarded all disagreement as mutinous, put Darwin under a constant obligation of self-restraint. For the most part he was successful in avoiding contentious subjects, such as the

Reform Bill which was being passionately debated at home. Distance, fortunately, lent support to his forbearance, and as the familiar scenes of England gave way to exotic tropical forests and volcanic mountains, his comments on political events at home became more and more laconic. When, finally, the momentous news arrived of the passage of the bill, he merely passed on to his family, with more indulgence than indignation, FitzRoy's remark that it remained to be seen whether England would continue to be a monarchy or would become a republic.

Slavery, on the other hand, was too obtrusive a fact to be ignored. For a good part of the voyage it was an ever-present affront to Darwin, and his feelings about it were a constant source of exasperation to FitzRoy. When FitzRoy, returning to the ship after a visit to one of the great slave-holders of Brazil, triumphantly reported that the owner had called together his slaves and asked them whether they wished to be freed, to which the slaves had unanimously replied "No," Darwin could not resist asking whether such an answer made in the presence of the owner was worth anything. In a rage, FitzRoy informed him that since he did not credit his word, they could no longer live together. For a time it seemed as though Darwin would have to leave the ship (although the officers extended to him an invitation to share their mess), but FitzRoy finally sent a formal apology. In this case it may have been the captain's dislike of contradiction even more than his dislike of Whiggism that had provoked him, for in his journal of the voyage he took an entirely moderate and humane view of slavery. There he deplored the extension of slavery as an evil which no amount of private benevolence could remove, and he recommended both the abolition of the slave trade and the emancipation of the slaves.[24] And later still, as Governor of New Zealand, he was so zealous in defending the natives against the white settlers that he had to be recalled. To Darwin, however, he appeared as an unregenerate Tory, ready to defend slavery with all its abominations.

When the Beagle finally sailed from Brazil on the homeward journey, Darwin thanked God that he would never again visit a slave country. "To this day," he wrote a decade later, "if I hear a distant scream, it recalls with painful vividness my feelings when, passing a house near Pernambuco, I heard the most pitiable moans, and could not but suspect that some poor slave was being tortured, yet knew that I was as powerless as a child even to remonstrate."[25] Yet he did not believe either that slavery was an absolute evil or that liberation would be an absolute good. He admitted that "the actual state of by far the greater part of the slave population is far happier than one would be previously inclined to believe";[26] the practical interests of the slave-owner, fortified by whatever good feelings he might possess, saw to this. Yet the misery of the slaves was none the less appalling, and their condition of servitude none the less intolerable, for its being an evil often more spiritual than physical. Eventually, with the growth of the black population, liberation would become inevitable. When that day came, Darwin hoped that the Negroes would be content to assert their rights without avenging their wrongs.[27]

What might be regarded as an even more serious cause for conflict were FitzRoy's intense religious beliefs. Yet these did not, in fact, occasion any open quarrel or even, apparently, any sharp difference of opinion. FitzRoy admitted as much in his journal. In his final chapter, "A Very Few Remarks with Reference to the Deluge," he affirmed the compatibility of geology, rightly understood, with the Bible and warned his readers not to be led astray, as he had been, by geologists "who contradict, by implication, if not in plain terms, the authenticity of the Scriptures." He confessed that he himself once suffered from a "disposition to doubt, if not disbelieve, the inspired History written by Moses." And he recalled how once, during the voyage, while accompanying a friend (surely Darwin) across a plain composed of rolled stones bedded in "diluvial detritus" some hundred feet in depth, it occurred to him that "this could never have been effected by a forty days' flood."[28]

Two decades later, in the shock of reading the *Origin*,

this confession and the memory of his own lapse of faith were forgotten. He then spoke only of having often in the past expostulated with his old comrade of the Beagle for entertaining views contrary to Scripture. The expostulation, however, must have taken place not on the voyage itself, as has been assumed, but on one of their few meetings afterward, when FitzRoy's religious scruples clashed with Darwin's theory of evolution; it is probably to this episode that Darwin referred in his autobiography when he spoke of having once, years after the voyage, offended FitzRoy "almost beyond mutual reconciliation."[29]

During the voyage Darwin was less a heretic than Fitz-Roy, his "disposition to doubt" being as little violent as his disposition to believe. His was the indifference of orthodoxy. He may have been taken aback by the passion Fitz-Roy could bring to bear on such questions as the precise measurements of Noah's ark; but he was even more shocked when one of his shipmates flatly denied the fact of the flood.[30] Indeed, he was remembered by them as being naively orthodox in his beliefs. Several of the officers (though themselves orthodox) were amused once when he unhesitatingly gave the Bible as a final authority on a debated point of morality; and another recalled how he and Darwin had requested a chaplain at Buenos Aires to administer the sacrament of the Lord's Supper to them before they ventured into the wilds of Tierra del Fuego.[31] He was even able to enter with some enthusiasm into Fitz-Roy's proselytizing schemes; and one of the curiosities of Darwin's intellectual history is the fact that his first work intended for publication was a pamphlet, signed jointly with FitzRoy, on "The Moral State of Tahiti," which was an appeal for the support of the missionaries in the Pacific.[32]

Several times during the voyage he alluded to the vision of a quiet English parsonage glimpsed through a grove of tropical palms. To a college friend already installed in a country parish he wrote: "I hope my wanderings will not unfit me for a quiet life and that on some future day I may be fortunate enough to be qualified to become like you a country clergyman. And then we will work together at Natural History."³³ And again the following year: "I often conjecture what will become of me; my wishes certainly would make me a country clergyman."³⁴ If this vision slowly faded during the course of the long trip, it was not because heresy intervened but because another career did. What had originally been intended as a parson's innocent hobby had become a respectable, independent profession.³⁵

It was FitzRoy's proselytizing zeal that was responsible for one of the most entertaining episodes of the trip. Officially the purpose of the voyage was to survey the southern coasts of South America, draw up detailed navigation charts, and carry a line of meridians around the earth. Incidentally it also had the purpose, approved by the Admiralty, of repatriating three Fuegians whom FitzRoy had taken as hostages on the earlier trip and who now weighed heavily on his conscience. In 1828, soon after he had succeeded to the command of the Beagle, he had taken these hostages in reprisal for the theft of one of His Majesty's whaleboats (and for the indignity suffered by his first mate, who had had to paddle his way back to the ship in a reed basket). Unfortunately, such disciplinary measures were lost on the Fuegians, who proceeded to make a farce of the affair when the adult prisoners, after eating the best meal of their lives, jumped overboard and swam home, leaving the captain with an eight-year-old girl and three bawling infants. The girl remained, while the infants were forced upon some reluctant natives (their parents having refused to claim them). Later two young men casually attached themselves to the ship, and one boy was acquired from his father at the cost of a mother-of-pearl button. This singular assortment of volunteer hostages were christened by the sailors Fuegia Basket (the little girl), Jemmy Button, Boat Memory, and York Minster. Returning to England, FitzRoy had the embarrassing task of informing his superior of the complement of extra passengers carried by his man-of-war and suggesting that they now be turned over to the government, with the view of putting them to "some public advantage." This could best be done, he thought, by having them educated in England for two or three years and then returning them to their land with a stock of equipment designed to improve the condition of their countrymen, "who are now scarcely superior to the brute creation."[36]

The Fuegians, unfortunately, proved to be no less immune to civilization than to discipline. In spite of several attempts at inoculation, Boat Memory, the most promising of the hostages, died of smallpox soon after the ship's landing. The others were boarded in the house of a schoolmaster whose duty it was to teach them English, gardening, husbandry, and "the plainer truths of Christianity."[37] They seemed to be making satisfactory progress, behaving creditably when presented before the Queen, and accepting their many gifts graciously, when one day, about a year after their arrival, York Minster was surprised in sexual embrace with the ten-year-old Fuegia Basket. It was little consolation to FitzRoy to be told that heathen women mature early; it must have occurred to him that it would be safer and less scandalous if they matured somewhere far away from England. When he was assigned to a second trip to the Tierra del Fuego, he saw his opportunity. Still hoping to turn the affair to some advantage, he took with him on this voyage, besides his precocious natives, a young missionary to settle in their country and disseminate the arts and morals of Christian civilization. Accompanying him were several boatloads of gifts which were to be the instruments of his mission; when this bulky cargo was unpacked, it was found to contain such amenities of civilization as wine glasses in assorted sizes, soup toureens, beaver hats, and a mahogany dressing case.

After a year's voyage down the eastern coast of South America, the Beagle reached the Tierra del Fuego. In his diary Darwin recorded his first impressions of the naked, yelling savages who ran alongside the ship, their appearance so bizarre that it was "scarcely like that of earthly inhabitants."[38] He did not know whether to be more astonished by their bland disregard of firearms or by their nonchalant manner of greeting their long-lost relatives. The meeting of Jemmy Button and his parents, he thought, "was not so interesting as that of two horses in a field."[39] Even more shocking than the sullen suspiciousness with which Jemmy's parents approached him was the indecent haste, once the ship had departed, with which they proceeded to rob him. If Jemmy was not safe from their plundering zeal, the pathetic missionary, Matthew, was even less so, and, after a week of being teased, threatened, and mauled about, he decided that either they were not convertible or he was not the person to convert them. The garden, hopefully designed by FitzRoy and laboriously laid out by the crew as the first step in the material advance of the Fuegians, had been willfully trampled, and everything visible and portable in Matthew's tent had been stolen. Even the captain was persuaded to abandon the experiment temporarily and restore the missionary to the safety of the ship. Further infamies were discovered when the Beagle returned a month later and found that York Minster, the oldest of the hostages who was to have been the spearhead of the civilizing mission, and Fuegia Basket, the little girl who had captivated the Queen (to say nothing of York), had collaborated in an elaborate scheme to rob Jemmy of everything his less efficient family had left him, including the very clothes he was wearing. Still FitzRoy did not despair, continuing to hope that they would be a Christian inspiration to their countrymen: "Perhaps a ship-wrecked seaman may hereafter receive help and kind treatment from Jemmy Button's children; prompted, as they can hardly fail to be, by the traditions they will have heard of men of other lands; and by an idea, however faint, of their duty to God as well as their neighbour."[40]

Darwin, in spite of his missionary sympathies, was less optimistic. With a surprisingly fresh insight, he reflected that it was the absolute equality of the Fuegians that prevented their advance. If only all the plunder remained in the hands of one family or tribe, instead of being aimlessly passed from one to another, there might be some hope of progress. Otherwise there was none. Individuals, he conceded, might be redeemed: Jemmy, York, and Fuegia had been transformed, at least physically, into a replica of civilized youth. Yet even this was not an unmixed good, for while preferring, as all sensible men must, civilized habits to uncivilized ones, they were doomed to the latter. Whatever ends had been served by their excursion to England, their own personal happiness, he feared, had not been one of them. He did not see how any Englishman could enter into the minds and souls of creatures more animal than human, with their stunted bodies and hideous faces, their greasy skins and voices barely articulated, who lived on a diet of fish and berries and slept naked on the ground. Their condition would seem to be one of utter misery; yet, judging by the fact that their numbers were not decreasing, he presumed that they must enjoy a happiness of some kind to make their lives worth living. "Nature," he reflected, "by making habit omnipotent, has fitted the Fuegian to the climate and productions of his country."[41]

Darwin spoke more wisely than he knew in cautioning against the attempt to impose the values of civilization upon people in a state of barbarity. Years later it appeared that the captain's prediction of the services his protégés might perform for needy seamen had been fulfilled in a fashion never intended by that proper gentleman: some sailors returning from South America told Darwin that an English-speaking Fuegian woman had spent several days aboard their ship. This was not, presumably, the "help and kind treatment" FitzRoy had hoped for from little Fuegia Basket. A more serious travesty of his intentions was enacted in 1859, when the Fuegians massacred most of the missionaries in the area—the massacre being led by Jemmy Button's son. Darwin then concluded that it was useless and dangerous to send missionaries to a people so savage as the Fuegians, and he refused to support the society set up for that purpose. Several years after the massacre, however, he was prevailed upon, by reports testifying to the great advances made among the natives, to subscribe to the mission, and he continued to do so until the end of his life. Had he known more of conditions there, he might have been reinforced in his first opinion, for the glowing reports of the society were based upon the discovery that the land was perfectly suited for sheep grazing and that the saving of souls could be profitably combined with the growing of sheep. Unfortunately, the new civilization brought with it not only the benefits of Christianity but also new and fatal diseases. In thirty years the population declined from ten thousand to one thousand, and by the end of the century both the mission and the industry were dying for lack of human material.

A far graver historical irony—as FitzRoy would have thought it, tragic irony—was that the Beagle should later have become notorious as the laboratory or testing ground of the *Origin of Species*. The irony must have been all the more painful to FitzRoy because the discrepancy between cause and effect was so gross. In inviting Darwin to accompany him on the trip, he had simply wanted to make certain that "no opportunity of collecting useful information during the voyage should be lost."[42] And by "useful information" he meant just that: not grand theories deriving from obscure facts, the age of a continent deduced from the fissures in its rocks or the origin of living things from the eye structure of a bat. What FitzRoy wanted to know was whether, for example, there was metallic ore in the Fuegian mountains. If a naturalist could confirm his suspicions about the presence of such ore, then FitzRoy would have felt himself repaid for the inconvenience of having on board yet another "supernumerary," as Darwin figured on the ship's register.

As it happened, Darwin found no metal of commercial value in the Fuegian mountains.[43] He did find, however, a wealth of natural history, and he discovered in himself an unexpected fund of enterprise. Even the martinet FitzRoy, after only a few weeks at sea, felt called upon to pay tribute to Darwin's exceptional industriousness. And with this sudden access of energy, there also came upon Darwin an accession of self-confidence that transformed him, almost overnight, from an enthusiastic amateur into a dedicated professional. Suddenly the world appeared to be wide open, his for the taking. Within a fortnight he was proposing to write a book on geology; a week later it was zoology that was to be his subject. "I think if I can so soon

judge," he wrote his father, "I shall be able to do some
original work in natural history. I find there is so little
known about many of the tropical animals."[44] The less
that had already been done in the natural sciences, the
more there remained for him to do. "It is a new and
pleasant thing for me," he confided in his diary, "to be
conscious that naturalizing is doing my duty, and that if I
neglected that duty I should at the same time neglect
what has for some years given me so much pleasure."[45]
The idea of a quiet parsonage at home lingered on, but it
is doubtful whether Darwin ever again saw it as anything
but a convenient base from which to conduct his geological
and zoological expeditions. Even hunting was sacrificed
to his new passion. After two years at sea, he practically
gave up shooting the birds and animals for his collections,
and finally he turned over his gun entirely to his servant,
complaining that shooting took time away from his real
work. His family must have been startled to read: "There
is nothing like geology; the pleasure of the first day's partridge shooting or first day's hunting cannot be compared
to finding a fine group of fossil bones, which tell their
story of former times with almost a living tongue."[46]

It was at St. Jago, one of the Cape Verde islands and the
first port of call (Tenerife they had only glimpsed from
the distance), that the idea came to him of writing a book
on the geology of the countries he was to visit. Taking
shelter from the sun under a cliff, he gloried in the sight of
the exotic desert plants blooming in the arid soil around
him and the living corals growing in the pools at his feet.
Not even the familiar descriptions of Humboldt had prepared him for the rich colors of tropical vegetation, the
irregular shapes of volcanic mountains and boulders, the
unknown insects and birds fluttering about the equally unknown flowers. "It has been for me," he exulted, "a glorious
day, like giving to a blind man eyes; he is overwhelmed
with what he sees and cannot justly comprehend it. Such
are my feelings and such may they remain."[47] They did
not, however, remain quite such. He soon discovered that
his pleasure came less from the passive, sensual act of seeing than from the effort of comprehending and analyzing.

The very next day, his second day on shore, he wrote of the "memorable epoch" passed through by the geologist examining for the first time volcanic rock.[48] Geology was the most interesting natural feature of St. Jago, and geologizing soon became his chief occupation, although he could not ignore the tempting array of rare zoological and botanical specimens. After three days he was able to look back upon that first overwhelming, uncomprehending day as "a period long gone by,"[49] and he could abandon himself to the pleasurable anxiety that there might be no one in England willing to take on the task of examining the formidable collection he was amassing.

By the time the Beagle was ready to sail from St. Jago three weeks later, the novelty of that island had begun to pall, and Darwin was impatient for the more exotic creatures to be found farther south. Just above the equator, the Beagle stopped at the small island of St. Paul's, where Darwin was delighted to find birds so unaccustomed to men that they did not move when approached and could be brought down by so unconventional an instrument as the geologist's hammer. The crossing of the equator also proved to be a memorable, if less instructive or pleasurable experience, as the crew carried out the prescribed initiation ceremonies, energetically lathering the faces of the novices with pitch and paint, shaving them with a roughened iron hoop, and then conducting them along a plank leading to a bath of icy water.

In Brazil, with the ship anchored first at Bahia and then at Rio de Janeiro, Darwin made his first acquaintance with truly tropical life. Even as he was lost in raptures at the richness of the spectacle, he found himself thinking how fine a feast it would provide for the scientist. If he tried to follow the flight of a strange butterfly, his eye was arrested by the even stranger flower or fruit upon which it alighted; if he tried to fix his mind on the splendor of the scenery, he found his attention being diverted to some curious detail in the foreground. He collected enough brilliantly colored flowers, he boasted, to make a florist go wild and enough insects to make an English entomologist madly jealous. In arranging for an expedition into the interior, he met with

the kind of bureaucratic insolence that generally follows a
revolution such as Brazil had recently experienced, but he
submitted to it with the thought that "the prospect of wild
forests tenanted by beautiful birds, monkeys and sloths,
and lakes by cavies and alligators, will make any naturalist
lick the dust even from the foot of a Brazilian."[50] He
calculated that it took more time to sort and preserve his
specimens than to acquire them, and he pretended to envy
the naturalist in England who, in the course of a day's
walk, might find some one thing worthy of his attention,
while the naturalist in the tropics suffered the inconvenience of not being able to walk a hundred yards without
encountering a rare and wondrous species.

For nine months the Beagle sailed along the eastern
coast of South America, while Darwin made inland expeditions geologizing, zoologizing, and botanizing. He witnessed a revolution in Montevideo, which he thought ludicrous until it interfered with his naturalizing activities. He
hunted Gaucho-fashion in Argentina, lassoing the animals
by the legs with two heavy balls suspended from a long
thong. And he came upon a cache of fossils which he
triumphantly identified as belonging to "the antediluvial
animal," the Megatherium.[51] He complained of the mo-
notony of the sand hillocks along the coastline and of the
endless green plain bounded by the interminable brackish
river that made up the interior. Even the tablelands of
Patagonia, that wretched stretch of land in South Argentina without houses, water, trees, or mountains, he found
preferable to the level, green, and fertile pampas, so rich
and benign and boring. He did not know precisely why
the somber south appealed to him more. Perhaps it was
because, being of no practical use, it was unknown and
hence boundless. And not only boundless but also ageless,
its past being as limitless as its future. By December
1832, when the Beagle was making its way along the coast
of the Tierra del Fuego, he was looking forward to the
even wilder country that awaited him.

In the Tierra del Fuego the three natives were de-
posited, the captain and crew set to their navigation in
earnest, and Darwin became acquainted with a wilderness

that, to anyone but a determined scientist, would have been appallingly barren. Even the forest, filled with decayed and fallen trees, gave off an aroma of death. On the nearby Falkland Islands zoology and botany very nearly deserted him. The few living productions which the islands boasted were soon gathered, and it was depressing to find again the symbols of death in the wrecks of four large ships lying in a single harbor. The islands redeemed themselves only by the presence of a rock abounding with shells of a particularly interesting geological era. Here, as on the mainland, it was the geologist who carried the day; to him death offered up a rich harvest.

During the following year and a half the Beagle retraced its course as far north as Montevideo and south again to the Tierra del Fuego, giving Darwin the opportunity for leisurely expeditions along the Rio Negro and into the interior of South America—riding, shooting, climbing mountains, incessantly exploring and collecting—the impressively mysterious title of "El Naturalisto" on his passport having gained him a hospitable reception among even the most ignorant and suspicious. Several adventures that he modestly refrained from mentioning have been hidden in the obscurity of FitzRoy's *Narrative*. One such was his rescue of the boats after a fall of ice threatened to sweep them away; at another time, in great fatigue and thirst, he persisted in a grueling search for water. Nor did he mention the christening of Darwin Mountain and Darwin Sound in the Tierra del Fuego, in testimony, FitzRoy stated, of his exertions beyond the call of duty. Other episodes Darwin could recount without immodesty. At Buenos Aires, for example, he again found himself frustrated by a revolution—a revolution, he indignantly observed, that was "nothing more or less than a downright rebellion"[52] and that was frivolous enough to succeed by the simple expedient of stopping the supply of meat to the insatiably carnivorous populace.

He speculated about the people as he did about the flora and fauna. Where the Fuegians were so barbarous as to be almost a different species from the human, the townsmen of Buenos Aires were so debased morally as to be a different species from the civilized Europeans. They were profligate sensualists, deriders of religion, entirely wanting in principle, honor, chivalry, or any generous feeling; the best thing that could happen to them, Darwin judged, would be to fall under the iron hand of a dictator. Once they showed themselves so deficient in sensibility as to mistake the Beagle for a smuggling ship. "A person," Darwin despaired, "who could possibly mistake Captain Fitz-Roy for a smuggler, would never perceive any difference between a Lord Chesterfield and his valet."[53] As for the Indians, he predicted that their linguistic facility would inevitably lead to both their civilization and their demoralization, "as these two steps seem to go hand in hand."[54] Taken aback by the violence of these sentiments, he recalled how, only a year earlier, he would have accused himself of illiberality for harboring such opinions.

In June 1834 the Beagle rounded the Cape and proceeded to explore the western coast of South America. While visiting a gold mine near Valparaiso, Darwin drank some native-made wine that made him violently ill and confined him to bed for a month. When he resumed work, it was the mountain ranges that principally occupied him: volcanoes belching smoke, mountain peaks whose ascent was so steep that the trees served as ladders, woods too thickly tangled to be penetrated except on hands and knees, solid masses of granite looking "as if they had been coeval with the very beginning of the world."[55] The granite delighted him as being the "classic ground" of the geologist —extensive in its range, beautifully compact in texture, one of the earliest stones to have been recognized by the ancients, and the deepest, most fundamental layer in the crust of the globe to which man can penetrate. It was here, at the point where knowledge had to give way to imagination, that geology most fascinated him.

No sooner had he arrived at this idea of a geological bedrock, basic both in fact and in conception, than he was dramatically confronted with its exact negation, the transient and ephemeral. On February 20, 1835, while lying peacefully in a wood, resting and speculating about such questions as how long it would take for the last remnants of a fallen tree to vanish, he felt the shock of what turned out to be one of the worst earthquakes in the living memory of Chile. In Valdivia, where he experienced it, the impact of the quake was great, but the damage was relatively mild. When he arrived at Concepción, however, he was shocked to find not a house standing, and in nearby Talcuhano a tremendous wave had washed away even the ruins of the buildings. The earth was traversed by great cracks, solid buttresses six to ten feet thick were shattered into fragments like so many biscuits, massive stones formerly lying in deep water were thrown high on the beach, and a schooner in the bay was carried well within the town. He promptly added the earthquake to the two other most memorable spectacles he had witnessed on the trip: the Fuegian savage and the tropical vegetation. The moral of the granite was replaced by the moral of the earthquake: "An earthquake like this at once destroys the oldest associations; the world, the very emblem of all that is solid, moves beneath our feet like a crust over a fluid; one second of time conveys to the mind a strange idea of insecurity, which hours of reflection would never create."[56]

During most of the voyage along the west coast of South America, geology was Darwin's main occupation, not only for its own interest but also for lack of another. Vegetable and animal life were often so scarce that by the time he reached Peru he confessed to being weary of repeating the epithets "barren and sterile."[57] By contrast, the Galapagos Islands, reached in the autumn of 1835, were a paradise of reptiles. The hordes of clumsy, black, hideous creatures swarming among the black rocks, leafless brushwood, and stunted trees resembled nothing so much, he fancied, as the cultivated parts of the infernal regions. The tortoises were so huge that they could be barely moved by one man, and so numerous that one ship's company caught well over five hundred of them in a short time. The lizards were two or three feet long, as black as the lava rocks among which they crawled, and fittingly called, Darwin thought, the "imps of darkness."[58] He wondered what "centre of creation"[59] could have produced this strange combination of tropical creatures and Arctic plants, giant reptiles and volcanic craters.

From the Galapagos the Beagle sailed west to circle the globe, a journey of one year, during which time Darwin alternated between gratitude for his great fortune in having more than realized his most extravagant dreams and impatience for his return to England, for the undulating green fields and shady lanes that often seemed more desirable than the most exotic tropical sight. Tahiti, "that fallen paradise,"[60] was as charmingly picturesque as any traveler's tale would have it. New Zealand, on the other hand, was correspondingly disagreeable, and Australia was only a little better—New Zealand because of its primitive bestiality, and Australia because of its prosperous barbarity. The Coral Islands were gratifying because they gave him the occasion for testing his theories about the formation of these curious structures. Some desultory exploring of the Cape Colony and a memorable meeting in Cape Town with Sir John Herschel punctuated the long stretches of months at sea, months that were no less agonizing for the homesickness from which he now suffered than for his perennial seasickness. On the return voyage, after four and a half years at sea, FitzRoy, experiencing one of his frequent crises of conscience, decided that the Beagle had to revisit the coast of South America to check on some discrepancies of longitude. Darwin's impatience could no longer contain itself, and he swore that he loathed the sea and all the ships that sailed upon it. Having landed once again in Argentina, however, he found the tropical scenery as irresistible as ever and the geology as absorbing.

For the last time he deplored the impossibility of expressing in words the physical appearance of a natural scene, let alone the sensations aroused by natural beauty. The scientist's display of specimens, pinned, tagged, and arranged, was as far removed from reality as his catalogue of identifying marks. "Who, from seeing choice plants in a hot house, can multiply some into the dimensions of forest trees, or crowd others into an entangled mass? Who, when examining in a cabinet the gay butterflies, or singular cicadas, will associate with these objects the ceaseless harsh music of the latter, or the lazy flight of the former—the sure accompaniments of the still glowing noon day of the tropics."61 In his last week in South America he tried to fix forever in his mind the tropical scene. But he knew, even as he exerted himself to this feat of memory, that although the shapes and colors of the orange tree, the coconut, the palm, and the mango would remain with him, the thousand details that made up the texture of the living scene would perish. He hoped only that there would remain, like a tale told in childhood, a picture full of beautiful though shadowy figures.

On the last stretch of the homeward journey he composed his retrospect of the trip, a calculus of the pains and pleasures of five years' wandering. The pains were obvious: the absence of family and friends, the sacrifice of physical well-being, the privation of luxuries such as music and comforts such as solitude and rest. In addition, there was the disproportionate time spent on the sea compared to the days in harbor—disproportionate, he was surprised to learn, even for the taste of professional sailors, most of whom did not really enjoy the sea itself. The boasted glories of the illimitable ocean he declared to be but "a tedious waste, a desert of water." No doubt there were delightful sights: "a moonlight night, with the clear heavens, the dark glittering sea, the white sails filled by the soft air of a gently blowing trade wind; a dead calm, the heaving surface polished like a mirror, and all quite still excepting the occasional flapping of the sails." And it was exciting to witness one's first squall at sea, with its rising arch and coming fury, or a gale with its mountainous waves. But for the most part these were more dramatically seen elsewhere—on the canvas of a Vandervelde, he suggested, or, better yet, from the shore, "where the waving trees, the wild flight of the birds, the dark shadows and bright lights, the rushing torrents, all proclaim the strife of the unloosed elements." The deck of a ship was both more uncomfortable and less revealing: "At sea, the albatross and petrel fly as if the storm was their proper sphere, the water rises and sinks as if performing its usual task, the ship alone and its inhabitants seem the object of wrath."[62]

If the pains of such a voyage were great, he found the pleasures to be no less so. The first of these was the contemplation of the varieties of nature: the sublimity of primeval forests, the awesomeness of sterile wastes, the sweeping perspective from lofty mountains, the sight of savage man in his native haunts. This vision of nature in all its primitive virility, he declared, arouses man to an awareness of his own spiritual potency; the sight of a plain with an unfathomable past and an infinite future provokes him to a frenzy of imagination, to an assault upon the traditional boundaries of knowledge. Yet, both the instincts and the sensations of the explorer are confused. For while he is urged forward by the civilized motives of comprehension and conquest, he is also driven by a primitive passion for the chase; he is the savage returning to his wild and native habits. A journey to distant countries thus sharpens and partly also allays that want and craving which, as Herschel had said, "a man experiences although every corporeal sense is fully satisfied."[63]

It was to the young naturalist that Darwin was addressing his remarks, for in this calculus of pains and pleasures only the naturalist found the balance to be on the side of the pleasures. As in music, he suggested, the listener who understands the notes more thoroughly enjoys the concert, so in travel the man with an informed interest in natural history profits most from the experience.

For Darwin himself, the harvest proved to be even richer than he knew, for he reaped the fruits not only of professional mastery but also of personal maturity. On October 2, 1836, five years and two days after he had left Shrewsbury, he returned home. The original two-year voyage, which his father had come so near disallowing, had been more than doubled, yet he had felt independent enough not to solicit his permission or even accede to his request that he leave the Beagle and return by some speedier means. To be sure, the paternal authority continued to be felt, particularly in the matter of money; hardly a letter home did not contain a reference to a draft of money he had been obliged to issue, with its precise amount, the reason for the expenditure, and an abject apology. (These communications were always addressed to his sisters, the fountainhead of authority being rarely approached directly.) Apart from this relic of dependence, however—less trivial than it may seem to a generation lacking the Victorian respect for either parents or money—it was clear to both the father and the son that Charles had, in fact, entered upon his majority when he boarded the Beagle, that the sailing date was indeed, as he had predicted, the birthday of a second life. He had left England a man somewhat younger than his twenty-two years, with too many enthusiasms to be convicted of idleness but not enough perseverance or knowledge to be acquitted of aimlessness. He returned a man somewhat older than his twenty-seven years, physically more debilitated and mentally more invigorated than might have been expected from an absence of only five years. "Why the shape of his head is quite altered," his father at once observed, apparently with satisfaction.[64]

That the voyage of the Beagle was a turning point in his life he had realized almost as soon as the ship dropped anchor at its first port of call and he dared to think that he might write a book on geology. And his superiors instantly knew it when they received his letters and the cases of specimens which arrived in such alarming quantities. Henslow was gratifyingly flattering about the value of the collection, particularly the fossil bones, and he was so enthusiastic about Darwin's letters that he took it upon himself to have excerpts from them printed.[65] Sedgwick called upon Dr. Darwin to say that Charles would take a place among the leading scientific men. To Dr. Butler, Charles' former headmaster, Sedgwick wrote with greater frankness but no less enthusiasm: "[Charles] is doing admirable work in South America, and has already sent home a collection above all price. It was the best thing in the world for him that he went out on the voyage of discovery. There was some risk of his turning out an idle man, but his character will now be fixed, and if God spares his life, he will have a great name among the naturalists of Europe."[66]

Professionally, the voyage proved to be more productive than Sedgwick or even Darwin suspected. During the trip he had industriously kept a variety of records: personal and scientific notes hastily jotted down in small pocket notebooks; large notebooks containing more extensive scientific memoranda composed during the quiet intervals at sea; and a diary, portions of which were sent home periodically for safekeeping. Originally the diary was intended simply for his own edification, as an exercise in self-discipline and a corrective for the memory. It occurred to him that it might be of wider interest when FitzRoy, himself engaged in writing the official account of the voyage, asked to read it and liked it well enough to want parts of it incorporated into his own book. Darwin readily agreed, surprised only that FitzRoy should want to preserve chitchat about countries so frequently and adequately described. His sisters loyally hastened to assure him that not only was his diary worth publishing, but it was worth publishing on its own, rather than being anonymously buried in a book that even the tolerant Charles suspected would be a bore. They submitted the matter to Henslow and Sedgwick, who declared themselves enthusiastically for its independent publication. It was then decided to issue it under the title "Journal and Remarks," as Volume III of the *Narrative of the Surveying Voyages of H.M.S. Adventure and Beagle,* Volume I being on the initial voyage of 1826–30 by Captain King, and Volume II the official report of the 1831–36 voyage by FitzRoy.

Upon his return Darwin set to work to convert the diary into a journal suitable for publication by deleting the personal passages and adding scientific material culled from his notebooks. He worked at it for over six months, complaining of the difficulty of writing and the tedium of proofreading. By the summer of 1837 he had completed a volume that was more radically different from its original than might be suspected from a mere comparison of lengths—224,000 words, or 600 pages, compared with the 189,000 words, or 425 pages, of the original manuscript.

The book was set in type, and page proofs were corrected, although FitzRoy's dilatoriness delayed publication until 1839.

Darwin's book may have been fettered, as his friends suspected, by its association with the two other volumes in the series, but enough people saw its superiority to warrant a second printing of his alone almost immediately and a third the following year. These reprints appeared with his name on the cover for the first time, under the title, *Journal of Researches into the Natural History and Geology of the Countries Visited during the Voyage of H.M.S. Beagle Round the World*.[67] In 1845 a second and revised edition was published by John Murray, this time with some material added but more subtracted, making it some 10,000 words shorter than the first published edition. (To distinguish these several versions, the original manuscript has been referred to as the "Diary," the first published edition as the "Journal," and the edition of 1845 as the "Revised Journal."[68]) This revised edition, of which all the later ones are reprintings, was the first to provide for royalty payments. When the publisher of the first edition had submitted a statement in 1842, it was to the effect that 1337 copies had been sold and that the author owed him twenty-one pounds and ten shillings, the cost of presentation copies.

Yet Darwin was far from displeased, and toward the end of his life he confessed that he was more delighted with the success of the *Journal* than with that of any of his other works. Eventually, the voyage of the Beagle was to provide material for several more ponderous tomes. But until the *Origin* itself, none displaced the *Journal* in his affections. And none, not even the *Origin*, is more revealing as an exhibit of the workings of his mind.

CHAPTER 4

GEOLOGY: METAPHYSICS AND METHOD

THAT the voyage of the Beagle was not simply a trial run for the *Origin* is suggested by the far greater emphasis upon geology in the documents of the voyage than in the *Origin*. To produce a book on geology was then the height of Darwin's ambition; and in fact he produced three such books: *The Structure and Distribution of Coral Reefs* (1842), *Geological Observations on the Volcanic Islands* (1844), and *Geological Observations on South America* (1846). Even when his subject was more general, geology predominated: the first edition of the *Journal* had so many long geological disquisitions that he was obliged to delete and abbreviate them for the more popular revised edition; and in the notebooks his geological interests were indulged even more freely.

In all branches of natural history, except perhaps entomology, Darwin was very much an amateur when he started on the voyage, but he was nowhere so crass an amateur as in geology. Several weeks of geologizing in North Wales would hardly seem much of an introduction to the geology of a volcanic island off Africa. Indeed, when he first started to explore St. Jago, his geological observations were chaotically entered into his little pocketbook with the comment: "What confusion for the geologist."[1] Several months later he was writing of this science, which he had once found to be "incredibly dull," as the most exciting of all subjects: "Geology carries the day: it is like the pleasure of gambling. Speculating, on first arriving, what the rocks may be, I often mentally cry out three to one tertiary against primitive, but the latter have hitherto won all the bets."[2]

What evidently impressed Darwin immediately, even before he could order his impressions sufficiently to understand the exact meaning of the geological formations of St. Jago, was the simple idea suggested by the mere presence of volcanic rocks—the idea of change. "Everything betrays marks of extreme violence," he recorded in his diary.[3] Change, or violence, as the basic theme of geology would seem to be so elementary as not to call for special attention. But it was seeing the idea embodied in the volcano, witnessing the dramatic evidence of the upheaval and subsidence of the earth, that made Darwin realize that geology was not the dull, academic enterprise he had once thought it, but an exciting, living science. It also transformed what theories of geology he had brought with him from Cambridge, particularly from Henslow and Sedgwick. St. Jago, he later declared, convinced him of "the wonderful superiority of Lyell's manner of treating geology, compared with that of any other author."[4] The sight of the volcanoes had plunged Darwin not only into the study of geology but also into one of the great scientific controversies of the nineteenth century.

In the brief span that has been assigned as its "heroic age"—from about 1790 to 1820—geology had gone through several battles, the first of which was waged under the banners of Neptune and Vulcan.[5] The Neptunists, more formally known as the Wernerians after their German leader (their chief English representative was Robert Jameson, Darwin's soporific professor of geology at Edinburgh), had held the principal fact in geological history to be the envelopment of the earth by the sea, with the rocks now found in the earth's crust the result of the mechanical and chemical action of water. They were opposed by the Vulcanists, led by James Hutton and his successor John Playfair, who emphasized in place of the action of water the effect of the internal heat of the globe in the formation of the earth's surface. By 1820 this conflict had been largely decided in favor of the Vulcanists. But the victory of Hutton was only partial, for his secondary thesis, later known as "uniformitarianism," was not so much disputed as ignored. This was the idea that no extraordinary powers were to be allowed in nature, no uncommon events alleged in order to explain the appearance of the earth; nature was assumed to be pursuing an orderly, regular, lawful course, uniform with that observed in our own time and experience.

In the 1820's uniformitarianism was an unfashionable doctrine, held by few reputable scientists. Most geologists, including the converts from Wernerianism (one of whom was Sedgwick), were "catastrophists." Catastrophism assumed, in opposition to uniformitarianism, that the history of the earth testified to violent catastrophes or cataclysms which interrupted its regular order and radically altered its surface, one of these being the flood commemorated in the Bible. If the catastrophists were in the ascendant, it was not only because what they believed seemed to be attested by common sense and the consensus of geological experience, but also because theirs seemed to be the simpler, more economical view. Uniformitarianism, in contrast, imposed upon geologists too heavy a burden of abstract theory and speculation. Fresh from the controversy of Neptunism versus Vulcanism, and wearied with the fanciful and elaborate cosmogonies that had engaged them in the preceding decades, most geologists determined to confine themselves to the empirical and descriptive aspects of their subject and to eschew the theoretical and general. Catastrophism was for them not so much a theory as a lack of theory. Thus William Smith, celebrated as the father of English geology, who professed to regard geology simply as an aid to engineering and manufacturing, and abstained from the Neptunist-Vulcanist controversy on the grounds that it was irrelevant to these practical concerns, was a catastrophist by instinct rather than by dogma.

Catastrophism was inflated into a general theory only when it came into conflict with the new and invigorated uniformitarianism of Charles Lyell. In 1830 Lyell published the first volume of his *Principles of Geology*, aptly subtitled "An attempt to explain the former changes of the earth's surface by reference to causes now in operation." The *Principles* instantly became a popular and professional success, and new editions of the first two volumes were called for even before the third and last volume was published in 1833. This work, which remained the classic of geology for half a century, was written by a young man (Lyell was thirty-two when it was published) who was by profession a lawyer; who studied geology during the holidays when the courts were not in session and when he was not pursuing his side hobbies of conchology, zoology, and botany; and whose eyesight was so bad that he had had to interrupt for a time both his professional and amateur activities. Partly because of the overwhelming confidence of the author, reflected in the grand scope and assured tone of the book, and partly because geology was in so feeble a state that it could not resist this determined assault, the *Principles* soon dominated geological discussion, and uniformitarianism became a serious and ultimately successful contender against the entrenched catastrophism.

From the perspective of a later century, the triumph of uniformitarianism has been interpreted as one episode in the progressive ascendancy of scientific enlightenment over obscurantist religion. Since the catastrophists allowed a wider latitude for the action of God in nature, they have become known as the party of religious orthodoxy; while the uniformitarians, preferring to rely on "the undeviating uniformity of secondary causes,"[6] have been represented as naturalists, theists, and even atheists. The protagonists themselves, however, did not see the issue so simply. The catastrophists were, in fact, less catastrophic and the uniformitarians less uniformist than has been thought. Sedgwick, one of the leading catastrophists, had repudiated the diluvian theory—the idea of the flood as a primary and universal geological agency—even before Lyell's book appeared, and Whewell, the philosopher of catastrophism, did so afterward. Geologists, including the most enthusiastic of catastrophists, had long been in the habit of amending and liberalizing the text of the Bible by a process of exegesis familiar for centuries. Referring to one geologist who had found as many as three deluges before Noah's and to another who had discovered an untold number of catastrophes not mentioned in the Bible, Lyell concluded: "We have driven them out of the Mosaic record fairly."[7] For his own part, Lyell did not deny the fact of the Biblical flood, but only the idea that it was a universal and decisive geological event. Those who indulged in elaborate calculations to determine the precise hour of the creation or the exact quantity of water required to fulfill the Biblical account of the deluge were neither catastrophists nor uniformitarians but theologians and pseudo-theologians. It was the young John Newman, not Sedgwick, who declared that if the bones of antediluvian men were not recovered, it must have been because they had inhabited the darkest regions of Africa or because the deluge had washed them away into the depths of the ocean.

The catastrophists, no less than the uniformitarians, were jealous of their reputations as scientists. This was true even of William Buckland, the most theological of the catastrophists, whose appointment as Dean of Westminster in 1845 came as an appropriate sequel to his twenty-five years as Professor of Geology at Oxford. After listening to one of his geological lectures, one German professor commented on the Englishman's "peculiar love of regarding nature from a theological point of view";[8] he had been provoked by Buckland's discourse on the wisdom of Providence in bringing together coal and iron deposits in the neighborhood of Birmingham for the express purpose of making Britain the richest nation of the earth. The same comment might also have been suggested by a superficial acquaintance with Buckland's works: his inaugural lecture, "Vindiciae Geologicae, or the Connexion of Geology with Religion Explained," and his more famous work, published in 1823, *Reliquiae Diluvianae* or "Observations on the Organic Remains Contained in Caves . . . Attesting the Action of an Universal Deluge." A closer examination of these two, however, shows them to be primarily scientific in their intention and theological only by derivation. Buckland did not try to deduce the facts of geology or paleontology from the Bible but sought to use those facts, arrived at by the then conventional scientific methods, to confirm the Bible. Even the *Edinburgh Review,* a vigorous defender of Hutton's uniformitarianism long before Lyell brought it back into fashion, greeted Buckland's book with enthusiasm; in spite of its inauspicious title, the *Review* was happy to report that the subject had been treated scientifically and without religious bias.

If Buckland was not quite the obscurantist he is generally made out to be, the other catastrophists were much less so. Daubény, Professor of Chemistry at Oxford, who was sufficiently catastrophist to use the classifications of ante- and post-diluvian, was nevertheless so determined to maintain his independence from religion that he refused even to speculate as to whether the deluge attested to by geology was the same as that of the Bible. Sedgwick was less of a purist, conceding that since truth was ultimately indivisible, the conclusions of science might be compared and found to be at one with those of revelation. But he too insisted that science had to pursue its investigations and arrive at truths independent of the Bible. Scientific truth, he said, could not be deduced from the "records of the moral doctrines of mankind." And in his presidential address to the Geological Society in 1830 he attacked "the unnatural union of things so utterly incongruous" as science and revelation.[9] Similarly, Whewell, in an entirely amicable review of Lyell's book, criticized its uniformitarian thesis as being speculative and untrue, at the same time praising the dignified way in which the scientist treated his subject. It was time, he urged, that men ceased to look with suspicion upon such speculations, and that "the condition and history of the earth, so far as they are independent of the condition and history of man, are left where they ought to be, in the hands of the natural philosopher."[10]

Since Lyell was as little inclined as Whewell himself to say anything heretical about the history of man, this advice was directed not at him but rather at those religious zealots who failed to distinguish between uniformitarians and catastrophists, flaying both indiscriminately. To Sedgwick, Whewell, and even Buckland, these, not Lyell, were the main enemy. The year before Buckland became Dean of Westminster, he had to listen to his orthodoxy being impugned by the Dean of York. And Sedgwick, having been accused of subverting Scripture, was driven to the same refuge Lyell so often found himself in: "Geology introduces some tender topics which require delicate handling. I must speak truth, but by all means avoid offence if I can."[11] As for Lyell, it was not only to avoid offense but also out of genuine conviction that he denied harboring any irreligious sentiments. To argue for the "undeviating uniformity of secondary causes," he insisted, was to affirm rather than deny the supernatural as a primary cause. Later it was to appear how inconstant was this distinction between secondary and primary causes. Inconstant and yet seemingly indispensable, for however far back the primary cause might be pushed, it always hovered in the background for the consolation of theologians and the reconciliation of science and religion.

If religion did not speak unambiguously for one party, neither did science for the other. Lyell, of course, thought it did, representing his doctrine as part of the inexorable march of scientific progress. Where catastrophism, he said, induced indolence and despair, throwing geologists back upon the deadening assumption of inexplicable acts of violence, uniformitarianism for the first time encouraged them to the laborious but fruitful task of examining those minute changes that were visibly in operation. In fact, however, neither the virtue nor the novelty of uniformitarianism was as certain as Lyell believed and as later historians and biographers have almost unanimously assumed. Contemporary critics did not share this illusion, charging that uniformitarianism, so far from being the vanguard of scientific thought, was actually a throwback to the earlier, rejected scheme of Hutton. Lyell might sometimes be pressed to admit this, but he never conceded their main point: that uniformitarianism was retrogressive in bringing back those sweeping and speculative generalizations which had so befogged geology in its early days. What was needed, they felt, were new and reliable facts rather than new and unproved theories. And they were aware that Lyell himself was a theorizer rather than a discoverer; as one of his young disciples put it: "We collect the data, and Lyell teaches us to comprehend the meaning of them."[12] (This was literally true, for Lyell's eyesight was often so weak as to prevent him from even examining the data; by 1850 he was quite incapable of making a geological map, and his near-blindness was often a serious physical danger.)

Even after the publication of the *Principles*, the catastrophists continued to do more for the prosaic advance of geology than the uniformitarians. Just as the Neptunists, however inferior to the Vulcanists in theory, had been their superiors in such practical fields as mineralogy, petrography, and even paleontology, with the result that their defeat had a crippling effect on some fields of geology,[13] so the catastrophists, whatever might be said of their theory, did the more fruitful practical work. It could not have been sheer perversity, or even accident, that caused so many of the most productive geologists of the time to be numbered among the catastrophists: Sedgwick, Buckland, Conybeare, Murchison, Agassiz, Cuvier, de Beaumont, and others. Today the uniformitarian doctrine seems simple and obvious. Then it was catastrophism that was simple and obvious. Common sense and the empirical evidence combined in favor of the idea that violence and catastrophe, rather than an undeviating uniformity, was the message engraved upon the rocks. The catastrophists of the 1830's may be forgiven for not conceding defeat, for not realizing that the judgment of posterity would be against them. In the vicissitudes of ideological warfare, victory is uncertain, defeat unacknowledged, and the verdict of history unpredictable.

Not only as a scientific theory but even as a philosophical doctrine, uniformitarianism failed to be entirely persuasive. It suffered from the same logical difficulty that was to plague Darwin and all later evolutionary thought—the confusion between naturalism and gradualism, the assumption that the method of naturalism necessarily involved a theory of gradualism. In his *Principles*, Lyell not only put forward the proper claim of naturalism, that nothing be admitted into science that implied a breach in the natural order; he went further in arguing that nothing be assumed to have taken place in the past that was not taking place in the present. It was from this second proposition that he derived the theory that all changes were the product of a slow, gradual, and cumulative process such as was observable—if he was right in his observations—in the present. At the time, the catastrophists obscurely sensed that Lyell's argument was not all of a piece, that even a naturalist might find it objectionable. Sedgwick charged the *Principles* with exceeding the bounds of the evidence:

> It assumes, that in the laboratory of nature, no elements have ever been brought together which we ourselves have not seen combined; that no forces have been developed by their combination, of which we have not witnessed their effects. And what is this but to limit the riches of the kingdom of nature by the poverty of our own knowledge; and to surrender ourselves to a mischievous, but not uncommon philosophical scepticism, which makes us deny the reality of what we have not seen, and doubt the truth of what we do not perfectly comprehend?[14]

Unfortunately, neither Sedgwick nor any of the other catastrophists was in a position to exploit the confusion in Lyell's thinking. They could not defend naturalism against the usurpation of gradualism because they themselves were not naturalists. It remained for a confirmed naturalist to try to separate naturalism from gradualism, to try to save naturalism if gradualism should prove unsound. Thus fifty years after the appearance of the *Principles*, Thomas Huxley could be found arguing that uniformitarianism did not necessarily imply the "extreme slowness of all geological changes." Lyell, he said, had not excluded the possibility of leaps or even catastrophes if the term was understood as a natural event; he had only excluded those changes, whether violent or slow, which implied a breach in the order of nature: "There is no antagonism whatever, and there never was, between the belief in the views which had their chief and unwearied advocate in Lyell and the belief in the occurrence of catastrophes."[15] Yet, Huxley notwithstanding and in spite of the occasional concessions he was able to quote from the *Principles,* there is no doubt that it was precisely Lyell's intention to exclude the likelihood of catastrophes by insisting upon the principle of the slowness of change.

It may be that Lyell himself was not unaware of the objections to this principle. Certainly by 1872, when the eleventh edition appeared, he was sufficiently aroused to alter the subtitle from its original, ambitious form—"An attempt to explain the former changes of the earth's surface by reference to causes now in operation"—to the more modest and unobjectionable form, "The modern changes of the earth and its inhabitants considered as illustrative of geology." The change came too late to undo what had been done; by then naturalism and gradualism had been irrevocably identified with each other. Perhaps that is why Lyell agreed to the change. A supreme tactician in these matters, he was acutely aware that however logically independent the two concepts might be, psychologically they were intimately related.

The strategy of the *Principles* had been to create a new conception of time that would reinforce the new conception of nature. The catastrophists, for whom violence was a "mysterious and extraordinary agency,"[16] had envisaged nature as being "parsimonious of time and prodigal of violence"[17]—prodigal of violence *because* it was parsimonious of time. Lyell had to reverse the formula: Nature was parsimonious of violence and prodigal of time—parsimonious of violence *because* it was prodigal of time. The extension of past time provided a predisposition in favor of naturalism, as its diminution had acted as a "bias" or "prejudice" against it.[18] So long as the age of the world was taken to be only several thousand years, men would persist in relying upon supernatural intervention to account for the great changes that had occurred. It was as if, Lyell wrote, a historian, studying a two-thousand-year-old nation, were under the illusion that it was only a hundred years old, so that all of its history had to be telescoped into one-twentieth of its real time. Incidents would follow each other in rapid and bewildering succession, armies and fleets would no sooner be assembled than they would be destroyed, cities no sooner built than fall into ruin, and war and peace would alternate in a violent and incredible manner. Such a history would have the air of a romance, resembling nothing in the present course of human affairs and inexplicable except on the assumption of a super-human agency. Conceived under a similar delusion of time, Lyell argued, the history of the natural world presented no less grotesque a picture of violence and supernatural intervention:

> One consequence of undervaluing greatly the quantity of past time is the apparent coincidence which it occasions of events necessarily disconnected, or which are so unusual, that it would be inconsistent with all calculation of chances to suppose them to happen at one and the same time. When the unlooked-for association of such rare phenomena is witnessed in the present course of nature, it scarcely ever fails to excite a suspicion of the preternatural in those minds which are not firmly convinced of the uniform agency of secondary causes;—as if the death of some individual in whose fate they are interested happens to be accompanied by the appearance of a luminous meteor, or a comet, or the shock of an earthquake. It would be only necessary to multiply such coincidences indefinitely, and the mind of every philosopher would be disturbed.[19]

"The mind of every philosopher would be disturbed"— this was Lyell's fear. And it was to make sure that the philosopher did not go astray that he tried to eliminate those leaps in nature which the weak-minded might mistake for gaps in the natural order. Huxley may have been the superior logician; there was, indeed, no reason why nature should not occasionally proceed by leaps. But Lyell was the superior ideologue. To remove temptation from

the path of the weak, nature had to appear as a continuum, with no loopholes, no avenues of escape.

This history of controversy and confusion was reflected
in the development of Darwin's geological theories. Before
the voyage, as a student of Henslow and Sedgwick, he
had naturally regarded himself as a catastrophist. In the
summer of 1831, while preparing for the expedition with
Sedgwick, he had written to Henslow: "As yet I have only
indulged in hypotheses, but they are such powerful ones
that I suppose, if they were put into action but for one day,
the world would come to an end."[20] The doctrines of
catastrophism would naturally appeal to a young man
confronted with the evidence of geological upheaval. The
testimony of past change is so conclusive, while the sense
of present changelessness, of the fixity of the earth, is so
instinctive, that the mind grasps at the idea of a tremendous catastrophe in some remote and turbulent past. The
imagination of man has never been known to balk at the
idea of a primeval cataclysm; the universality of the idea
of the deluge is witness to this. What does repel the imagination is the idea of a world changing without the
benefit of cataclysms. As Darwin confessed to Henslow, it
was no trouble to fancy the most gory cataclysm; the only
difficulty was how the world could possibly survive what
the imagination might conjure up for it.[21]

It is all the more curious, therefore, that Darwin should
have abandoned catastrophism and espoused Lyell's *Principles* at the very moment when, as he observed, "everything betrays marks of extreme violence."[22] At first
thought it might seem that the volcanoes, giving evidence
of violence in the present as in the past, confirmed Lyell's
principle of the uniformity of changes throughout all geological time. This would certainly have been a confirmation
of uniformitarianism as Huxley understood it: for all of
their violence, the volcanoes connoted no breach of the
laws of nature, no inexplicable or mysterious agencies. This
reasoning, however, while sufficient to convert Darwin to
Huxley's brand of uniformitarianism, would not have sufficed to convert him to Lyell's brand. In fact, it would not

have taken him far from catastrophism itself. For the catastrophists, as much as the uniformitarians, were aware that the earth was in a continual state of change and that volcanoes were a powerful agent in effecting that change. Nor did they regard the volcano as a mysterious and inexplicable agency. Like working geologists of all schools, it was their ambition to discover the laws of nature that governed such phenomena; and one of the laws which they deduced from volcanoes was that nature often moved by leaps and bounds. What Darwin saw in the volcanoes of St. Jago to convert him to Lyell's view was not only the fact of change but a mechanism of change that was gradual rather than sudden. Volcanic craters that might have appeared as the very epitome of violent change, and that were regarded as such by the catastrophists, were revealed to him—perhaps under the influence of the *Principles,* which he was then reading—as the cumulative result of small and repeated explosions and ejections. This had been the discovery of George Scrope and of Lyell after him, and it was this theory that impressed itself upon Darwin as the only plausible explanation for the peculiar formation of the craters. Later in the voyage this theory of gradualism was to be demonstrated again on the coasts of South America, where he saw evidence of the slow elevation and subsidence of the land; on the plains of Patagonia, where the gradual process of elevation had created the typical, steplike formations of shingle; and in the Andes, where the slow, cumulative action of earthquakes and volcanoes had raised up a vast range of mountains. The theme of gradual elevation and subsidence became the focus of all of Darwin's geological research during the voyage and the theme of all his geological writings afterward.

Yet the catastrophist vision was so primitive and compelling that he could not always shake it off, and under the stress of a dramatic experience it was likely to reassert itself. Such an experience was the earthquake that occurred while he was in Chile, three years after his conversion to uniformitarianism. His initial, instinctive reaction, recorded at the time in his diary, was more fitting for a catastrophist: "I believe this earthquake has done more in degrading or lessening the size of the island than a hundred years of ordinary wear and tear."[23] Several weeks later he wrote to Henslow of "the picture so plainly drawn of the great epochs of violence."[24] And a few months later he confided to his notebook: "Strong earthquakes useful to geologists— can believe any amount of violence has taken place."[25] Only after some reflection was the earthquake brought into line with the gradualist thesis of Lyell. Thus in the *Journal* it appears as "a paroxysmal movement, in a series of lesser and even insensible steps" by which the coast was elevated, the most violent volcanic explosion being "merely one in a series of lesser eruptions." As volcanic eruptions are caused by the ejection of lava, so mountains owe their origin to the injection of lava; and it is this continuous action that accounts for the "extremely gradual" elevation of mountain chains. Even at this later date, however, Darwin was not able entirely to forget his excitement in "finding that state of things produced in a moment of time, which one is accustomed to attribute to a succession of ages."[26]

The historian who is tempted to wonder why more geologists were not converted to what he has come to see as the true or progressive doctrine of uniformitarianism, and who suspects that only incompetence or obscurantist prejudice can account for their recalcitrance, may be enlightened by a reminder of Darwin's own vacillations. It is a tribute to his honesty that his most eloquent testimonial to uniformitarianism should also have been a confession of ambivalence:

It is not possible for the mind to comprehend, except by a slow process, any effect which is produced by a cause repeated so often, that the multiplier itself ceases to convey any more definite idea than the savage receives when he points to the hairs of his head. As often as I have seen beds of mud, sand, and shingle accumulated to the thickness of many thousand feet, I have felt inclined to exclaim that causes, such as the present rivers and the present beaches, could never have ground down such masses. But, on the other hand, when listening to the rattling noise

of those torrents, and calling to mind that whole races
of animals have passed away from the surface of the
globe, during the period throughout which, night and
day, these stones have gone rattling onwards in their
course, I have thought to myself, can any mountains,
any continent, withstand such waste.[27]

Darwin's conversion to uniformitarianism is revealing
not only for what it suggests about the theory itself but
also for what it tells us of his mind—how he thought and
reasoned. One of the most striking circumstances of his
conversion is that it took place not on the abstract philosophical level but on the lowest theoretical level—the interpretation of a particular geological phenomenon. While
most of the readers of the *Principles* were agitated by the
philosophical and religious issues that generally engage the
passions of men, Darwin's imagination was excited by a
particular theory about the elevation and subsidence of
land. That his conversion was not primarily to uniformitarianism as a philosophical system is apparent from the fact
that it had to wait until he arrived at St. Jago and confronted the actual geological evidence. He must have been
familiar with the general thesis for some time. The first
volume of the *Principles* had been out for more than a year
and a half and had been the subject of many critical essays,
scientific dissertations, and dinner-table conversations before Darwin finally bought a copy to take with him on the
Beagle. (He had bought it at the urging of Henslow, who
had also warned him against believing it.) During the long
wait in port and in the first weeks at sea he must have
read a good deal, if not all of it. Yet the moment of revelation came only with the sight of the volcanoes of St. Jago.

Even after his conversion, Darwin was not consciously
interested in the philosophy of uniformitarianism. He accepted it, to be sure, but only because it was related to the
particular geological theories that had impressed him. During the course of the entire voyage, in letters, diary,
and notebooks, Darwin never seems to have mentioned
the word "uniformitarianism"—just as he never alluded,
whether from innocence, indifference, or distaste, to its

religious implications. This unconcern for philosophy was to become even more pronounced later, when his theories were to assume such momentous significance and the disparity between his intentions and the effect of his theories was to be most conspicuous. At no time in the course of his many works did he ever make any explicit and sustained philosophical statements such as appeared in Lyell's first work.

Yet, if any philosophy were to be inferred from his work, it would come close to that of the *Principles*. In spite of the divergencies that were to separate Darwin and Lyell in later years, their contemporaries had no doubt of their basic philosophical identity. Thomas Huxley spoke for many of his generation when he said: "Lyell, for others as for myself, was the chief agent in smoothing the road for Darwin";[28] and "Darwin's greatest work is the unflinching application to biology of the leading idea and the method applied in the *Principles* to geology."[29] Darwin himself was aware of this. Toward the end of the voyage he announced to a friend that he had become a zealous disciple of Lyell, adding, prophetically, that the geology of South America sometimes tempted him to go even further than his master.[30] A decade later, when he was well on his way to establishing his own reputation, he dedicated the revised edition of the *Journal* to Lyell, explaining: "I always feel as if my books came half out of Lyell's brain, and that I never acknowledge this sufficiently . . . for I have always thought that the great merit of the *Principles* was that it altered the whole tone of one's mind, and therefore that, when seeing a thing never seen by Lyell, one yet saw it partially through his eyes."[31]

In spite of this acknowledged debt, the terms "master" and "disciple," so often applied to Lyell and Darwin, sit uneasily upon them even at this early date. Darwin's conversion revealed a boldness that spoke of independence more than discipleship. Confronted with tropical geology for the first time, and undaunted either by his own meager training or by the geological confusion which presented itself to the inexpert eye, he did not do what most amateurs would have done in the circumstances—that is, devote the first bewildering weeks to sorting out and analyzing the composition of the rocks. Instead, he immediately leapt to the highest stage of inquiry, the problem of the meaning of the geological phenomena. He then proceeded to identify himself with a theory which, until that time, he had been taught to distrust. And no sooner had he identified himself with it than he started to take liberties with it, suggesting, for example, an important amendment—which he promptly characterized as "fact" rather than theory— to the effect that the craters had been formed or enlarged by the subsidence of the floors after eruption.[32] And all this in the first weeks of his novitiate—his novitiate not only in uniformitarianism but in geology itself.

Most of Darwin's original discoveries were of the same nature as this conversion: specifically scientific rather than philosophical in intention (although their effects may have been otherwise), and bold and confident in character. His mind was extraordinarily fertile in devising scientific theories. "It was as though," his son observed, "he were charged with theorizing power ready to flow into any channel on the slightest disturbance, so that no fact, however small, could avoid releasing a stream of theory, and thus the fact became magnified into importance."[33] He was incapable of harboring facts unfettered by a theory. He domesticated facts as quickly as he acquired them, subduing their natural unruliness, submitting them to an orderly regimen, and arranging them in tidy habitats. Yet while he was bold and adept as a theorizer, he was timid and inept as a philosopher. He would have liked his theories to exist in a safe middle region, bounded by facts on the one side and by metaphysics on the other.

Unfortunately, theories are not the obedient creatures of their masters and tend to stray outside their bounds. When this happened to Darwin, it was generally Lyell who came to his rescue. It is interesting that during the voyage of the Beagle, when Darwin was least experienced and might have been most in need of an authority, he actually depended upon Lyell less than in his later years. The references to Lyell are far fewer in the documents dating from the voyage than in those of the years immediately following it, in spite of his avid reading of the *Principles* while on board ship. (The second and third volumes were sent to him, at his request, as they appeared.) The explanation seems to be that he needed Lyell more as he became more ambitious and as his theories tended to encroach more upon the neighboring provinces of facts and metaphysics. In the published version of the *Journal* and even more in the revised edition, Lyell was pressed into service, on the one hand to furnish empirical facts to support Darwin's theories—facts about fossils, the distribution of animals, or the movements of glaciers—and on the other to provide a general philosophical framework for those theories. For the theories themselves, Darwin had no need to call upon Lyell; he had all he could do to keep up with his own inventiveness.

One of his most important moments of creation on the Beagle came, significantly, early in the voyage and is revealing not only as a measure of Darwin's inventiveness and independence but also as a case study of his mental processes. This concerned the theory of coral islands, those singularly shaped lands rising out of the depths of the tropical ocean.

It had long been known that coral rocks were built up by the gradual accretion of the remains of polyps and other organisms. The question that was still in dispute was whether the coral polyps built up their structure from the bed of the ocean or from some other base. Toward the end of the second volume of the *Principles,* Lyell advanced the theory that coral atolls were built upon submerged volcanic craters. Only upon this assumption, he argued, could one account for the peculiar forms taken by these coral structures—the ring shape, the central lagoon, the sudden rising of an isolated mountain in a deep sea—as well as for the fact that they were to be found in those areas which had only the two kinds of rocks, coral limestone and volcanic. Darwin had not yet received this volume of the *Principles* when an alternative theory began to suggest itself to him: that it was not volcanic craters under the surface of the ocean that formed the base of the coral structure, but mountains that had once had their peaks above the surface and were gradually subsiding, providing a foundation, as they sank, for the growth of the polyps.

This theory began to suggest itself to Darwin not only before he had read Lyell's account but before he had so much as seen a coral island. Many years later, when writing his autobiography, he half-apologetically explained:

> No other work of mine was begun in so deductive a spirit as this, for the whole theory was thought out on the west coast of South America, before I had seen a true coral reef. I had therefore only to verify and extend my views by a careful examination of living reefs. But it should be observed that I had during the two previous years been incessantly attending to the effects on the shores of South America of the intermittent elevation of the land, together with denudation and the deposition of sediment. This necessarily led me to reflect much on the effects of subsidence, and it was easy to replace in imagination the continued deposition of sediment by the upward growth of corals. To do this was to form my theory of the formation of barrier-reefs and atolls.[34]

In fact, it is likely that the idea first came to him even before his examination of the shores of South America. It was at St. Jago that the white cliffs of lava rock first revealed to him the "new and important fact" of subsidence,[35] and it was while sitting beneath such a cliff, contemplating the living corals in the pools at his feet, that the thought came to him of writing a book on the geology of the various countries he was to visit. If one of the main themes of his proposed work was to be the idea of subsidence, it is not too fanciful to suppose that he already intended to apply it to the formation of coral islands.

The method by which Darwin worked out his theory is as revealing as the spirit in which he conceived it—and as deductive. The entire structure of reasoning was based on the principle of exclusion and was derived from two facts. The first was the known fact that corals could not live more than twenty or thirty feet below the surface of the water;

this meant that they could not build up their structures from the ocean bed but had to build on some agreeably warm base below the sea surface. The second was the fact of the peculiar shapes assumed by coral islands: the lagoon-island or atoll, a vast ring of coral rock; the barrier reef, a ring of coral encircling an island and separated from it by a deep channel of water; and the fringing reef lying near the shore of an island. Lyell's volcanic-crater theory was unsatisfactory because of the difficulty of supposing a chain of craters, all of them with their summits at precisely the optimum distance below the water level and many of them of very great dimensions. That there was no such chain of mountains beneath the sea Darwin deduced from the fact that there was no such chain on land.

The reasoning continues in this abstract fashion:

> If the data be thought insufficient, on which I have grounded my belief, respecting the depth at which the reef-building polypifers can exist, and it be assumed that they can flourish at a depth of even one hundred fathoms, yet the weight of the above argument is but little diminished, for it is almost equally improbable, that as many submarine mountains, as there are low islands in the several great and widely separated areas above specified, should all rise within six hundred feet of the surface of the sea and not one above it, as that they should be of the same height within the smaller limit of one or two hundred feet. So highly improbable is this supposition, that we are compelled to believe, that the bases of the many atolls did never at any one period all lie submerged within the depth of a few fathoms beneath the surface, but that they were brought into the requisite position or level, some at one period and some at another, through movements in the earth's crust. But this could not have been effected by elevation, for the belief that points so numerous and so widely separated were successively uplifted to a certain level, but that not one point was raised above that level, is quite as improbable as the former supposition, and indeed

differs little from it. . . . If, then, the foundations of
the many atolls were not uplifted into the requisite
position, they must of necessity have subsided into it;
and this at once solves every difficulty, for we may
safely infer, from the facts given in the last chapter,
that during a gradual subsidence the corals would be
favourably circumstanced for building up their solid
frame works and reaching the surface, as island after
island slowly disappeared. Thus areas of immense extent in the central and most profound parts of the
great oceans, might become interspersed with coral-
islets . . .[36]

Darwin admitted that there was no proof of the gradual
subsidence of land over a large area. But for this too he
had a reasonable explanation: such proof could not be
expected since it would obviously be difficult, except in
countries long since civilized, "to detect a movement, the
tendency of which is to conceal the part affected."[37] The
only "proof" he could point to was the fact that there
seemed to be no other way of accounting for the phenomenon. An additional virtue of his theory—to Darwin this was
almost presumptive evidence of its truth, although he did
not say so explicitly—was that it provided a single explanation for all three types of coral islands. And not only a
single explanation but, better yet, one which established
the three in a progressive relationship toward each other,
with one type evolving out of the other. As the land subsided, he reasoned, a fringing reef was formed; if it continued to subside, this would pass into the condition of an
encircling or barrier reef; and finally, with the complete
submergence of the land, the reef would be transformed
into an atoll. Thus might be explained not only the strange
and yet fixed shapes taken by the coral islands but the
even stranger circumstance that certain types should be
common in some areas and not in others.

Darwin's theory was carefully argued. Every eventuality
seemed to be accounted for, every alternative excluded.
It was so persuasive that Lyell himself was promptly converted—not without misgivings, however, for it meant that

he had to abandon his own theory to which he was publicly committed. After he had been informed of the new theory, Lyell wrote that he could think of nothing else for days. "It is all true," he warned Darwin, "but do not flatter yourself that you will be believed till you are growing bald like me, with hard work and vexation at the incredulity of the world."[38] Yet a paper by Darwin on this subject, although it by no means convinced the rival schools, was respectfully received at a meeting of the Geological Society in May 1837; and Lyell could only conclude that men were now prepared to accept ideas they would not have tolerated seven years earlier. (He also mentioned the more tolerant hearing given to Darwin's statement of the theory of the slow rise of the Andes than had been given to his own earlier presentation of the same theory.) By 1842, when Darwin's volume on *Coral Reefs* was published, the scientific public was even more receptive.

When the geologist Archibald Geikie read it years later, it had lost none of its charm:

> "How one watched the facts being marshalled into their places, nothing being ignored or passed lightly over; and how, step by step, one was led to the grand conclusion of wide oceanic subsidence. No more admirable example of scientific method was ever given to the world, and even if he had written nothing else, the treatise alone would have placed Darwin in the very front of investigators of nature."[39]

Yet it was Geikie himself who helped expose the fallaciousness of this "admirable example of scientific method" by demonstrating conclusively that some coral islands had undergone not subsidence but considerable elevation. "It has been to myself and to many other geologists a matter of keen regret," he confessed, "that this brilliant generalization of the great naturalist has been deprived of the wide application which for many years was attributed to it." But even as he was regretfully obliged to discredit the theory as such, he persisted in praising it as "a monument of his genius, which did good service by lifting geological speculation to a higher plane, and filling our minds with a more vivid conception of the gigantic scale on which the movements of the terrestrial crust may have been effected."[40] The idea that somehow Darwin's errors were more valuable—in some sense "truer"—than his opponents' truths was to play an important part in the later history of his theory; but there were few critics, apart from Geikie, frank enough or astute enough to see this.

The strength of Darwin's theory was also its weakness. Its novelty and its genius lay in replacing a fixed and static conception (this was as true of Lyell's volcanic-crater theory as of any other) by a dynamic, evolutionary one. The word "evolutionary" is not one that appeared in his own discussion or in the accounts of his contemporaries, or even in those of later, more evolution-minded generations. Nor is it now imported into this discussion as a portent of his future—although it was that, to be sure. It is used because it best describes the essential meaning and originality of his theory, which was the idea of a gradual and steady movement of the earth's surface, a movement to which all phenomena were subject. The genius—and the folly—of such a theory is that it can explain anything and everything. Any particular variety of coral island, indeed the very presence or absence of such islands, can be accounted for as being at some stage of its evolutionary development. Logically, the theory was irrefutable. At only one point could the armor of its logic be pierced; this was the empirical fact of subsidence. And it was at this point that the theory eventually gave way.

Late in life Darwin was obliged to make the admission that coral islands did not invariably result from subsidence, that they could result from the elevation of submarine mountains by means of an accumulation of minute oceanic organisms. Yet he did not renounce his theory, insisting that although elevation might explain some isolated examples of coral islands, it could not account for the majority of them. In spite of the fact that his theory collapsed with the admission of this exception, he would not abandon it because he did not believe that reason could so betray him. "It will be a strange fact," he protested, "if there has not been subsidence of the beds of the great oceans, and if this has not affected the forms of the coral reefs."[41] Two years later, as the evidence against him mounted, he could only reiterate his belief that the facts could not be other than he had made them out to be: "It still seems to me a marvellous thing that there should not have been much, and long continued, subsidence in the beds of the great oceans."[42]

The facility—and precipitancy—of Darwin's theories may also be exemplified in an episode occurring soon after his return from the voyage. In the summer of 1838 he made an eight-day expedition to the Scottish Highlands to examine the curious "parallel roads" of Glen Roy, which were the subject of many Highland legends and of much scientific speculation. His findings were presented in a paper before the Royal Society in 1839. His main conclusion was that the parallel roads were formed by the action of the sea and were raised to their present level by a gradual elevation of the district. He arrived at this theory by the same deductive method, the method of exclusion, that he had used with such telling effect in the case of the coral reefs. Since the waters that had formed the terraces could not possibly have been dammed back by barriers of rock, he saw no alternative, in spite of some unresolved difficulties, but to regard these terraces as the work of the sea. As he explained to Lyell: "I have fully convinced myself (after some doubting at first) that the shelves are sea-beaches, although I could not find the trace of a shell; and I think I can explain away most, if not all, the difficulties."[43]

A year after he had decided that his was the only possible theory, two other geologists, Agassiz and Buckland, discovered a more plausible explanation for Glen Roy in the idea of transient barriers of glacier ice serving to dam back the waters shaping the terraces. Although Darwin had earlier recognized the importance of glacial action in the geology of South America, it took him twenty more years before he announced his conversion to the theory as it applied to Glen Roy. Not until 1861 did he frankly confess his paper on Glen Roy to be "one long gigantic blunder from beginning to end."[44] He accounted for this blunder in his autobiography: "Because no other explanation was possible under our then state of knowledge, I argued in favour of sea-action; and my error has been a good lesson to me never to trust in science to the principle of exclusion."[45] The lesson was a long time in the learning, however; so long that his most important work, the *Origin of Species,* was completed before, by his own admission, it was brought home to him.

Darwin readily admitted that in his geological work he relied heavily upon deduction, generalization, and theorizing. Late in life, in the retrospect of his career, when he was looking for the source of his discoveries, it was natural that he should credit much of his success—for it was his success he was trying to account for and not his failures—to this facility for generalizing. (Also, he hastened to add, not entirely accurately, to his willingness to give up any hypothesis, "however much believed," as soon as the facts were shown to oppose it.[46]) Earlier in his career, however, when success was uncertain and the failures disconcertingly fresh in his mind, he was far less sanguine. Almost his last words in his diary and *Journal* were a reproach to himself and a caution to others:

> As a number of isolated facts soon become uninteresting, the habit of comparison leads to generalization; on the other hand, as the traveller stays but a short space of time in each place, his description must generally consist of mere sketches, instead of detailed observation. Hence arises, as I have found to my cost, a constant tendency to fill up the wide gaps of knowledge by inaccurate and superficial hypotheses.[47]

CHAPTER 5

INTIMATIONS OF THE *ORIGIN*

FROM the perspective of history, the Beagle appears as the laboratory in which were first distilled those ingredients later to be combined with such explosive effect in the *Origin*. This interpretation comes not only with all the compulsion of hindsight and with the universal assent of Darwin's biographers, but with the authority of Darwin himself. The opening sentences of the *Origin* read:

> When on board HMS Beagle as naturalist, I was much struck with certain facts in the distribution of the organic beings inhabiting South America, and in the geological relations of the present to the past inhabitants of that continent. These facts, as will be seen in the latter chapters of this volume, seemed to throw some light on the origin of species—that mystery of mysteries, as it has been called by one of our greatest philosophers.[1]

The autobiography is more specific:

> During the voyage of the Beagle I had been deeply impressed by discovering in the Pampean formation great fossil animals covered with armour like that on the existing armadillos; secondly, by the manner in which closely allied animals replace one another in proceeding southwards over the Continent; and

thirdly, by the South American character of most of the productions of the Galapagos archipelago, and more especially by the manner in which they differ slightly on each island of the group, none of the islands appearing to be very ancient in a geological sense.[2]

These statements, delivered with such authority and assurance, would seem to come very close not only to the mystery of mysteries, the origin of species, but to that lesser mystery, the origin of the *Origin*. Fortunately, it is possible to test Darwin's memory and to reconstruct—so far as it is ever given a biographer to "reconstruct"—the episodes that are said to have played so crucial a role in the history of his mind.

Of the experiences referred to in the autobiography, that of the fossil discoveries takes priority in both logic and time. The first reference to fossils occurs in the diary under the date of September 22, 1832. Its importance only barely emerges through the laconic prose: "Had a very pleasant cruise about the Bay with the Captain and Sullivan. We stayed some time on Punta Alta about ten miles from the ship; here I found some rocks. These are the first I have seen, and are very interesting from containing numerous shells and the bones of large animals"[3]—after which Darwin returned to his familiar complaints about the weather and scenery. He was, however, diligent enough to go back to his cache the following day, where after three hours of digging he finally managed to extract the head of a large animal which he tentatively identified as related to the rhinoceros. He returned twice more during the fortnight the Beagle was anchored at Bahia Blanca and was rewarded the last time with a jawbone containing a tooth of "the great antediluvial animal the Megatherium."[4] This was a particularly interesting find, he noted, since the only specimen in Europe was in the royal collection at Madrid, where it was as good as hidden in its primeval rock. A year later, on the second trip up the coast, the sight of the bones of this animal had become too familiar to evoke any excitement. He was disappointed when the promise of some "giant's bones" turned out to be those of the usual Megatherium and complained that he had to work too hard to extract fragments of them.[5] The monster only redeemed itself when, the following day, Darwin was able to purchase a part of a head in perfect condition for a few shillings.

Darwin responded to these finds as any naturalist—more particularly, as any geologist—would have done. Repeatedly, he cautioned his sisters and Henslow against detaching the identifying numbers from his specimens, for only by their numbers could their original locations be determined. "All the interest," he wrote his sister, "which I individually feel about these fossils is their connection with the geology of the Pampas, and this entirely rests on the safety of the numbers."[6] Again, informing her of a new find, he explained that it interested him "as connecting the geology of the different parts of the Pampas."[7] And when he commented, in passing, on the differences or similarities between fossil shells and existing forms, it was almost always to elucidate the geological age and structure of the area.[8] Even his gratification at finding fossils in the oldest rocks known to contain organic remains—"this has long been a great desideratum in geology, viz., the comparison of animals of equally remote epochs at different stations in the globe"[9]—was that of the geologist, who was interested in comparing not the animals themselves in different stations of the globe but rather the geological formations of different parts of the globe as they were revealed by the presence of these fossils. In his letters and diary there is no attempt to relate his fossil discoveries to his descriptions of the fauna, no attempt to relate "the present to the past inhabitants," the living to the dead. The geological enterprise was kept quite distinct from the zoological.

Only much later, after he had returned from the voyage and the fossils had been identified by Owen, did Darwin seriously begin to attend to them as zoological, rather than purely geological, phenomena. Thus the accounts of his discoveries in the *Journal* were far more weighty and portentous than the original diary reports of the same findings.

The "giant's bones" were now identified as those of the "Toxodon," and a long citation from a paper by Owen placed them in the order of rodents. Other finds were related to the existing armadillos, and still others, again on the authority of Owen, to the guanaco or llama. He also referred to discoveries made some years earlier in Australia of the remains of extinct species of kangaroo and other marsupial animals. All this evidence he took to be confirmation of the "law of the succession of types," the law that "existing animals have a close relation in form with extinct species."[10]

That Darwin's appreciation of the significance of his findings was belated is itself a trivial conclusion, although it is worth making if only to dispose of the impetuous and overdramatic accounts of those biographers who have him exclaiming "Eureka" on the plains of South America or on board the Beagle. Darwin himself confessed, and Huxley confirmed, that he could not have appreciated the significance of his findings while on the voyage because he lacked the necessary training in dissection and drawing and the knowledge of comparative anatomy; and that all his zeal and industry only resulted, as Huxley said, in "a vast accumulation of useless manuscript."[11] His appreciation of their significance had to await his return and the identification of the specimens by experts. More important, however, than the gap in time between his findings and his appreciation of them is the discrepancy in quality between those experiences and the theories later associated with them.

The history of paleontology, as well as Darwin's own admissions, attests to the fact that neither his fossil discoveries nor the laws he claimed to have deduced from them were as original as has been commonly supposed. Certain words are apt to have a titillating effect upon the modern ear. Such is the word "fossil." We respond to it as something daring, iconoclastic, brazenly modern, as if every appearance of the word holds promise of being the missing link between man and the lower animals. So it is that biographers of Darwin have pounced upon every mention of fossils in his early writings as if the mere word was an augury of the momentous theory to come. In fact, however, both fossil hunting and fossil theorizing had a long and respectable history when he came upon the scene. He himself was moved to complain of the invasion of fossil hunters that had preceded him in South America, carrying off the choice specimens and spoiling the natives by making them too knowing about the location and value of the caches. He wrote ruefully of the "confounded Frenchmen" who turned up everywhere, one of whom even appeared in person to present him with a copy of a monograph he had published in one of the scientific journals.[12] And he was amused, when he was drawn into a discussion with some South Americans on the meaning of fossils, "to find the same subject discussed here, as formerly amongst the learned of Europe, concerning the origin of these shells, whether they really were shells or were thus 'born by Nature.'" Pressed for his opinion, he had replied, ambiguously, "God made them"[13]—as the learned of Europe, with the same ambiguous intent, had once replied.

It was, indeed, a good many centuries since the learned of Europe had resorted to such evasions. As far back as Herodotus and Theophrastus, they had known about fossils and had properly understood them to be the remains of animals and plants. Occasionally, during the Middle Ages and the Renaissance, schoolmen might be found arguing that fossils were freaks or sports "born by nature," the abortive attempts of nature to produce life, or even stones fashioned by nature to imitate living things, possibly for the purpose of beautifying the world. But this was neither the prevalent nor the respectable opinion, and by the seventeenth century treatises on fossils were in substantial agreement with the modern view. It testifies not so much to the backward state of opinion as to the erratic course of intellectual progress that Voltaire, the disciple of Newton and the paragon of enlightenment, should have mocked his contemporary Buffon for supposing that fossils could actually have originated on the top of mountains; he thought it more plausible that the shells had been left there by pilgrims returning from the East. Except by such unbelievers as Voltaire, however, it was not only generally understood by the close of the eighteenth century what fossils were, but it was also known what scientific use could be made of them. The English geologist, William Smith, established paleontology as an instrument of geology by perfecting the procedure of identifying and dating the age of strata according to the fossils contained in them. At the same time, his French contemporary, Cuvier, gave a biological cast to paleontology when he treated the fossils as individual forms of life, zoological specimens that happened to be extinct. It remained for Lamarck to carry this biological or zoological interpretation still further by deliberately associating the investigation of existing species with the remains of extinct forms.

The first decades of the nineteenth century witnessed the rapid advance of paleontology along these zoological lines. Shortly before Darwin sailed on the Beagle, Buckland delivered his presidential address to the newly founded British Association on the subject of the Megatherium. Discoveries such as that of the Australian marsupial fossils, later cited by Darwin in his *Journal,* were reported, with a proper appreciation of their significance. And scientists began to systematize and synthesize their findings. While Darwin was occupied in South America, the American naturalist, Louis Agassiz (later to emerge as one of his most determined opponents), was devising a scheme by which fossils were divided into such categories as "progressive," "synthetic," "prophetic," and "embryonic"—terms implying a keen appreciation of that relation between the extinct and the living that Darwin later found to be crucial in the development of his own views. Nor was Lyell slow in taking up the new study; in the *Principles* he paid tribute to paleontology as an adjunct of zoology:

> The adoption of the same generic, and, in some cases, even the same specific names for the exuviae of fossil animals, and their living analogues, was an important step towards familiarizing the mind with the idea of the identity and unity of the system in distant eras. It was an acknowledgement, as it were,

that a considerable part of the ancient memorials of nature were written in a living language.[14]

Darwin himself never claimed that such principles as the succession of types or the relation between the living and the dead were original to him. Of one important find, perhaps the most important one, he said frankly that he himself "had no idea at the time to what kind of animal these remains belonged,"[15] and that it was entirely owing to the zoologist, Richard Owen, that their relation to the existing llama was disclosed; the result of this disclosure he took to be not the discovery but rather the "confirmation" of the law of the relation of the living to the dead.[16] And twenty years later, when he resurrected these phrases about the living and the dead and the succession of types, he made it clear that they had been inspired not only by his own findings in South America but also by his reading of the professional literature.[17] These qualifications have been largely ignored by his biographers, who have assumed that it was his unique fossil discoveries, or his unique deductions from those fossils, that gave rise to his theory. Actually, neither was unique, naturalists of conflicting schools having appreciated both the importance of fossils and their relation to living forms—a thought suggesting that the genesis of his theory was more complicated than either Darwin or his biographers have intimated.

If the discovery of fossils and their relation to living forms was not quite the revolutionary experience it has been thought, neither was the discovery of the "distribution of organic beings inhabiting South America," designated as the other main source of inspiration of the *Origin*. His autobiography specified more exactly what it was about geographical distribution that had so "deeply impressed" him on the Beagle: "the manner in which closely allied animals replace one another in proceeding southwards over the Continent," and "the South American character of most of the productions of the Galapagos archipelago, and more especially . . . the manner in which they differ slightly on each island of the group."[18]

Of the first point—the manner in which animals replace

one another on the continent—there is, in fact, little explicit mention in the diary or letters. The closest he came to it was an item in one of his notebooks: "It will be interesting to observe differences of species and proportionate numbers: what also appear characters of different habitations."[19] This, however, refers not to the progress southward on the continent but to a more obviously dramatic situation: animal life on the mainland compared with that on the neighboring islands, in this case the Falkland Islands. There are, in the notebooks, catalogues of species inhabiting different regions: "Falkland larks here. Double barred kelp birds here . . ."[20] But such observations, after all, were the normal routine of the naturalist, and there is nothing to suggest that Darwin's were distinguished by any special theory or principle. Certainly there is no hint that species "replace" one another, with the evolutionary implications conveyed by that word.

Nor did the Galapagos immediately inspire Darwin to any great flights of speculation. He had looked forward to visiting these volcanic islands mainly because of their interesting geological structure, and most of his time there seems to have been devoted to geology. Even from the vantage point of New Zealand, two months and three thousand miles away, when he thought of the Galapagos it was as "that land of craters."[21] To be sure, he also attended to the flora and fauna of the islands. But in writing of the first to Henslow, who had a special interest in the subject, he satisfied himself with the terse remark: "I shall be very curious to know whether the Flora belongs to America or is peculiar." And of the latter: "I paid also much attention to the birds which I suspect are very curious." Having delivered himself of these brevities, he returned to the more congenial topic of geology.[22] In his diary he made more of the lizards swarming about those "infernal regions" and the tortoises whom he fancied as the inhabitants of another planet or as "old-fashioned antediluvian animals."[23] Having industriously collected all the animals and plants of the islands, he observed: "It will be very interesting to find from future comparison to what district or 'centre of creation' the organised beings of this archipelago must be attached."[24] It is difficult to see how these speculations about "antediluvian animals" and "centres of creation" qualify as intimations of the *Origin.*[25]

Only after his return from the voyage, when he had discussed his findings with experts and had read the available literature, did Darwin begin to reflect more seriously on the problem of geographical distribution. In the *Journal* he made much of the curious similarity in structure and habits among allied species existing in opposite parts of the continent and in widely varying conditions of soil, climate, and longitude; and the equally curious circumstance that species separated only by a chain of mountains or a channel of water, but otherwise sharing the same physical conditions of life, should be so noticeably dissimilar. The Galapagos confronted him with the even odder situation that the species should be similar to those on the mainland and at the same time different in small yet distinct features. At times the archipelago seemed to be a "little world within itself," while at other times he thought that anyone familiar with the continent would "feel convinced that he was, as far as the organic world was concerned, on American ground."[26]

Yet this discussion, while sometimes tantalizingly reminiscent of the *Origin* in suggesting problems that were to feature so prominently in that later work, was also provokingly confused in its conclusions. In part, the confusion came from the inadequacy of his materials. Not knowing what to look for on the voyage, he had, as he later realized, let some of the most promising opportunities escape him. In the *Journal* he confessed to some of these lost opportunities. The English resident in charge of the islands, a Mr. Lawson, had boasted to him of being able to identify, by the size and shape of any tortoise, the particular island of its origin. And other inhabitants confirmed the fact that many species of plants and animals were to be found exclusively on one island. Yet Darwin took this information so little to heart that he partially mingled the collections from two islands (whether this was done before his talks with the islanders is not clear), and even later, although taking care to identify the source of his specimens, he failed to consider each island as a separate entity and so did not try to gather a complete series of specimens from any one island. It had never occurred to him, he confessed, "that the productions of islands only a few miles apart, and placed under the same physical conditions, would be dissimilar."[27] This failure still rankled many years later, when he told Henslow: "I need not say that I collected blindly and did not attempt to make complete series, but just took everything in flower blindly." But not everything, as it appeared, for he also admitted: "I was able to ascend into the high and damp region only in James and Charles Islands."[28] Thus, when he came to evaluate his experiences he was hampered by a lack of exact information on the distribution of species on the islands.

Even where the facts were adequate Darwin's explanations were often inconclusive and confusing. It was apparent, for example, that the tortoises, while differing from each other distinctly on each of the islands, were of a species familiar in many parts of the world. Darwin accounted for this latter fact by supposing that the species had originated in the Galapagos and had been distributed to the far reaches of the world by buccaneers who had carried them off alive and who then, presumably, instead of eating them, as was the custom on the islands, had let them loose. This explanation suggests more difficulties than it solves. His reason for assuming, in the first place, that the species had originated on the islands was simply that "nearly all the other land inhabitants seem to have had their birthplace here," so that if these did not, it would be a "remarkable anomaly."[29] Considering the fact that what he was called upon to explain was a whole series of anomalies, there seems no reason to balk at this one. Moreover, his explanation for their world-wide range runs counter to the fact of their singular appearance on each of the islands, for if so impressive a feat of world-wide distribution could have been accomplished by the occasional visits of buccaneers, surely the daily visits of the natives, continually transporting the tortoises from island to island, should have succeeded in thoroughly mingling the varieties among the several islands.

This difficulty in accounting at the same time for the similarity and the dissimilarity of species also mars his account of the other great oddity in the geographical distribution on the Galapagos: the fact that the mocking birds were of different forms on each of two islands, with a third form common to two other islands. Earlier he had observed that birds of a fairly long flying range were not affected by the barrier of a mountain range, so that while species of land animals differed on the two sides of the Andes, species of birds did not. Yet he did not seem to find it surprising that on the Galapagos one of his main examples of a species inhabiting only one island should happen to be a bird.

Since the problem of the geographical distribution of species, like that of fossils, later assumed such an important role in the *Origin*, it is tempting to attach great significance to its first appearance in Darwin's writings. It is salutary, therefore, to keep in mind not only the inadequacy of his explanations in the *Journal*, but also the fact that the problem itself had not originated with him. As in the case of fossils, he was working with the accepted tools of the trade. One of the chapters in Lyell's *Principles*, entitled "Laws which Regulate the Geographical Distribution of Species," opened with the statement: "Next to determining the question whether species have a real existence, the consideration of the laws which regulate their geographical distribution is a subject of primary importance to the geologist." Lyell himself did not pretend to any originality on this score: "That different regions of the globe are inhabited by entirely distinct animals and plants is a fact which has been familiar to all naturalists since Buffon first pointed out the want of specific identity between the land quadrupeds of America and those of the Old World." And he proceeded to cite Buffon, Humboldt, de Candolle, and others in proof of the fact that although the same physical conditions of life might give rise in different localities to different species, these species were often similar or analogous in form.[30] It is little wonder that Darwin, who had just adopted the *Principles* as his testament, should have taken advantage of the ideal conditions of the Galapagos to investigate this "subject of primary importance." Indeed, if there is any occasion for surprise, it is that he was not sufficiently impressed with the importance of the subject to make more of the opportunities presented by the Galapagos, and that his own conclusions should show so little advance over those of his predecessors.

Nor were other subjects, dealt with in the *Journal* and now regarded as a portent of the *Origin*, unique to him. There was the matter, for example, of transportation. Again and again in the *Journal* Darwin remarked upon the ways by which animals and plants were transported and distributed from one place to the next. He had found microscopic animals floating fifty miles off the coast of Chile, crustacea in the seas around the Tierra del Fuego, and vegetable matter off the coast of Brazil. He had seen seeds carried for vast distances by the heavy gales and sea currents of the Tierra del Fuego, tree trunks drifting to islands far to the west, and pebbles traveling more than four hundred miles from the Andes along the Rio Negro. And far out at sea he had caught beetles of a variety that rarely take wing, while a grasshopper had flown on board the Beagle when the nearest point of land not directly opposed to the trade winds was 370 miles away. That some of these curious examples of the means by which animals and plants might be transported were later to be cited in the *Origin* does not alter the fact that their appearance in the *Journal* signified no revolutionary discovery or even novel theory. This subject too had been amply discussed in the *Principles*, and Darwin himself, in the *Journal*, fortified his own examples by a host of still more remarkable cases culled from the writings of other travelers and naturalists.

Even the several allusions to adaptation were neither original nor specially suggestive of the *Origin*: the rat that one zoologist judged to be quite distinct from the English kind but that Darwin suspected was "only the same altered by the peculiar conditions of its new country";[31] the peculiarities of a lizard that another zoologist attributed to "a variation in habit, accompanying change in structure";[32] and the tameness of the birds remarked upon by so many travelers and commonly ascribed to the fact that they were unaccustomed to men and so had not developed the instinct of fear. (This latter case presented its own difficulties, however, because even after the birds had been in prolonged contact with men, they did not acquire the habit of caution.) These, as the *Journal* made clear, were the standard fare of discussion among naturalists, and Darwin could confidently draw upon the experiences and suggestions of others to eke out his own.[33]

It is one of the many ironies in the development of Darwin's thought that the question of the distribution of species, which was to figure so prominently in his theory of evolution, should also have loomed so large in the orthodox theory of creation. For it was the creationist, even more than the evolutionist, who had urgently to account for the transportation or distribution of a species from its original "centre of creation." Lyell understood this well, and in the second volume of the *Principles* he complained of the excesses of those naturalists who, having too narrow and rigid a conception of creation, were too speculative and fanciful on the subject of distribution. The first travelers, he explained, had assumed that they would find wherever they went the same species they were familiar with at home, and they took pleasure in giving them the same names. When this provinciality was at long last overcome and the ancient floras had fallen into disrepute, naturalists went to the opposite extreme and assumed that different species had been created for different areas. Every apparent exception to this rule was examined with the most scrupulous severity. If it withstood this examination, the naturalist then began to speculate on the means by which the seeds might have been transported from one country into the other, and tried to determine in which of the two countries the species was indigenous, the assumption being that a species, like an individual, could have only one birthplace.

That Darwin's early speculations about distribution were indeed inspired by this theory of creation is apparent from the few explicit remarks in his diary and *Journal*. In the diary the only reference to the distribution of species on the Galapagos was an inquiry about the "centre of crea-

tion" to which the species might be found to belong.[34]
That he was here using "creation" in an orthodox sense
appears in another remark: "It seems not a very improbable
conjecture that the want of animals may be owing to none
having been created since this country was raised from the
sea."[35] Nor did the *Journal*, at least in its first edition, introduce any novel conceptions of the origin of species. From
the absence of trees in a coastal region of Brazil he concluded that "herbaceous plants, instead of trees, were created to occupy that wide area, which within a period not
very remote, has been raised above the waters of the
sea."[36] The fact that the species on opposite sides of the
Andes were dissimilar he found not at all surprising—"unless we suppose the same species to have been created in
two different countries."[37] And in the Galapagos he speculated that a particular species of lizard must have been
"created in the centre of the Archipelago" and from there
dispersed over a limited area.[38]

If the origin of species was to be accounted for, as Darwin was at this time inclined to think, by acts of creation
taking place at given times and places, the extinction of
species might also reasonably be supposed to have occurred
in a similarly fixed and arbitrary fashion. Viewing the fossils of extinct animals, Darwin asked himself why these
species had died out. He could see no signs of violence, no
appreciable change in the physical conditions of their life.
It would be tempting, he speculated, "to believe in such
simple relations, as variation of climate or food, or introduction of enemies, or the increased number of other species, as the cause of the succession of races."[39] Unfortunately, not only did he find no evidence of "such simple
relations," but he thought it extremely improbable that
they could have had the effect of wiping out completely,
in a short period of time and over a vast range of space, a
species of elephant inhabiting the shores of Spain, the
plains of Siberia, and the interior of North America, while
not affecting the other animals in these regions. Nor could
a failure of pasture account for the death of the fossil
horse, when the same plains were afterward overrun by
hordes of fresh stock introduced by the Spaniards. It could

not be a simple matter of adaptation that explained why
some species died while others lived; for why should indigenous plants and animals, after thriving in a particular
country and presumably being well adapted there, be ex-
terminated by others of a different type, arbitrarily imported by man? He was reminded of certain fruit trees
which, though grafted on young stems under the most
varied conditions and fertilized by the richest manures,
withered away and died, as if "a fixed and determined
length of life has in such cases been given to thousands
and thousands of buds (or individual germs), although
produced in long succession." "All that at present can be
said with certainty," he concluded, "is that, as with the
individual, so with the species, the hour of life has run its
course, and is spent."[40]

If there were still any lingering temptation to see the
voyage as a preview of the *Origin*, it would be dispelled by
Darwin's own reflections recorded in his diary during the
last months of the trip and later reproduced, with the exception of the final sentence, in the *Journal:*

> A little time before this I had been lying on a sunny
> bank and was reflecting on the strange character of
> the animals of this country as compared to the rest of
> the world. An unbeliever in everything beyond his
> own reason might exclaim, "Surely two distinct Cre-
> ators must have been at work; their object, however,
> has been the same and certainly the end in each case
> is complete." Whilst thus thinking, I observed the
> conical pitfall of a lion-ant:—a fly fell in and im-
> mediately disappeared; then came a large but unwary
> ant. His struggles to escape being very violent, the
> little jets of sand described by Kirby were promptly
> directed against him. His fate, however, was better
> than that of the fly's. Without doubt the predacious
> larva belongs to the same genus but to a different
> species from the European kind. Now what would
> the Disbeliever say to this? Would any two workmen
> ever hit on so beautiful, so simple, and yet so arti-
> ficial a contrivance? It cannot be thought so. The one

hand has surely worked throughout the universe. A Geologist perhaps would suggest that the periods of Creation have been distinct and remote the one from the other; that the Creator rested in his labour.[41]

Darwin himself never claimed, as his more ambitious biographers do,[42] that the theory of the *Origin* was first conceived on board the Beagle. On the contrary, as he later remembered it, although "vague doubts occasionally flitted" across his mind (doubts such as must have occurred to any naturalist), he definitely believed, at the time of the voyage, in the doctrine of the permanence of species.[43] Yet in discussing the genesis of the *Origin*, he always invoked his experiences on the Beagle. It has generally been assumed that those experiences were the compelling reasons for the new theory, in confronting him with facts that either could not be assimilated into the old theory or that more dramatically suggested the new. Neither, as has been seen, is true. The facts about fossils and the geographical distribution of species had long been available to naturalists, without making them dissatisfied with the conventional theory of creation or impelling them to embrace a theory of evolution. These facts, it is apparent, could not have been the "sufficient cause" for his later ideas.

Yet, if Darwin's memory is to be trusted, his experiences on the Beagle must be held in some sense responsible for his later ideas. Intellectual responsibility is a delicate thing, not entirely to be comprehended in the ordinary categories of cause and effect. It may be that in this case his experiences were not so much the inspiration as the confirmation of his theories. Once the idea of evolution had suggested itself, the memory of fossils pried out of ancient rocks and resembling living animals, or of giant tortoises varying from island to island, may have assumed a new significance. It was not a significance irresistibly borne in upon him by the facts, but it was a possible significance. If this was so, it was not the Beagle that was responsible for the *Origin*, but rather the *Origin*, so to speak, that was responsible for the Beagle—or at least for a new interpretation of his findings on the Beagle.

There is, in fact, no real continuity between the Beagle and the *Origin*. Between the two there intervened an idea. It was in the light of that idea that the experiences on the Beagle were re-ordered and re-interpreted by Darwin until they were ready to stand witness for the idea. Later they became so firmly identified in his mind with his theory, indeed were made to occupy pride of place in the genesis of that theory, that he could not think of them apart from it. And later generations, knowing of these events only through Darwin's reconstruction of them, and hypnotized by history into believing that the antecedents of an event must be the cause of that event, have acquiesced in the myth that the *Origin* was the Beagle writ large.

EMERGENCE OF
THE THEORY

BOOK III

CHAPTER 6

PORTRAIT OF THE SCIENTIST

THE vision of Darwin as a country clergyman, alternating between the leisurely rites of the cleric and the more strenuous devotions of the sportsman-naturalist, did not survive the long trip on the Beagle. Instead there emerged an energetic young scientist whose sense of dedication was obscured only by the modesty of his manner and the unworldliness of his ambitions. Even before his return, his family had come to recognize the new Darwin. "Papa and me," wrote his sister during the last year of the voyage, "often cogitate over the fire what you will do when you return, and I fear there are but small hopes of you going into the church. I think you must turn Professor at Cambridge and marry a Miss Jenner if there is one to be had."[1] There may not have been available a Miss Jenner (of the famous family of scientists[2]), but a professorship probably would have been Darwin's for the asking.

Indeed, it was to Cambridge that Darwin returned, after brief family visits at Shrewsbury and Maer. First as a guest of Henslow and then in lodgings, he spent the winter of 1836–37 arranging his geological specimens and preparing his *Journal* for publication. In the spring he moved to London to embark upon what he later realized was the most active period of his life. For two years, living in bachelor lodgings in Great Marlborough Street, he worked on the Beagle volumes, prepared several papers for the Geological Society, and collected material on species. As Secretary of the Geological Society and a member of the Athenaeum, he met many of the scientists and literary men of London, and found himself mixing in society as he had not had the opportunity to do before and was not to have the energy to do again.

His most intimate friend at this time was Lyell, whom he relied upon for information, advice, and a sympathetic hearing. He also saw a good deal of the botanist Robert Brown, who was full of the minutiae of his science but could never be inveigled into theoretical speculation; was honored with a single and somewhat disillusioning meeting with Humboldt, an egotistical and garrulous old man; and had more frequent and satisfying meetings with Sir John Herschel. Historians, however, rather than scientists, seemed to be the staple of most of the dinner parties he attended, and he became acquainted with Henry Buckle (who favored him with a long monologue and, when Darwin finally escaped, was heard to remark: "Well, Mr. Darwin's books are much better than his conversation"[3]), Macaulay (to whom he was grateful for not having monopolized the conversation), Motley, Grote, and Carlyle. It was Carlyle, a friend of Charles' brother Erasmus (also living in London then), and so a more than passing acquaintance, who made the most profound impression upon him. Years later Darwin recalled his harangue, lasting the entire length of dinner, on the virtues of silence. He remembered him as a "depressed, almost despondent yet benevolent man"; and while he was fascinated by his vivid characterizations and moral fervor, he was appalled by his obtuseness in defending slavery and condemning science.[4]

It is only in retrospect, however, as several years of life are compressed in as many pages of memoirs, and by comparison with the later period of almost complete seclusion, that Darwin's London life takes on a somewhat frenetic appearance. In fact, the ill health that was later to make of him a recluse had set in almost as soon as the Beagle berthed. In a letter of 1837 appears the first reference to the chronic state of illness that was already beginning to curtail his activities. Characteristically, it occurs as one of a series of reasons why he felt obliged to refuse the secretaryship of the Geological Society. From this time on, Darwin never lost sight of the need to husband his time and strength for his own work. Secretaryships (although he was later prevailed upon to hold the post from 1838 to 1841), professorships, even friendships were relegated to the periphery of his life, where they subsisted, if at all, on the sufferance of a peculiarly inhospitable nervous system.

That his ill health should have had so decisive and, it is thought, fortunate an influence upon his career has led to a good deal of speculation, much of it ill-natured—his suffering apparently inspiring people more with envy and rancor than pity. For suffering was the motif of Darwin's life, as surely as science was its motive. It was, as his son later recalled, "a principal feature of his life, that for nearly forty years he never knew one day of the health of ordinary men, and that thus his life was one long struggle against the weariness and strain of sickness."[5] Actually, it was closer to forty-five than forty years that he suffered from that assortment of ailments which defied medical diagnosis and cure. The symptoms were already acute in October 1837, when he complained that the least flurry of excitement "completely knocks me up afterwards, and brings on a violent palpitation of the heart."[6] Upon a doctor's advice, he gave up writing and even the correcting of proofs for some weeks. The spells of sickness, characterized by stomach upset, nausea, headache, and insomnia, returned in even more serious form, and the following May he confided to his diary his suspicion of a general physical breakdown. During the next few years the attacks became more severe and frequent, until his father warned him that he could hold no quick expectation of relief.

His invalidism did have the effect, as so many have remarked, of protecting him from social and even professional distractions and isolating him with his work. The implication, however, that it was mere hypochondria, designed to give him an unfair competitive advantage over his more able-bodied and social-minded colleagues, is one

of the perversities of reasoning by which failure would im-
pugn success; there was nothing, after all, to prevent even
the most robust scientist from reproducing those hermetic
conditions which Darwin had imposed upon himself. And
surely his unconscious could have contrived some better
means for protecting and nourishing his genius[7] that would
have left him more rather than less time in which to work.
He himself felt, and with reason, that his professional prospects were the worse, not the better, for his illness. In 1841
he said: "It has been a bitter mortification for me to digest
the conclusion that the 'race is for the strong,' and that I
shall probably do little more, but be content to admire the
strides others make in science."[8] He did not often complain,
but to his good friend Hooker, pressing him for information about his health, he wrote: "I well remember when I
thought it utter nonsense to talk of eight hours being too
much, but I now find two more than I can stand";[9] and
somewhat later he reported that he was seldom able to
take more than one hour's scientific reading—a complaint
that his most harassed colleagues could not fairly make.

After one particularly bad spell in 1848–49, he decided
upon the desperate measure—desperate because it meant
leaving home—of taking the water cure at Malvern; for a
time hydropathy seemed to help him, and he went so far
as to build a douche at home and press the butler into
service as bathman. But soon this too proved of no avail,
and he had to reconcile himself to living with a sensitive
stomach, spells of faintness, twitching muscles, spinning
head, spots before the eyes, and the other nervous disorders that assailed him. A week's visit to his brother was
a rare event, social dinners became an extravagant indulgence, and even the arrival of a friend for an afternoon of
scientific talk and gossip was an occasion for delighted
planning, only to be followed by a sick headache that laid
him up for the whole of the following day. In 1863 he experienced a particularly severe attack during which he
suffered a loss of memory; his wife was warned to be prepared for an epileptic fit, but it is not known whether this
followed.[10] It is no wonder that once, reporting that for
three years he had not known a whole day without suf-

fering from stomach disorder and most days from "great prostration of strength," he should have been betrayed into bitterness by the thought that "many of my friends, I believe, think me a hypochondriac."[11]

What was hypochondria to the cynics of Darwin's generation has become neurosis to the sophisticates of a later generation, so that today it is the fashion to ascribe his illness not only to an unconscious wish to shield his genius from a predatory world, but also to a neurotic fixation upon the father image and feelings of inadequacy about himself. His father's cold and autocratic nature, it is said, produced a household of "disagreeable and unfulfilled daughters, and neurotic sons," all morbidly craving an affection that was never given.[12] Others have suggested that his work was calculated to defy authority, a defiance subconsciously directed against his father and so fearful as to produce the "phobic anxiety state"[13] which was his real malady: "If Darwin did not slay his father in the flesh, then in his *Origin of Species, Descent of Man,* etc., he certainly slew the Heavenly Father in the realm of natural history"—the identification between the earthly and the heavenly fathers being facilitated by the fact that Robert Darwin had been "inimicably disposed" to such heterodox views as those of Erasmus Darwin.[14] (Actually, of course, Robert Darwin had been as little orthodox as Charles himself, and there is no evidence of any antipathy to Erasmus' theories.) The most forthright of these psychoanalytic commentators has had the grace to add that we must be thankful for a neurosis that could produce an *Origin of Species,* and that the last thing a civilized person, even a psychoanalyst, should desire is the cure of such neurotics as Darwin: "It is a terrifying thought that the Darwins of today may be known to posterity only in the case books of the psychiatrist."[15] Not even this disarming concession, however, can atone for the arbitrariness of the psychoanalytic interpretation—or rather interpretations, for they disagree among themselves.

There have been purely medical attempts at diagnosis. One ophthalmologist saw all of Darwin's troubles as the result of eyestrain, a theory that would merit more serious consideration had he not presumed every one else with any record of bad health (and almost every celebrity of the time did seem to be sickly) to be suffering from the same cause.[16] Another doctor, diagnosing Darwin as an "asthenic grade IV," suggested that his disorder was functional, the result of "an inherited peculiarity of the nervous system" reflected in the history of suicide, melancholia, stammering, and marked eccentricity in the family.[17] These medical and psychological interpretations are not, of course, necessarily contradictory, since they may complement and reinforce each other. What makes them all suspect is precisely this variety of explanation, the endless permutations and combinations that can be played upon them. It is hard to give credence to explanations that can presume to go so deep and in so many directions at once, on the basis of so few reliable facts.

If Darwin's ill health was not, as some seem to think, a pretext to isolate himself with his work, neither was it, as Darwin had right to fear, an insuperable obstacle to work. One reason why it did not prove fatal to his ambitions was the devotion and sympathy of his wife.

In November 1838, two years after his return from the voyage, he became engaged to his cousin, Emma Wedgwood. Emma, then thirty-one, a few months older than Charles, was the youngest daughter of his uncle Josiah, and their childhood ties had recently been drawn closer with the marriage of her brother to Charles' sister Caroline. It was so much a matter of course that a Darwin should marry a Wedgwood that the only thing in doubt was which Darwin and which Wedgwood it would be. For a time it appeared to be Erasmus who had designs upon Emma,[18] while Charles seemed equally well disposed toward all the Maer girls. After the engagement was announced, Emma said that the proposal had come as a surprise, and that she thought they might have gone on as friends for years "and very likely nothing come of it after all."[19] Charles, however, had convinced himself that marriage was a good thing—"only picture to yourself a nice soft wife on a sofa, with good fire and books and music perhaps; compare this vision with the dingy reality of

Great Marlborough Street . . . Marry, marry, marry. Q.E.D."—and was only worried lest Emma find him "repellently plain."[20] For her part, Emma declared herself delighted with his good nature and affectionate temper; even his habit of not drinking wine pleased her. The only threat of domestic dissension she anticipated was his dislike of the theater, but she consoled herself that "he stands concerts very well."[21]

Charles had more serious misgivings. Against the vision of the "nice soft wife on a sofa," he had to consider the disagreeable possibility that having many children might oblige him to earn a living and that a wife might seek to entice him into society. He warned Emma of the austere life she might expect with him:

> I was thinking this morning how it came, that I, who am fond of talking and am scarcely ever out of spirits, should so entirely rest my notions of happiness on quietness, and a good deal of solitude: but I believe the explanation is very simple and I mention it because it will give you hopes, that I shall gradually grow less of a brute; it is that during the five years of my voyage (and indeed I may add these two last) which from the active manner in which they have been passed, may be said to be the commencement of my real life, the whole of my pleasure was derived from what passed in my mind, while admiring views by myself, travelling across the wild deserts or glorious forests or pacing the deck of the poor little Beagle at night. Excuse this much egotism,—I give it you because I think you will humanise me, and soon teach me there is greater happiness than building theories and accumulating facts in silence and solitude.[22]

Letters to fiancées written on the eve of marriage, like the resolutions to reform which they often resemble, need not be taken at face value; and it may be supposed that Darwin did not really feel the need to be "humanised," that in fact he thought happiness to lie precisely in "building theories and accumulating facts in silence and solitude."

The marriage took place on January 29, 1839. Almost

immediately Emma was initiated into the task of caring for an invalid, Charles having been ailing all winter. Many years later she was to write to her daughter, then on her honeymoon, that "nothing marries one so completely as sickness."[23] Whatever neurotic significance may be read into this, sickness certainly married them more completely than health normally does. And more happily— for Emma was the perfect nurse as Charles was the perfect patient. (On the occasions of Emma's headaches they reversed the roles, and both ministered devotedly to their ailing children.) It does not reflect upon the genuineness of Darwin's illness to say that they enjoyed catering to their own and each other's ailments more than might be thought proper, and that they made a norm out of sickness as others did out of health; indeed, this may be the happiest way to adjust to a chronic sickness such as Darwin's. Nor is there any suspicion of sacrifice or resentment on Emma's part, no sense of her playing a part. She was, in fact, a forthright, outspoken, and notably unsentimental woman who happened to enjoy a quiet, entirely domestic life. And so she provided for Darwin that solitude and tranquility he so desperately needed—or as much solitude and tranquility as could be had in a large and happy family.[24]

If it was entirely with his person and not at all with his work that she busied herself, this was also to Darwin's liking. While she was still his fiancée he had discouraged her from trying to read Lyell's book, with which she was nobly struggling under the impression that it was one of the duties of a loyal wife to converse intelligently about her husband's interests. Later she discarded such illusions so completely that she admitted she could not become interested in his experiments or books. It was perhaps just as well that she did not pretend to any knowledge of his work; when he was editing the volume on the *Zoology of the Voyage of the Beagle,* she casually described it as a book about the animals of Australia.

A devoted wife and financial security—it is hard to say which serves the thinker better. Fortunately, Darwin had both in great measure. When the American philosopher, John Fiske, came to write an obituary of him, he said that the one thing a man of genius should not be called upon to do in a well-ordered world was to earn a living. From this odious necessity Darwin was free. When he married, his father settled upon him over 13,000 pounds, and Emma brought a dowry of 5000 pounds and an annual allowance of 400 pounds. The income from these investments, plus some smaller amounts which his father had earlier turned over to him (and without including the interest on his wife's dowry), came to 1250 pounds in the first year of his marriage. By the time he died, an additional bequest of 50,000 pounds in his father's will, and the natural inflation of capital, brought the value of his estate to 282,000 pounds and his annual income from investments alone (apart from royalties on his books) to 8000 pounds (on which he had to pay 40 pounds income tax).[25] Some part of this increase must be attributed to his sound financial judgment and management. To the question of whether he had any "special talents," apart from his professional ones, Darwin once replied: "None, except for business as evinced by keeping accounts, replies to correspondence, and investing money very well."[26]

The habit of acquisition, with Darwin as with many of his contemporaries, was largely the habit of methodical economy. In twenty-four carefully preserved books may be found an account of almost every penny received and spent from the day of his marriage until his death forty-three years later. These were not the records of a professional accountant but the personal, handwritten journals of a man for whom it was an act of honor to keep an account of every financial transaction, no matter how lowly. He even managed to impress his methodical habits upon his more casual wife, so that every cab or omnibus fare, every theater ticket and charitable donation, to say nothing of every item of every tradesman's bill, was punctiliously recorded; to be sure, Emma's accounts did not always balance, but this was a failure of arithmetic, not of will.

As in most such cases, the habit of orderliness was

enhanced by the instinct for economy. And economy, in turn, was prescribed not only as a means of preserving and augmenting the fortune held as a sacred trust for one's children but also as a virtue in its own right. Like so many Victorian gentlemen who did not have to work for a living, Darwin was reluctant to dispose of what he had not earned. He practiced such typical economies as saving the spills of paper that had been used to light the candles, and writing notes and even whole chapters of books on the backs of letters and odd remnants of paper. If nineteenth century letters are often so difficult to decipher, it is because many shared Darwin's habit of writing in the margins and, when those gave out, of scrawling cross-wise over the already filled page in order to save the postage on another sheet of paper. And if the original manuscripts of their books are not easier to come by, it is because the virgin backs of the sheets were so tempting a prey to these rapacious gentlemen. The palimpsest is an old and respectable device, and although paper in Victorian England was not nearly so precious as parchment in ancient Greece, the virtue of economy was more so.

For the other tools of his trade, Darwin showed as little respect and as little self-indulgence. He rarely bought a book he could borrow, either from circulating libraries or from friends. And he urged his own books upon others with the same solicitude for their money that he had for his own. He often felt called upon to restrain his friends from purchasing for him books they thought might interest him. "It would be extravagant," he reproached Hooker, "to buy Flinders without it being very cheap, as upon reflection I remember it so easily borrowed from public libraries."[27] By such devices he was able to keep the cost of new books down to twenty pounds or less a year. When Hooker had to return the precious manuscript of part of the *Origin*—this was before its publication—Darwin instructed him on how to tie and label it so that its postage would not be more than fourpence. And just before the *Origin* was published, he confided to Huxley that he would like to send out a few presentation copies, but could not tell how many he could afford until he found out the selling price of the book—this at a time when his annual income from investments alone came to over five thousand pounds.

His son tried to account for these excesses of economy by suggesting that Dr. Darwin had led Charles to believe that he was poorer than he was. This may, indeed, explain his schoolboy anxieties about money, but it can hardly account for the persistence of those anxieties long after he discovered that he would have an independent income, and still more after his return from the voyage when his father settled upon him a substantial amount of stock and Charles took over the management of his own affairs. No rational explanation can account for the letter to his sister in 1845, relating that, in spite of improvements he was making in his house, he had managed to save 400 pounds out of an income of 1400: "So I hope the Shrewsbury conclave will not condemn me for extreme extravagance, though now that we are reading about Sir Walter Scott's life, I sometimes think that we are following his road to ruin at a snail-like pace."[28]

The recurrent fear of ruin which assaulted him whenever he was confronted with any small but unforeseen expenditure was undoubtedly exacerbated by the Puritan-Victorian tenet that waste was sinful, however trivial the amount. The origin of this fear, however, was probably more personal: a sense of guilt at not earning his keep, and an acute sense of inadequacy at not being able to do so. This sense of inadequacy was even projected to his children. The business of earning a living seemed so formidable, so unthinkable, that he could not imagine anyone dear to him capable of it, least of all a Darwin who must have inherited his own constitutional weakness. To an old school friend he sometimes reminisced about the carefree days when there were no worries about inflation, invasion, the ill health of children or the choice of professions for sons: "How paramount the future is to the present when one is surrounded by children. My dread is hereditary ill-health. Even death is better for them."[29] Many years later, when his son Frank decided to give up medicine and become Charles' assistant, Erasmus made the typical comment: "After all he is a Darwin, and the chances are against any of our unfortunate family being fit for continuous work."[30]

The dread of hereditary ill health was not entirely illusory. With his own ailments resembling in so many respects those of both his grandfather and his brother, his mother-in-law's (and aunt's) medical history of epilepsy and invalidism, and his wife's chronic headaches, he had good reason to fear for his children's health. And as it happened, of his ten children one girl died shortly after birth, another, the much-beloved Annie, died in childhood, his youngest son Charles was a mental defective who lived only two years, Henrietta had a serious and prolonged breakdown at fifteen, and three sons suffered such frequent illness that Darwin regarded them as semi-invalids. Even if all these ailments were not the constitutional disorders he took them to be, they were real enough to warrant some amount of anxiety. In this, as in other matters, he was, of course, excessive. It hardly needed his repeated urgings to convince his sons, then in boarding school, not to overwork. And his ejaculation of relief when he discovered, after going through his accounts, that he would be able to provide for the whole of his ailing, helpless brood —"Thank God, you'll have bread and cheese"—was hardly calculated to reassure the little boy who took him literally.[31]

For all of his economy-mindedness, however, Darwin was by no means miserly. He was certainly prey to neurotic fears about the future, but it was an exaggerated sympathy and sense of responsibility that aggravated these fears. Indeed, toward others he was conspicuously generous. His personal expenditure for one year might be as little as six pounds, eleven shillings, including all his clothing, hairdressing, and snuff bills; but his children's dress, education, and personal allowance were ample, and, perhaps in memory of his own guilt-ridden youth, he was invariably generous and uncomplaining in meeting his sons' college debts. Nor was there anything mean about the conduct of his household. When he was first married, the staff comprised cook, maid, and manservant; toward the end of his life this had been expanded to cook, four maids, and three menservants. By the standard of the day, this meant if not luxury or elegance at least security and comfort.

That his concern was for security and comfort rather than luxury and elegance was evident in Darwin's choice of a home. For the first several years of their marriage he and his wife lived in London in a small house in Bloomsbury which he christened "Macaw Cottage," in testimony of the discordant colors of the furnishings. Here they learned to do without not only the aesthetic amenities but also the social ones. For this was one of Darwin's worse periods, when he was sick more often than well; to a friend resting in the country, he wrote that even Delamere was a hotbed of dissipation compared with his own retreat at 12 Upper Gower Street. After the birth of his first child, William, at the end of 1839 and his second, Annie, in 1841, the pressure of a growing family, as well as the determination to live a still more secluded existence (even the occasional meetings of the scientific societies were becoming too much for him), persuaded him to seek a house in the country.

Darwin explained that he chose Down[32] because of the extreme quiet and rusticity of the place and the diversity of its vegetation. More fanciful explanations have been suggested by others. The psychoanalyst who interpreted his devotion to science as a form of religious consecration found it significant that it should have been in his thirty-third year, which was also the year of the crucifixion, that he chose to retire from London and seclude himself in so isolated a village as Down.[33] A more prosaic consideration was economy. The anguish of house hunting, negotiating, and bargaining is revealed in a series of unpublished letters to his sisters, from which it appears that he was determined not to pay more than 2200 pounds for the house, while the owner was asking 2500. And although Darwin admitted to his family that the house was well worth the asking price and that living in the country would be cheaper than in London, he apparently did not feel justified in paying the additional 300 pounds, so that for some time it seemed as if the house would not be his.

And even after the owner had capitulated, the subject of money continued to dominate his letters, his father and sisters animatedly discussing such matters as how much should be paid for a wall of specified dimensions. His father was profusely thanked for lending him the 2500 pounds for the purchase of the house and the cost of the alterations, and his sisters received an accounting of his cash assets: 429 pounds in the bank and 84 pounds in his great iron chest. At one point, in justification of his purchase, he quoted Sir John Lubbock on the desirability of having some part of one's property in land—as if it needed a Lubbock to argue the point with a Darwin-Wedgwood.

In September 1844 the family moved to Down. Darwin described it fairly when he called it a "good, very ugly" house.[34] It was not the tawdry ugliness of gingerbread and gimcrack but the unpretentious ugliness of a dull, ungraceful, but solid and eminently utilitarian structure. It was a square, whitewashed brick building of three stories, situated on a high tableland of eighteen acres. The view was as bleak as the house itself, with no shrubbery or walls to soften the harsh lines of the surrounding chalk country. Later, with the landscaping of the gardens and the addition of a large bow to the house which eventually was covered by a tangle of creepers, the building became less severe in its proportions and less formidable in appearance.

The house commended itself to Darwin by its solidity and even more by its location. Down was a small village in Kent consisting of some forty houses, Darwin's being a quarter of a mile from the village on a small lane leading to the Westerham highroad. When he first moved there, a coach drive of twenty miles was the only means of access to London, but twenty miles hardly gives a fair notion of the remoteness and ruralness of the area. Even later, when the railroads began to approach, the nearest station was eight and a half miles away, which meant a wearisome drive up and down hills with the old and exceedingly cautious gardener who also doubled as coachman. Since there were chalk hills on two sides and the house was connected with the main road only by stony lanes, Darwin could boast that he had never been in a "more perfectly quiet country."[35] If it was not as retired a place as one German had it, who said that the house could only be approached by a mule track, it was isolated enough to make the two-hour trip into London, when Darwin's health was at its best, a monthly or, at most, fortnightly event.

What rescued Darwin from the indolence that might so easily have settled upon a man with a good wife, an ample income, and a chronic illness were the daily discipline and the methodical habit of work—virtues he said had been instilled in him on the Beagle. Darwin must take his place alongside Anthony Trollope and the other great Victorians whose creativity has been impugned by their methodicalness, on the supposition that genuine creation can only be erratic. Unlike Trollope, however, who was methodical in work so as to be prodigal of leisure, Darwin's methodicalness had no other intention than to extract from the day a few good hours of work.

By a careful alternation of work and play, he hoped to pacify his delicate nerves. His day started early with a walk and solitary breakfast, and by eight he was in his study where he worked until nine-thirty, considering that hour and a half his most productive time. He then joined the family in the drawing room to look over his mail; family letters and perhaps part of a novel would be read aloud while he relaxed on the sofa. At ten-thirty he went back to work until noon, often emerging from his study to announce with satisfaction that he had put in a good day's work. Another walk took him past the greenhouse, where he might inspect some experimental plants, and then on to the "sand-walk," a narrow strip of land an acre and a half in extent encircled by a gravel walk. Here he took his specified number of turns (the set number was abandoned when he grew older), watching his children at play, observing some bird or animal, or examining a grass or flower. For a time, on his doctor's advice, he took a canter on the oldest, gentlest horse that could be found, marveling at the memory of his youthful riding

exploits; this practice was discontinued after several unnerving spills.

Lunch was followed by the reading of the newspaper
(the only non-scientific reading he did himself) and the
writing of letters. If these were numerous or lengthy, they
were dictated from a rough draft which he jotted down;
a printed form of reply to cranks—this was after he had
become famous—was used. About three o'clock, with his
letters finished, he went to rest in his bedroom, lying on
the sofa, smoking a cigarette (during working hours he
preferred the greater stimulation of snuff), and listening
to a novel or some other non-scientific book read aloud
by his wife. The reading often put him to sleep, and he
used to complain that he missed whole chunks of books
because his wife went on reading, not wanting the cessation
of the sound to awaken him. At four he had another half-
hour's walk, followed by an hour of work. A session of
novel reading preceded dinner, and two games of back-
gammon with his wife invariably followed it. In the evening he read a scientific book until fatigue overtook him,
when he would lie back to listen to his wife at the piano
or reading aloud. He retired at ten or ten-thirty, often to
lie awake in pain for hours.

The whole of this elaborate schedule was devised to
give him a total of four hours of work a day—at most,
because the schedule prevailed only on his good days.
On his bad days, which might seize him as often as twice
a week, all pretense of work was abandoned. Looking
over his diary for the 1840's and '50's, he calculated that
of the eight years he was employed on his study of the
Cirripedia he had lost a total of two years in sickness.
At this time he complained that his normally high complexion deceived acquaintances into thinking him healthier
than he was, and that only his family and friends knew
of his misery. It may have been unconscious pique more
than hypochondria that prompted him once to describe
his complexion as "rather sallow," when it was, in fact,
distinctly ruddy.[36]

It is to his children that we are indebted not only for
the account of Darwin's normal working day but also for

an idea of his personality. Without their testimony, it would be easy to fancy him as a grim, dyspeptic, domineering father—the very image of the Victorian *paterfamilias*. In fact, he was anything but that. However suspicious we may be of that pious formula—the father respected and beloved—in the case of Darwin we must credit its truth. His children were not, as in the fashion of their class, confined to the nursery and servants' quarters, instructed to tiptoe past their father's study, and brought into the drawing room after tea for a brief and well-groomed appearance. On the contrary, they spent a good deal of time, and informally, with their parents. Their favorite playing grounds were the drawing room and garden. It was their father's study they invaded when they wanted the sticking plaster, string, or ruler that could be found on his desk—when they had not been misplaced in play. An ailing child would be tucked up on a sofa in the study so that he might not be lonely, and a romping child might offer his father the magnificent bribe of sixpence if only he would come out and play during working hours. He was, if anything, too much moved by their childhood sorrows, too much impressed by their youthful aptitudes. And when they were older he was full of gratitude for their assistance and too easily persuaded of the superiority of their judgments. The most cynical reader of biographies would be hard put to it to dispute the genuineness of the love and respect borne him by his family, and his most determined enemies were unable to call into question his gentleness, modesty, and good nature. There may be much in his work and mind to criticize, but little in his character.

Yet his character and mind were all of a piece, and what was admirable in the one was not necessarily so in the other, tenderness of character sometimes showing itself as softness of mind. His esthetic sensibility, for example, was soft and homely rather than firm and cultivated. He himself deplored the "curious and lamentable loss of the higher esthetic tastes" that seemed to afflict him sometime after the age of thirty.[37] He could no longer endure the poets who had moved him in his youth, and his one attempt to reread Shakespeare bored him to the point of physical nausea. His favorite books were popular, sentimental novels, his only requirement being that they have a happy ending; if they also had at least one thoroughly lovable character, preferably a pretty woman, so much the better. Walter Scott, Jane Austen, and Mrs. Gaskell he found most satisfying. Of non-fiction he still enjoyed history, biography, travels, and a variety of essays. He learned French and German in order to follow the scientific monographs, but he frankly regarded such reading as a disagreeable chore, never looked at anything but a scientific work in either language, and persisted in pronouncing foreign words exactly as if they were English. He did not even try to understand more in them than he had to for his immediate purposes. His son remembered his once asking, "Where is this place Wien where such lots of books are published?"[38] (He had visited the continent for a fortnight once, when he was eighteen.) Even pictures, which had once delighted him, no longer gave him pleasure. And although he still liked a vigorous tune, he was generally unable to identify or reproduce even the most familiar one.

Such were Darwin's habits and tastes. He indulged them frankly, making no pretense of greater sophistication than he felt. At the same time, he was sufficiently modest and self-critical to realize that they were not of a high order. He had none of the belligerency, and therefore none of the philistinism, of those whose tastes he happened to share. Later he concluded that his mind was not so well organized or finely constituted as it might have been, else he would not have suffered "the atrophy of that part of the brain . . . on which the higher tastes depend." He also suspected that the loss of those tastes was "a loss of happiness, and may possibly be injurious to the intellect, and more probably to the moral character, by enfeebling the emotional part of our nature."[39]

Darwin himself seemed to think that his esthetic faculty was so weak because his scientific faculty was so strong. It might be argued, instead, that his scientific imagination exhibited something of the same tender-mindedness evident in his esthetic tastes. His attitude toward his experiments and instruments was a curious compound of romanticism and common sense—in both respects far removed from the modern ideal of the impersonal, white-frocked precisionist of the laboratory. His sons used to laugh at him for personalizing the subjects of his experiments, for abusing the seedlings who were defying his wishes—"the little beggars are doing just what I don't want them to"[40]— or admiring the ingenuity of a leaf in forcing its way out of a bowl in which he had tried to fix it. Even counting, that most mechanical of occupations, kept him alert and almost excited, as if, it seemed to his son, he was personifying each seed as "a small demon trying to elude him by getting into the wrong heap."[41] For certain flowers he felt a personal affection, gently stroking them as a child might or praising the beauty of their color and form. The most unlikely objects could provoke him to admiration. His family was fond of taunting him with a sentence in the *Origin* describing a particular barnacle with "six pairs of beautifully constructed natatory legs, a pair of magnificent compound eyes, and extremely complex antennae."[42]

Even his instruments and tools were endowed with personality, very different from the standardized equipment the modern technician insists upon. There was some excuse for the elaborate improvisation he had resorted to on the Beagle, where the weight of one rat was entered in his notebook as being equivalent to his "flask, with water, without bottom, 2 bullets, 4 pellets," compared with another equal to "flask without top or bottom, big scissors"; where an object was noted to be the "length one handkerchief and half"; or the size of a river as being larger or smaller than the Severn at home.[43] But at Down, where there was no such excuse, he extemporized in the same way. His three-foot rule was old and battered, the common property of the household; the seven-foot deal rod used in measuring plants had been roughly calibrated by the village carpenter; while for millimeter measurements he used paper rules. His weighing scales were faulty, and his chemical balance dated from his childhood experiments with his brother in the garden shed. For liquid-capacity measurements he used an apothecary's measuring glass, roughly and unevenly graduated. He had two micrometers which gave differing results, and took his equivalence of inches and millimeters from an old book where, as one of his children later discovered, it was incorrectly given. When he wrote enthusiastically about the Beagle's elegant mahogany fittings and admirable scientific equipment, he never thought it a deficiency that it should have only a simple microscope; and at home, where he had both, he preferred the simple to the compound.

It is almost as if he sought out the rough and homespun in preference to the standard and precise, for it was not one or two but almost all of his instruments that were obviously and needlessly faulty. It could not have been simply as an economy measure that he put up with these makeshifts, although this may partly explain his reluctance to have proper containers made up for his experiments, so that he had to hunt up odd boxes and blacken the inside with shoe polish if he wanted them dark, or find bits of broken glass to cover his tumblers containing seeds. Such improvisations, to be sure, were often entirely adequate for his purpose. Jagged pieces of glass made inelegant but none the less effective covers, and a measurement requiring a seven-foot rule probably did not have to be precise within an eighth of an inch. It is refreshing to find a scientist not afflicted with that occupational neurosis, the fetishism of precision, which compels him to be precise whether or not precision is called for. The fact is, however, that it was not from so liberated a motive that Darwin acted as he did. The simple truth seems to be, as his son found, that it apparently never occurred to him that his instruments were anything but precise. He thought the trade of instrument making, when he thought of it at all, rather mysterious, and he had implicit faith in all instruments *per se*. It may be that many of his experiments required no great precision, but it was not on this assumption that he acted. On the contrary, he took great pains to be precise, to obtain perfect measurements with his imperfect seven-foot ruler, and to get the fluid line to correspond exactly with the inexact graduations on his beaker.

That Darwin's equipment and attitude were different from those of the scientist today is not at issue here. Only in recent years has science reached that stage of self-consciousness and self-doubt in which it loses faith not only in the perfection of its instruments but also in the objectivity of the scientist himself. It is not expected of Darwin that he should have been troubled by thoughts of fallibility, relativity, or indeterminacy; but only that he should have observed the standards of his own time. And it was by those standards that he was in arrears. Nineteenth-century science was sufficiently aware of the desirability of precision and standardization to make Darwin's tool chest seem distinctly unprofessional. In this, as in other respects, he gives the appearance of an amateur, an amateur even for his own day. And sometimes the impression is more of a child earnestly and meticulously performing some primitive experiment in the sanctum of the garden shed. There is something attractive in all this —in Darwin's capacity to retain, as a mature and distinguished scientist, the unassuming ways of the youth and amateur. It was part of his charm that age and fame should have so little altered him.

CHAPTER 7

GENESIS OF THE THEORY

If Darwin ever found himself, on the Beagle, thinking about the problem of species, it was to confirm his original impression that species, being immutable, had originated in special acts of creation. Only after his return did he seriously begin to consider the possibility that species were not immutable and had gradually changed and evolved in the course of time.[1] Having once considered this theory, he was so taken with it that he promptly made it the frame of reference for all his work. For all practical purposes, this may be taken as the moment of his conversion.

The date of this conversion is almost certain. The first notebook embodying his new views, dated July 1837 to February 1838, contains the note: "In July opened first notebook on Transmutation of Species. Had been greatly struck from about the previous March on character of South American fossils, and species on Galapagos Archipelago. These facts (especially latter), origin of all my views."[2] Although this note was probably not, as is generally assumed, contemporaneous with the rest of the notebook (its portentous tone suggesting that it was added at a later date), it was certainly very near the truth. It must have been about this time, in the spring or early summer of 1837, that Darwin's ideas took this turn. Had the change occurred earlier, it would have been reflected in the *Journal*, which, more than half completed by March, shows no trace of it. His notebook, on the other hand, started in July, proclaims it explicitly and insistently.[3]

It is the theory of mutability that informs almost the whole of this first notebook. The questions to which Darwin addressed himself were those later to appear in the *Origin*. Why are individual plants and animals sufficiently alike to group themselves into species, and yet sufficiently different to vary among themselves within species? Why do modern species resemble the extinct ones in any area, and why have the extinct ones become extinct? Why are some species born while others die? And finally, the ultimate question: Why are species related to each other, and what is the chain of their relationship? His answers to these questions were groping, tentative, highly speculative, yet invariably derived from the hypothesis of mutability. The notebook deserves to be quoted at length, partly because much of it is not available in print,[4] and also because it is only by extensive quotation that it can be seen how readily Darwin took to his theory and how bold and elaborate his speculations were, even at this early stage.

He speculated about the most fundamental facts—the facts of life and death, of generation and variation:

> Why is life short. Why such high object generation. [i.e., Why is generation so high an object?] We *know* world subject to cycle of change, temperature and all circumstances which influence living beings. We see the young of living beings become permanently changed or subject to variety, according to circumstances . . . Hence we see generation here seems a means to vary or adaptation . . . There may be unknown difficulties with *full grown* individual with fixed organization thus being modified. Therefore generation [is designed] to adapt and alter the race to *changing* world. On other hand, generation destroys the effect of accidental injuries, which if animals lived for ever would be endless (that is with our

present system of body and universe). Therefore final cause of life.

Why does individual die? To perpetuate certain peculiarities (therefore adaptation), and obliterate accidental varieties, and to accommodate itself to change (for, of course, change, even in varieties, is accommodation). Now this argument applies to species.

If individual cannot propagate he has no issue— so with species.

If *species* generate other *species*, their race is not utterly cut off:—like golden pippins, if produced by seed, go on—otherwise all die.

The fossil horse generated, in South Africa, zebra —and continued—perished in America.

There is nothing stranger in death of species than individuals.

Absolute knowledge that species die and others replace them.

With respect to extinction, we can easily see that variety of ostrich (Petise), may not be well adapted, and thus perish out; or, on the other hand, like Orpheus, being favourable, many might be produced. This requires principle that the permanent variations produced by confined breeding and changing circumstances are continued and produced according to the adaptation of such circumstances, and therefore that death of species is a consequence (contrary to what would appear from America) of non-adaptation of circumstances.

With this tendency to vary of generations, why are species constant over whole country. Beautiful law of intermarriages partaking of characters of both parents, and these infinite in number. . . . According to this view animals on separate islands right to become different if kept long enough apart . . . Now Galapagos tortoises . . .

A species, as soon as once formed by separation or change in part of country, repugnance to intermarriage settles it.

Propagation explains why modern animals same type as extinct, which is law almost proved. We can see why structure is common in certain countries when we can hardly believe necessary, but if it was necessary to one forefather, the result would be as it is.

This view supposes that in course of ages and therefore changes, every animal has tendency to change. This difficult to prove.

From the variability of individuals and the mutability of species, Darwin went on to speculate about the relations of species and the great chain of being—"the tree of life"—of which they are part:

If we choose to let conjecture run wild, then animals, our fellow brethren in pain, disease, death, suffering and famine—our slaves in the most laborious works, our companions in our amusements—they may partake our origin in one common ancestor—we may be all melted together.

The different intellects of man and animals not so great as between living things without thought (plants), and living things with thought (animals).

Organized   beings   represent   tree   irregularly branched.

The tree of life should perhaps be called the coral of life, base of branches dead; so that passages cannot be seen.

The bottom of the tree of life is utterly rotten and obliterated in the course of ages.

It leads you to believe the world older than geologists think.

There never may have been grade between pig and tapir, yet from some common progenitor. Now if the intermediate ranks had produced infinite species, probably the species would have been more perfect.

Cuvier objects to propagation of species by saying, why have not some intermediate forms been discovered between Palaeotherium, Megalonyx, Mastodon, and the species now living? Now according to my view (in S. America) parent of all Armadilloes might be brother to Megatherium—uncle now dead.

Species according to Lamarck disappear as collections made perfect.

We have not the slightest right to say there never was common progenitor between Mammalia and fish when there now exist such strange forms as . . .

If all men were dead, then monkeys make men, men make angels.

Finally, he defended the utility and legitimacy of a general law such as that of evolution:

Before the attraction of gravity discovered it might have been said it was as great a difficulty to account for the movement of all by one law, as to account for each separate one; so to say that all mammalia were born from one stock, and since distributed by such means as we can recognize, may be thought to explain nothing.

How does it come wandering birds such as sandpipers not new [originated] at Galapagos. Did the creative force know that this species could arrive. Did it only create those kinds not so likely to wander . . . Astronomers might formerly have said that God fore-ordered each planet to move in its particular destiny. In the same manner God orders each animal created with certain forms in certain countries; but how much more simple and sublime power—let attraction act according to certain law, such are inevitable consequences. Let animals be created, then by the fixed laws of generation, such will be their successors.

Let the powers of transportal be such, and so will be the forms of one country to another. Let geological changes go at such a rate, so will be the number and distribution of the species!!

It is remarkable how much of the *Origin*—with the notable exception of the theory of natural selection, which was to come later—is anticipated here. This is revealed not only in his instant and total grasp of mutability as a general explanation of the life course of species and their relations to one another, but even in such details as why some genera have few species compared with others or the meaning of vestigial organs:

In a decreasing population at any one moment fewer closely related (few species of genera); ultimately few genera (for otherwise the relationship would converge sooner), and lastly, perhaps, some one single one. Will not this account for the odd genera with few species which stand between great groups which we are bound to consider the increasing ones?

When one sees nipple on man's breast, one does not say some use, but sex not having been determined. So with useless wings under elytra of beetles. Born from beetles with wings, and modified. If simple creation merely, would have been born without them.

The whole of this chain of reasoning may be found in his first notebook, covering a period of no more than seven or eight months. If the two other volumes, up to October 1838, are included, the extracts make an even more impressive record. In the first volume his suspicions of the kinship of man with the animals were couched in the most tentative, apologetic terms: "If we choose to let conjecture run wild, then . . ."; or in the most speculative way, so as almost to defy credence: "If all men were dead, then

monkeys make men, men make angels." The last entry
came late in the first volume. Early in the next one, and
probably not more than a couple of weeks later, his tone
had become considerably more assured:

> Let man visit orang-outang in domestication, hear
> expressive whine, see its intelligence when spoken, as
> if it understood every word . . . see its affection to
> those it knows, see its passion and rage, sulkiness and
> . . . despair; let him look at savage, roasting his par-
> ent, naked, artless, not improving yet improvable;
> and then let him dare to boast of his proud pre-
> eminence.

> Man in his arrogance thinks himself a great work,
> worthy the interposition of a deity. More humble and
> I believe true to consider him created from animals.

Darwin had come a long way in a short time. By the
second of these volumes it is clear that emotionally, if not
by the strict canons of scientific inquiry, he was entirely
committed to evolution. Resolved to make of this his life-
work, he braced himself for the ambitious task that lay
ahead. And he indulged himself in the gratifying fancy
that, like the intellectual pioneers of old—Copernicus, Kepler, Galileo—he had to be prepared to brave persecution in
the service of the truth:

> This multiplication of little means and bringing the
> mind to grapple with great effects produced is a most
> laborious and painful effort. . . . Will never be con-
> quered by anyone . . . who just takes up and lays
> down the subject without long meditation.

> Mention persecution of early astronomers. Then add
> chief good of individual scientific men is to push their
> science a few years in advance of their age (differently
> from literary men). Must remember that if they *be-
> lieve* and do not openly avow their belief, they do as
> much to retard, as those whose opinions they . . .

It is a long way from these speculations about birth and
death, varieties and species, the tree of life and the laws of

creation, even the role of the noble heretic, to Darwin's later description of his state of mind at this time. In the introduction to the *Origin*, he wrote:

> On my return home, it occurred to me, in 1837, that something might perhaps be made out on this question [of the origin of species] by patiently accumulating and reflecting on all sorts of facts which could possibly have any bearing on it. After five years' work [i.e., in 1842] I allowed myself to speculate on the subject, and drew up some short notes. . . .[5]

The autobiography reaffirmed this interpretation in the much-quoted statement:

> After my return to England it appeared to me that by following the example of Lyell in Geology, and by collecting all facts which bore in any way on the variation of animals and plants under domestication and nature, some light might perhaps be thrown on the whole subject. My first note-book was opened in July 1837. I worked on true Baconian principles, and without any theory collected facts on a wholesale scale, more especially with respect to domesticated productions, by pointed enquiries, by conversation with skilful breeders and gardeners, and by extensive reading.[6]

Before inquiring into the general truth of this "Baconian" account, it would be interesting to examine some of the facts which he patiently and, as he said, indiscriminately collected. Many years later he declared himself surprised and impressed by the industry he had displayed, particularly as evidenced by the quantity of books and journals he had managed to read at this time. His industry was indeed impressive. But sometimes his "facts" and his sources for them are more surprising than impressive. What he prided himself on, and what is most original in his work, was not the abstractions from the professional journals but the information he received from "skilful breeders and gardeners," the practical men of affairs who supposedly harbored untapped resources of knowledge. All too often, how-

ever, the practical wisdom of these men is revealed to be
the fruit more of prejudice than of mature experience. And
unfortunately Darwin could not always distinguish between the two, particularly when his source was so un-
exceptionable as his father, uncle, or friend. The tolerance
he displayed toward his instruments was extended to his
informants, and often with the same sad results. On the
subject of races, for example, he received such confident
firsthand information as the following:

> Strong odour of negroes—a point of real repugnance.
> Parasites of negroes different from Europeans.

> Dr. Smith says he is certain that when white men
> and Hottentots or Negroes cross at Cape of Good
> Hope, the children cannot be made intermediate. The
> first children partake more of the mother, the later
> ones of the father. [Added later: Is not this owing to
> each copulation producing its effect; as [?] bitches
> puppies are less pure bred owing to having once been
> mongrels.] He has thus seen the black blood come out
> from the grandfather (when the mother was nearly
> quite white) in the two first children.

> My father says, on authority of Mr. Wynne, the
> bitch's offspring is affected by previous marriages with
> impure breed.

> The cat had its tail cut off at Shrewsbury and its
> kittens had all short tails; but one a little longer than
> the rest; they all died. She had kittens before and
> afterwards with tails.

In the light of this "amiable credulity," as one biologist
has described Darwin's reception of information,[7] one can
appreciate Darwin's statement in his autobiography that he
was not very skeptical—"a frame of mind [meaning, apparently, an excess of skepticism] which I believe to be
injurious to the progress of science."[8]

If Darwin's facts were not all they might be, neither were
his claims to have worked on "true Baconian principles,"
collecting facts for five years without any theory and with-

out any speculation. As the notebooks amply demonstrate, he was speculating boldly from the very beginning of this period, and his speculations were all directed to a particular theory—that of mutability. What is impressive about these early notebooks is not the patient marshaling of the evidence, which in fact was conspicuously absent, but rather the bold and spirited character of his thought. What clearly urged him on was a theory capable of the widest extension and a mind willing to entertain any idea, however extravagant. It is this combination that permitted him to go so far in so short a time, not only anticipating, in these first several months, a good part of his mature theory, but even pondering its difficulties and assessing its scientific status.

Darwin himself at other times admitted that he did not much respect those who pretended to work on the principle of gathering facts without reference to any theory: "How odd it is that any one should not see that all observation must be for or against some view if it is to be of any service."[9] And elsewhere: "No one could be a good observer unless he was an active theoriser."[10] He was, in fact, so active a theorizer that not only the facts he observed but also the facts he sought were those required by his theory. It was essential to his theory, for example, that there not be "any limit to the possible and even probable migration of plants," and when he solicited information on such migration, it was on the understanding that even if the desired evidence was not forthcoming, his theory would not be affected: "If you can show that many of the Fuegian plants, common to Europe, are found in intermediate points, it will be grand argument in favour of the actuality of migration, but not finding them, will not in my eyes much diminish the probability of their having thus migrated."[11] And later, when he undertook experiments to demonstrate the floating and surviving powers of seeds, again to show the infinite possibilities of migration, he was undismayed by failure. When his fresh seeds sank, he tried salting them. When these sank as well, he tried feeding them to fish in the hope that they would carry them hundreds of miles before voiding them onto some hospitable shore. And when his fish refused to oblige him, ejecting them instantly, he was forced as a last resource to believe that pods and even whole plants and branches were continually being washed into the sea and thus transported. "Float they must and shall," he vowed.[12]

Later, justifying the *Origin*, he defended the procedure of "inventing a theory and seeing how many classes of facts the theory would explain"[13]—a more apt description of his method and of that of most scientists than the ritualistic cant about "Baconian principles."

Darwin's ideas were speculative not only in the conjectural sense but also in the meditative or philosophical sense. He was, in fact, engaged in that traditional philosophical quest: the answer to the question of "why."

Why is life short? Why does the individual die, and why do species die? Why does nature put so high a premium on generation? And why does generation have the twofold character of perpetuation and variation? Why are species of one kind to be found in one place and time, and of a different kind in another place and time? Or, if they are of the same kind, why is that? Why do some genera have few species and others many? And the smaller "whys" of nature: Why are there nipples on men's breasts and useless wings on beetles?

In view of this pre-eminently teleological inquiry, it is strange to find Darwin hailed by such eminent philosophers as John Dewey as the father of the anti-teleological movement in modern thought. For more than two thousand years, the pragmatists tell us, men were enthralled by the primitive and vain quest for ultimate truths, the whys of existence. Only within the past century have they been liberated from their ancient folly and set free to pursue less ambitious and less chimerical ideals—the "what" and the "how" in place of the "why." And it is the scientific revolution inaugurated by Darwin that is reputed to have brought about this emancipation. Philosophers and scientists are still endorsing the tribute paid by Dewey to the new logic of Darwin which "forswears inquiry after absolute origins and absolute finalities in order to explore specific values and the specific conditions that generate them."[14]

Whether the *Origin* itself was as anti-teleological as has been made out, or whether it seemed so to its readers, will be the subject of later discussion. What is interesting now is that its inspiration was just this kind of metaphysical, teleological curiosity that recent pragmatic and positivist philosophy so deplores. There is no doubt that Darwin's own point of departure was an examination of "absolute origins and absolute finalities," the ultimate "why." Nor is it by accident that this was so. Darwin was able to give ultimate answers because he asked ultimate questions. His colleagues, the systematizers, knew more than he about particular species and varieties, comparative anatomy and morphology. But they had deliberately eschewed such ultimate questions as the pattern of creation, or the reasons for any particular form, on the grounds that these were not the proper subject of science. Darwin, uninhibited by these restrictions, could range more widely and deeply into the mysteries of nature. By breaking through the shell in which they had immured science—a shell made up of a myriad of "whats" and "hows"—he could make a fresh attack upon the old and traditional problems. And the instrument of his assault could only have been the question of "why."

Almost as curious as this anomaly by which Darwinism is made to play a part so out of keeping with its own origins is the anomaly in the person and mind of Darwin himself. It seems strange that Darwin, the least philosophical of men, should have fastened upon that most philosophical and metaphysical of all questions. Yet perhaps it is not so strange after all. It was with the sharp eyes of the primitive, the open mind of the innocent that he looked at his subject, daring to ask questions that his more learned and sophisticated colleagues would not have thought to ask. And thus his questions, while fundamental, were never abstract. Such professional philosophical problems as the concept of nature or the principles of naturalism never troubled him, just as he had not troubled, when becoming

a convert to Lyell, to inquire into the principles of uniformitarianism.

This fixation upon the "whys" of existence may account
for the strange lack of speculation in these notebooks about
the obvious "hows" of evolution—the actual mechanism of
the origin of species. Apart from an occasional half-hearted
reference to "adaptation," Darwin seems almost deliberately to have avoided at this time all the contemporary
theories of the mechanics of evolution. It was as if he was
determined not to be dissuaded from his general theory by
the unsatisfactoriness of the available mechanical explanations, as if he was determined to hang on to his theory
at all costs and against all odds.

Sooner or later, however, the question of "how" had to
be confronted. By September 1838 he was alternating between moods of satisfaction with his theory as it had progressed so far and moods of dissatisfaction that it had progressed no further. To Lyell he congratulated himself:
"Notebook after notebook has been filled with facts which
begin to group themselves *clearly* under sublaws."[15] But
the very next day he complained in his private diary:
"Frittered these foregoing days away in working on Transmutation theories."[16] It was a restless time, whether because of discontent with his scientific progress, thought of
his impending engagement, or ill health. "All September,"
his diary recorded, "read a good deal on many subjects;
thought much upon religion. Beginning of October ditto."[17]

Among the other books he read aimlessly, "for amuse-
ment," as he said,[18] was Malthus' *Essay on the Principle
of Population*. By this chance encounter, his theory was
provided with a rationale and mechanics that distinguished
it from other evolutionary theories, and the "how" of evolution came to supplement the "why." As Darwin later told
it, Malthus' description of the "struggle for existence" in
human society immediately suggested to him that under
the competitive conditions of animal and plant life, "favourable variations would tend to be preserved, and unfa-
vourable ones to be destroyed," the result being the formation of new species. "Here then," he explained, "I had at

last got a theory by which to work."[19] Almost all biographers have been content to accept this account of the matter and inquire no further. Yet the relationship between Darwin and Malthus is one of the most curious and misunderstood in the history of ideas.

What is generally taken to be the chief oddity of the affair turns out not to be odd at all. This is the circumstance that a purely economic and political tract should have inspired a purely scientific theory. In fact, Malthus' theory was itself derived from natural history, and Malthus himself, although a political economist by profession, was a devotee of the natural sciences, a member of the Geological Society, Fellow of the Royal Society, and regular follower of the meetings of the British Association. His addiction to science was so pronounced while still at college that he had had to reassure his father that while he hoped to become a "decent natural philosopher," he did not intend to study only the abstract principles of science but rather to "apply these principles in a variety of useful problems."[20] And it was the application of one of these natural principles that gave him the idea for his *Essay on Population* and Darwin the idea for the *Origin of Species*.

The first page of Malthus' *Essay* put the problem as the natural scientist saw it:

> It is observed by Dr. Franklin that there is no bound to the prolific nature of plants or animals but what is made by their crowding and interfering with each other's means of subsistence. Were the face of the earth, he says, vacant of other plants, it might be gradually sowed and overspread with one kind only, as for instance with fennel: and were it empty of other inhabitants, it might in a few ages be replenished from one nation only, as for instance with Englishmen.

> This is incontrovertibly true. Through the animal and vegetable kingdoms Nature has scattered the seeds of life abroad with the most profuse and liberal hand; but has been comparatively sparing in the room and the nourishment necessary to rear them.

The germs of existence contained in this earth, if they could freely develop themselves, would fill millions of worlds in the course of a few thousand years. Necessity, that imperious, all pervading law of nature, restrains them within the prescribed bounds. The race of plants and the race of animals shrink under this great restrictive law; and man cannot by any efforts of reason escape from it.[21]

The rest of the book was based upon this phenomenon of plant and animal life as it affected human life: the natural sparsity of plants and animals, upon which human nourishment depends, being the main check upon the natural fertility of mankind. Not all human beings who could be born are born; and not all who are born, live. In primitive tribes, such as the American Indians, "such a proportion of the whole number perishes under the rigorous treatment which must be their lot in the savage state, that probably none of those who labour under any original weakness or infirmity can attain the age of manhood."[22] More civilized people often responded by leaving their overcrowded communities to explore fresh regions; the resulting contests with the natives upon whom they encroached and with others acting from the same motives as themselves were "so many struggles for existence, and would be fought with a desperate courage, inspired by the reflection that death would be the punishment of defeat, and life the prize of victory."[23]

This concept of a "struggle for existence" (Darwin generalized and abstracted it by putting it in the singular) represents the high point of Darwin's debt to Malthus.[24] For the rest, however—the interpretation put upon it and the use he made of it—Malthus bears no responsibility. It was neither Malthus' intention to argue nor the effect of his argument that in these "struggles for existence" the strong triumphed and the weak succumbed. The closest he came to this was in the example quoted above of the American Indians. In general, what Malthus was concerned with was not how the struggle for existence affected the quality of the population but simply how it limited its numbers.

Indeed, it was precisely to deny the possibility of an improvement in quality that he had written his essay. The essay had been conceived in rebuttal to Godwin and Condorcet, both of whom had argued that the human species, under conditions of equality, was capable of infinite progress and perfectibility. Malthus' "principle of population" was intended to refute this idea. Human society could never progress toward perfectibility because the population inevitably tends to increase beyond the means of subsistence and is kept within the bounds of its resources only by misery, vice, and moral restraint.

At one point, the rebuttal of Condorcet verged perilously close to a rebuttal, by anticipation, of Darwin. This was when Malthus was inquiring into Condorcet's theory that the "organic perfectibility or degeneration of the race of plants and animals" was a general law of nature.[25] Malthus chose to refute this theory with precisely the same facts later used by Darwin to defend it: the facts of plant and animal breeding. He had been told, he wrote, that cattle breeders thought they could improve their breeds as much as they liked, on the assumption that some offspring would always show in exaggerated form the desirable qualities of their parents. Thus they had bred a variety of sheep with small head and small legs. But what they could never produce, he mockingly pointed out, was a breed in which heads and legs were "evanescent quantities."[26] For even if it was not known in advance just where the limit of improvement might be, it was known that there is such a limit, that however systematic the breeding, the heads and legs of sheep would never be reduced to those of a rat. "It cannot be true, therefore," he concluded, "that, among animals, some of the offspring will possess the desirable qualities of the parents in a greater degree; or that animals are indefinitely perfectible."[27] And the same was true of plants. Nothing was more striking than the development of a wild plant into a garden flower. Yet even here the limits were obvious, and it could be confidently supposed that no amount of cultivation would ever produce a carnation even the size of a large cabbage, to say nothing of the infinite range beyond the cabbage.

If the theory of infinite progress was questionable in the case of animals and plants, Malthus thought it was still more questionable in the case of human beings. During the whole of recorded history, he asserted, there had been no suspicion of organic improvement in the human frame. To be sure, if men were bred in the same spirit as domestic animals, some small degree of improvement might result. He quoted the *Tatler* to show how the ancient family of the Bickerstaffs had contrived to assume a more aristocratic appearance by lightening their complexion and increasing their height as a result of a series of judicious marriages. But such calculated breeding was both rare and limited. Certainly it did not encourage men to suppose that they might lengthen their life spans indefinitely, as Condorcet thought. Such wild speculations Malthus took to be a form of mental intoxication, "arising perhaps from the great and unexpected discoveries which had been made in various branches of science."[28]

It is likely that Malthus would have been much distressed had he lived to see what use Darwin made of him. For what Darwin did, in effect, was to use Malthus to prove the case of Condorcet. The principle of population, which Malthus had proposed as the decisive objection to all utopias, the inevitable frustration of all attempts at improvement and progress, Darwin took to be the very mechanism of improvement and progress. The struggle for existence, which to Malthus meant that vice and hardship and misery were the permanent qualities of human life, to Darwin meant that the strong survive, the weak perish, and the species was in a constant state of change and progress. It is rarely appreciated how thoroughgoing a reversal in the history of thought Darwin had effected. Neither he nor later commentators saw how ill used Malthus was, how completely ignored were his strictures on perfectionists and evolutionists, and how utterly perverted were his intentions. As surely as Marx stood Hegel on his head, so Darwin did to Malthus.

There was another sense in which Darwin stood Malthus on his head, and it was Marx, interestingly enough, who was the first to see this—Marx, who despised Malthus, revered Darwin, and was therefore quick to sense the opposition between the two. "What amuses me in Darwin," he wrote to Engels, a few years after the publication of the *Origin*, "is his assertion that he applied the theory of Malthus to plants and animals alike, whereas the whole joke in Malthus was that he applied the theory to men alone, with the geometrical progression, in opposition to plants and animals."[29]

What Marx discerned was not only the opposition between Darwin and Malthus but also the internal contradiction in Malthus' theory which Darwin, like so many others, failed to recognize. Malthus' principle of population was based on the supposed discrepancy between the reproduction rate of human beings and that of the animals and plants who were their means of sustenance, the first increasing geometrically and the second only arithmetically. Darwin, shifting the center of attention from mankind to the animal and plant kingdoms, was impressed by the enormous natural fertility of these animals and plants, which was kept in check only by their own limited means of sustenance. In this shift of perspective, he unwittingly revealed the basic fallacy of Malthus. For if human beings tended to increase geometrically, so did animals and plants —and perhaps even more than geometrically, their natural rate of reproduction being, if anything, higher than that of man. (In practice, of course, neither men nor animals nor plants increased geometrically, the rate of increase depending upon the respective checks upon their expansion.) Thus Malthus' distinction between the geometrical and arithmetical ratios, which was supposed to account for the pressures and checks upon the human population, proves to be entirely spurious, as is the elaborate set of calculations purporting to give the precise proportions of population (if unchecked) to food supply at the end of one, two, or three centuries.

This is not the place to enter into an extended critique of Malthus.[30] It need only be said that what he passed off as empirical evidence was as little empirical and as far from the point as his reasoning. It is all the more startling, therefore, to find Darwin and his friends eulogizing Malthus as master of facts and logic alike. When a critic of the *Origin* also ventured to criticize Malthus, Darwin wrote: "It consoles me that——sneers at Malthus, for that clearly shows, mathematician though he may be, he cannot understand common reasoning. By the way, what a discouraging example Malthus is, to show during what long years the plainest case may be misrepresented and misunderstood."[31] Alfred Russel Wallace, who came to the theory of natural selection independently of Darwin, and also by way of Malthus' *Essay*, said it was the most important book he had read in his youth, and praised it for its "masterly summary of facts and logical induction to conclusions."[32] And Hooker, one of Darwin's first confidants and converts, agreed that Malthus' arguments were "incontrovertible."[33]

Yet as Marx remained a Hegelian, so Darwin was essentially a Malthusian. For all his misreading of Malthus' intentions and theories, Darwin was indebted to Malthus perhaps even more deeply than he knew. For what he took from Malthus was more than a simple mechanism for the origin of species. It was a principle governing all the processes of nature: a natural, mechanical principle operating without the conscious intervention of either human or divine agents, a principle that was self-explanatory, self-sufficient, and self-regulating. As Malthus assumed that population was always adjusting itself, without the knowledge or even against the will of men, to the inexorable pressures of space and food, so Darwin saw men, animals, and plants, down to the lowliest organisms, responding involuntarily to the competitive conditions of life and hence involuntarily contributing to the evolutionary process, the creation of new species.

Neither Malthus nor Darwin, to be sure, saw the matter in just this light. Malthus, a pious and orthodox man, professed his entire faith in Paley's scheme of providence and design. And Darwin, acknowledging his debt to Malthus, spoke only of the idea of the struggle for existence which suggested to him how favorable variations would be perpetuated and unfavorable ones destroyed. Wallace, philosophically more sophisticated, saw more deeply into the problem. When he read Malthus, it immediately occurred to him that this struggle for existence—with its corollary, the survival of the fittest—provided him with the "self-acting process" which was the "long-sought-for law of nature that solved the problem of the origin of species."[34]

Another point of affinity between Malthus and Darwin is indicated by the similar career of their ideas. The first edition of Malthus' *Essay* was entirely *a priori* and abstract in its argument; only later, in preparing a second edition, did he trouble to amass the body of material which, although an afterthought and largely irrelevant to his thesis, earned him the title, by such otherwise discerning critics as Alfred Marshall and Lord Keynes, of a pioneer of inductive method and sociological history. Darwin was a more patient and cautious man than Malthus. It was not in a first edition but in a first private sketch, written in 1842, that he laid down almost the complete theoretical structure of the doctrine that was later, buttressed by a multitude of facts, to appear as the *Origin*. Since Darwin's theory, like Malthus', has been hailed as a triumph of induction, it is instructive to recall that its genesis was as little empirical as Malthus'.

Darwin was only twenty-nine and barely out of his apprenticeship, so to speak, when, by this second leap of imagination, his theory took full shape. If this chance reading—or misreading—of Malthus, like his first general speculations about evolution, seems too fortuitous a mode of inspiration, the fault may lie not with Darwin but with the conventional notion of scientific discovery. The image of the passionless, painstaking scientist following his data blindly, and provoked to a new theory only when the facts can no longer accommodate the old, turns out to be, in the case of Darwin as of others, largely mythical. It has recently been shown that the conventional account of Einstein's discovery of relativity is similarly fictitious, that the discovery was not occasioned, as is said, by experiments invalidating the old conception of time and space, but rather antedated those experiments, having originated in an act of pure thought "unaided by any observation that had not been available for at least fifty years."[35] As William James described it, the process of discovery is a series of random flashes visiting the investigator's mind, all the flashes being "on an exact equality in respect of their origin."[36]

Were the image not so incongruous with his person, it might be said of Darwin, as Freud once said of himself: "I am not really a man of science, not an observer, not an experimenter, and not a thinker. I am nothing but by temperament a *conquistador* . . . with the curiosity, the boldness, and the tenacity that belongs to that type of being."[37]

CHAPTER 8

PROGENITORS OF THE
THEORY

If it was as an innocent that Darwin embraced evolution, he was not alone in his innocence, any more than primitive man was alone in his primitivism. Indeed innocence, like primitivism, is itself a convention. Thus the idea of the evolution or transmutation of species, while not the accepted theory among most professionals, had a long and respectable history. This fact of its history may help explain how it happened that, once the right questions were asked, the answers should have emerged and matured so rapidly in Darwin's mind. For it was not so much a case of Athena emerging full blown from the brow of Zeus as of the Phoenix rising once again from its ashes.

From at least the sixth century B.C. until Darwin's own time, evolution had been a set piece of philosophical dispute. Anaximander, Empedocles, and others played with variations of the theory that all animals, of which man was one, had their beginnings in the water, discarding their shells and otherwise adapting themselves to their new circumstances when they came on land. Similar speculations, including even intimations of natural selection, became common enough to provoke Aristotle to refute them. The medieval philosophers were so enamored of the idea of a "scale of nature" linking minerals, plants, animals and men, that some of them even proposed to extend it from man to God; one school of anatomists contended that since

Galen had represented men with simian features, the men of antiquity must have looked like apes, an assumption they found confirmed in the human-like structure and behavior of apes. This kinship of ape and man remained a favorite theme of philosophers throughout the centuries, until La Mettrie actually devised a method for teaching apes how to speak based upon the deaf and dumb language.

Although the "scale of nature" was generally intended as a static rather than evolutionary concept, denoting the fixed relationships obtaining among the various creations of God, it sometimes tempted men to speculate about the possibility of one species having evolved from another. Newton gave his authority to the idea that nature was "delighted with transmutations,"[1] and Leibnitz found evidence of the continuity of nature in fish that had wings and lived out of water and birds that were cold-blooded and lived in water. Other links in the great "chain of being" were provided by the French philosopher de Maillet, who found that all land animals had their origin in marine forms, birds deriving from flying fish, lions from sea lions, and man from *l'homme marin* (presumably the husband of the mermaid).[2] Less spectacular but more substantial was the contribution of Kant, who reasoned, like Locke before him, that species were images contrived in the minds of men rather than realities of nature, and that the tree of nature would eventually show all living creatures, from polyps to man, to be of common descent. With the German romantics Goethe and Schelling, the history of evolutionary ideas reached Darwin's time—a history which any industrious scholar could spin out to almost any length.

Such industriousness would be wasted on Darwin, for while the history of evolution is relevant to the controversy that was to shake Victorian England, most of it has little bearing on the development of his thought, Darwin being equally indifferent to the philosophy of his own time and of antiquity. Even Aristotle, for whom he professed "unbounded admiration," praising him as "one of the greatest, if not the greatest, observers that ever lived,"[3] he read only in meager extracts—either so meagerly or so uncritically that he failed to realize that Aristotle was criticizing, not propounding, the evolutionary thesis. With the philosophers of the French Enlightenment and German Romanticism he was even less acquainted.

What Darwin could not have ignored, however, was the more recent scientific, rather than philosophic, history of evolution—notably the controversy between Linnaeus and Buffon which was still, in Darwin's time, lively matter for dispute among scientists. It is one of the ironies of history that Linnaeus' classification of species should in large part prevail even today, surviving so long the assumption upon which it was based, that of the immutability of species. It is also ironic that his famous dictum—"the number of species is the same as the number of forms created from the beginning"—which has come down as the classical expression of immutability, was retracted by Linnaeus himself within a decade of its publication. Confronted with the evidence of a hybrid form that could reproduce itself, Linnaeus was finally forced to concede that "it is possible for new species to rise within the plant world," and thus "the basis for all botanical science, and the natural classification of plants [is] exploded."[4] The explosion he anticipated, however, was long delayed. Although later editions of his work spoke of species as "the daughter of time,"[5] and although Linnaeus himself became suspect to the orthodox, his system survived together with the conception of species that had originally inspired it.

When Buffon came along to challenge Linnaeus' system, it was to challenge immutability as well. For Buffon, not only were there no absolute boundaries between species; there was not even a fixed line separating the animal from the vegetable kingdoms. Species, he said, progressed and degenerated by the favors and disfavors of nature; by changes in land and sea, food, climate, and the other prolonged influences to which they were subject. Formed upon no "original, special and perfect plan," they were a compound of other species. Thus the pig had toes for which it had no use; the camel had acquired his hump through domestication; and man and the ape, like the horse and ass, shared a common ancestor. By a rhetorical stratagem

familiar to the Encyclopedists, Buffon hoped to avoid ecclesiastical censure:

> In almost every part, as well externally and inter-
> nally, there is so perfect a resemblance [of the parts of
> the orang] to those of the human species, that we
> cannot compare them without expressing our wonder
> and admiration, that from such a similar conformation
> and organization the same effects are not produced.
> For example, the tongue, and all the organs of the
> voice, are exactly the same as in man, and yet this
> animal does not speak; the brain is absolutely of the
> same form and proportion, and yet it does not think.
> Can there be a more convincing proof that matter
> alone, however perfectly organized, cannot produce
> either speech or thought, unless animated by a su-
> perior principle? Or, in other words, by a soul to di-
> rect its operation?[6]

The church, however, was not taken in by this pretended
invocation of the soul, particularly since Buffon could not
resist adding: "If there were a step by which we could
descend from human nature to that of the brutes, and if
the essence of this nature consisted entirely in the form of
the body, and depended on its organization, the orang-
outang would approach nearer to man than to any other
animal."[7] Buffon was finally obliged to make a formal recantation of everything in his works that might be taken
to contradict the Biblical account of creation.

Since the authorities were not deceived by this subterfuge, it is odd that Darwin should have been. In his "Historical Sketch" appended to the third edition of the *Origin,*
Darwin dismissed Buffon shortly: "The first author who in
modern times has treated [the subject] in a scientific spirit
was Buffon. But as his opinions fluctuated greatly at different periods, and as he does not enter on the causes or
means of the transformation of species, I need not here
enter on details."[8] This cavalier dismissal may be explained
by the fact that Darwin had probably not read more than
extracts from the ten volumes of Buffon's *Natural History,*
with the result that he was merely passing on the con-

ventional disparaging opinion of Cuvier and other anti-evolutionists. Later, when Huxley brought to his attention the similarity between his own theory of heredity and Buffon's, he confessed ignorance of Buffon's views on this subject, and when he then read parts of the *History*, he admitted that "whole pages are laughably like mine."[9]

Erasmus Darwin is another name that is more conspicuous in the historian's account of Darwin's predecessors than in Darwin's own. In the "Historical Sketch" Darwin relegated his grandfather to a footnote as having "anticipated the views and erroneous grounds of opinion" of Lamarck.[10] And the autobiography remarked that although he had twice in his youth read the *Zoonomia*, which he had then "admired greatly," he was neither influenced nor persuaded by it, "the proportion of speculation being so large to the facts given."[11]

It is true that Erasmus Darwin, by his own admission, dealt largely in speculation,[12] which is perhaps why verse was so congenial a medium. Yet it was speculation which Charles might have found provocative at a time when he himself was being so speculative. A catalogue of evolutionary suggestions, anticipating not only Lamarckism but even the theory of natural selection may be culled from the *Zoonomia* and the notes to the *Botanic Garden*:

> The three great objects of desire, which have changed the forms of many animals by their exertions to gratify them, are those of lust, hunger and security. A great want of one part of the animal world has consisted in the desire of the exclusive possession of the females; and these have acquired weapons to combat each other for this purpose. . . . The final cause of this contest among the males seems to be, that the strongest and most active animal should propagate the species, which should thence become improved.[13]

> Another great want consists in the means of procuring food, which has diversified the forms of all species of animals. Thus the nose of the swine has become hard for the purpose of turning up the soil in search

of insects and of roots. The trunk of the elephant is an elongation of the nose for the purpose of pulling down the branches of trees for his food, and for taking up water without bending his knees. Beasts of prey have acquired strong jaws or talons. Cattle have acquired a rough tongue and a rough palate to pull off the blades of grass. . . . [The beaks of birds] seem to have been gradually produced during many generations by the perpetual endeavour of the creatures to supply the want of food, and to have been delivered to their posterity with constant improvement of them for the purposes required.[14]

The third great want amongst animals is that of security, which seems much to have diversified the forms of their bodies, and the colour of them; these consist in the means of escaping other animals more powerful than themselves.[15]

From thus meditating on the great similarity of the structure of the warm-blooded animals, and at the same time of the great changes they undergo both before and after their nativity; and by considering in how minute a portion of time many of the changes of animals above described have been produced; would it be too bold to imagine, that in the great length of time, since the earth began to exist, perhaps millions of ages before the commencement of the history of mankind . . . that all warm-blooded animals have arisen from one filament, which the Great First Cause endued with animality, with the power of acquiring new parts, attended with new propensities, directed by irritations, sensations, volitions and associations, and thus possessing the faculty of continuing to improve by its own inherent activity, and of delivering down those improvements by generation to its posterity, world without end?[16]

If this gradual production of the species and genera of animals be assented to, a contrary circumstance may be supposed to have occurred; namely, that some kinds by the great changes of the elements may have been destroyed. This idea is shown to our senses by contemplating the petrifactions of shells, and of vegetables, which may be said, like busts and medals, to record the history of remote times.[17]

From having observed the gradual evolution of the young animal or plant from its egg or feed; and afterwards its successive advances to its more perfect state, or maturity; philosophers of all ages seem to have imagined that the great world itself had likewise its infancy and its gradual progress to maturity. . . . The external crust of the earth, as far as it has been exposed to our view in mines or mountains, countenances this opinion; since these have evidently for the most part had their origin from the shell of fishes, the decomposition of vegetables, and the recrements of other animal materials, and must therefore have been formed progressively from small beginnings. There are likewise some apparently useless or incomplete appendages to plants and animals, which seem to show that they have gradually undergone changes from their original state. . . . Perhaps all the supposed monstrous births of Nature are remains of their habits of production in their former less perfect state, or attempts towards greater perfection.[18]

[The stronger locomotive animals] devour the weaker ones without mercy. Such is the condition of organic nature! whose first law might be expressed in the words, "eat or be eaten," and which would seem to be one great slaughterhouse, one universal scene of rapacity and injustice.[19]

If there was much in Erasmus Darwin to try the patience and respect of a modern scientist, there was less to offend the sensibilities of Charles, who was himself at the time engaged in a fair amount of speculation. The modern scientist might be put off by Erasmus' theory of "epigenesis," explaining the origin of life in terms of the "irritability" of fibers, and heredity in terms of a "filament" from which the embryo developed and to which the parents contributed in different and unequal ways. But contemporary scientists took the idea seriously; and Charles, whose own theory of "pangenesis" was only slightly more probable but no more demonstrable, had not much cause for contempt. Eramus is also, and properly, criticized for being too inventive in details and too chaotic in the whole. Thus he failed to realize the potentialities of his first law of nature, "eat or be eaten," lapsing into a paean to the benevolent deity who had so felicitously contrived to eliminate the aged and infirm and thus increase the stock of "pleasurable sensation" in the world.[20] Yet Malthus was no less obtuse (from Darwin's point of view) than Erasmus, and if Charles managed to extract the vital hint from the one, he might have been expected to find it in the other. In Erasmus, indeed, it was more amenable to the interpretation Charles later put upon it, apart from the advantage of being in a more familiar and homely context.

For all his vagaries, Erasmus Darwin may have been more influential than his grandson knew in preparing the way for Natural Selection, more influential than might be supposed from the casual admission: "It is probable that the hearing rather early in life such views maintained and praised may have favoured my upholding them under a different form in my *Origin of Species*."[21]

If, as Darwin implied, his grandfather's main distinction was in having anticipated Lamarck, it is generally Lamarck who is considered to have anticipated Charles Darwin. Lamarck was a promising botanist when the French Revolutionary Government, seeking to promote science while reforming society, appointed him to a new professorship of zoology; by the same strange workings of revolutionary logic, the other chair of zoology was given to the mineralogist Geoffroy Saint-Hilaire, while the zoologist Cuvier, who had never dissected a human body, became associated with the chair of comparative anatomy. By a triumph of personality over logic, the appointments proved eminently successful, producing the three main figures in the pre-Darwinian evolutionary controversy.

As in the case of Erasmus Darwin, there was much in Lamarck to repel the sober, workaday scientists of even his own generation. His wide range over botany, zoology, meteorology, mineralogy, paleontology, chemistry, and a variety of other subjects would not alone have alienated them, many of whom were equally versatile. They were, however, suspicious of the extravagant imagination that made him host to so many novel theories: chemical theories suggestive of alchemy; the denial that oxygen was a part of air and water on the grounds that no one had seen it and it went against reason; the concept of a primary fluid of fire-like substance, a *feu éthère*, penetrating all organic and inorganic matter and producing such diverse phenomena as color, sound and chemical change; or the notion that rocks and chemical compounds originated in organic life and that, like organic matter, they too were mutable, having no distinct "species." It would be unjust, however, to dismiss him too quickly, for while his work was often exceedingly speculative, it had a consistency and thoughtfulness that distinguished it from the more casual excogitations of an Erasmus Darwin.

Like many others, Lamarck conceived of living beings as forming a hierarchy, the highest orders having the greatest specialization of organs and the lowest orders the least specialization. This was familiar enough. More original was his theory of how the specialization of organs and the resulting hierarchy of orders arose:

> It is not the organs, that is to say, the nature and shape of the parts of an animal's body, that have given rise to its special habits and faculties; but it is, on the contrary, its habits, mode of life and environment that have in the course of time controlled the shape of its body, the number and state of its organs and, lastly, the faculties which it possesses.[22]

The animal's organs were thus shaped by its needs, which in turn were determined by the circumstances in which it might find itself. And whatever changes were so acquired would be inherited, as long as the external circumstances remained the same. Thus in the classic examples, the giraffe obtained his long neck by browsing on the tall branches of trees, moles became blind as a result of living underground, ant bears lost their teeth through their habit of swallowing their food whole, and birds who lived in water acquired webbed feet.

This theory of adaptation and evolution has been Lamarck's chief claim to fame. Yet there was much more in him that Darwin might have found suggestive. His conception of the course of evolutionary development was by no means the straight and simple ladder of most of his predecessors. On the contrary, he clearly envisaged a "branching series irregularly graded,"[23] reminiscent of Darwin's image in his 1837 notebook of a "tree irregularly branched." And like Darwin he looked to the practices of domestication for the principles of nature:

> What nature does in the course of long periods we do every day when we suddenly change the environment in which some species of living plants is situated. . . . Where in nature do we find our cabbages, lettuces, etc., in the same state as in our kitchen gardens? And is not the case the same with regard to many animals which have been altered or greatly modified by domestication?[24]

There were also clear intimations of a struggle for existence and the idea of the survival of the fittest:

> As a result of the rapid multiplication of the small species, and particularly of the more imperfect animals, the multiplicity of individuals might have injurious effects upon the preservation of races, upon the progress made in perfection of organisation, in short, upon the general order, if nature had not taken precautions to restrain that multiplication within limits that can never be exceeded.
>
> Animals eat each other, except those which live only on plants; but these are liable to be devoured by carnivorous animals.
>
> We know that it is the stronger and the better

equipped that eat the weaker, and that the larger species devour the smaller.[25]

Other of his theories, at the time equally unorthodox, are now little remembered, for the insufficient but usual reason that acceptance has made them commonplace. He clearly expressed, for example, those uniformitarian principles which Lyell, three decades later, was to propound as the new geological dispensation; and he used these principles, as Lyell did not, to support the theory of evolution:

> Why are we to assume without proof a universal catastrophe, when the better known procedures of nature suffices to account for all the facts which we can observe?

> Consider on the one hand that in all nature's work nothing is done abruptly, but that she acts everywhere slowly and by successive stages; and on the other hand that the special or local causes of disorders, commotions, displacements, etc., can account for everything that we observe on the surface of the earth, while still remaining subject to nature's laws and general procedures. It will then be recognized that there is no necessity whatever to imagine that a universal catastrophe came to overthrow everything, and destroy a great part of nature's own works.[26]

> Time and a favourable environment are as I have already said nature's two chief methods of bringing all her productions into existence; for her, time has no limits and can be drawn upon to any extent.[27]

> Losing trace of what has once existed, we can hardly believe nor even conceive the immensity of our planet's age. Yet how much vaster still will this antiquity appear to man when he shall have been able to form a just conception of the origin of living creatures, as well as of the causes of their gradual development and improvement, and above all when he shall perceive that time and the requisite conditions having been necessary to bring into existence all the living species now actually to be seen, he

himself is the final result and actual climax of this development of which the ultimate limit, if such there be, can never be known.[28]

In the light of all this, it would be churlish to deny to Lamarck, as so many Darwinians are today inclined to do, a prominent place among those who anticipated and even influenced Darwin. Darwin himself was not always so ungracious. His notebooks of 1837 and 1838 testify that Lamarck was an important source of inspiration, both for the theory of evolution in general and for suggestions of specific evolutionary processes and mechanisms. Even then, to be sure, there were reservations: "I am sorry to find Mr. Lamarck's evidence about varieties is reduced to scarcely anything—almost all imagination."[29] But he appears in Darwin's list of predecessors and anticipators; his authority is cited on such matters as the formation of the bat's wings; and there is a striking tribute to him as a scientist and thinker:

> Lamarck was the Hutton of geology [*sic*]. He had few clear facts but so bold and many [?] profound judgments that [?] Was endowed with what may be called the prophetic spirit in science, the highest endowment of lofty genius.[30]

Unfortunately, this early generous estimate of Lamarck is little known, although it has far more bearing upon the formative period of Darwin's life than his somewhat later, better-known, and more critical judgment. Today Lamarck is remembered in the harsh accents of the Darwin of the 1840's and '50's, who, occupied in developing his own theory and distinguishing it from that of his most famous predecessor, sought to assert his individuality and originality. It is the pronouncements of this middle period that have gained widest circulation: the declaration that the *Philosophie Zoologique* was "veritable rubbish . . . absurd though clever," doing the subject "great harm."[31] Heaven forfend him, Darwin protested, from Lamarck's nonsense of progression and adaptation, at the same time admitting that "the conclusions I am led to are not widely different from his, though the means of change are wholly so."[32] In the *Origin* even this admission was forgotten as he spoke of "my theory" to describe not only natural selection but also evolution in its most general sense. When Lyell, reading the proofs of the *Origin*, objected that it was not fair to Lamarck to say that the most eminent naturalists had rejected the theory of mutability, Darwin hastily amended the passage to read "living naturalists,"[33] insisting privately, however, that he had got "not a fact or idea" from Lamarck.[34] He even used Erasmus Darwin to belittle Lamarck, saying that Erasmus had anticipated Lamarck's theory—or rather error—"exactly and accurately."[35] In the "Historical Sketch" he was more judicious, describing Lamarck as "this justly-celebrated naturalist" who "first did the eminent service of arousing attention to the probability of all change in the organic, as well as in the inorganic, world, being the result of law."[36] (Later still, as will be seen, Darwin became more tolerant of Lamarck, perhaps because his own theory was by then securely established.)

It is generally assumed today that Lamarckism has been thoroughly discredited. Some reputable scientists, however, dispute this verdict, pointing out that some of the experiments which were presumed to be the definitive disproof of his theory are, in fact, indecisive. When Weismann cut off the tails of generations of mice and discovered that each generation nevertheless persisted in being born with tails, it was thought he had disposed of Lamarck. But the experiment failed to meet the essential condition of the theory, which was that the changes were responses to the changed needs and habits of the organism—deliberate mutilation hardly qualifying as a need or habit.[37] It is instructive that so reputable a scientist should have laboriously performed such irrelevant experiments and that even today they should be cited in evidence.

In part, the disrepute in which Lamarck is today held is a reflection of the high repute enjoyed by Darwin. In his own time, however, it was the prestige of the anti-evolutionists that condemned him to disfavor. His most formidable opponent was his colleague Cuvier, the eminent paleontologist. It is ironic that it should have been Cuvier, the arch anti-evolutionist, who perfected the paleontological technique later used so liberally by the evolutionists: the method of deducing from a single bone or remnant of bone the identity and structure of an extinct animal. Although the technique was the same, however, the conclusions were quite different, Cuvier's study of fossils proving to him that the extinct species were entirely unrelated to existing forms:

> There is nothing . . . that could, in the slightest degree, give support to the opinion, that the new genera which I have discovered or established among the fossil remains of animals might have been the sources of some of the present race of animals, which have only differed from them through the influence of time or climate.[38]

From the evidence of geological stratification, he deduced that catastrophes had been both violent and sudden—a conclusion all the more convincing because the particular evidence on which it was based, the geological formations of the Alps, is, in fact, difficult to interpret on any other assumption (as has been admitted by a reputable modern scientist, himself neither a disciple of Cuvier nor an opponent of evolution[39]). From this, Cuvier went on to reason that each such catastrophe would have obliterated most of the existing species, leaving in some oasis of life a species that would perpetuate itself and become dominant in the following era. There was in all this no talk of creation or creator; Lamarck, surprisingly, more often invoked the "Sublime Author of Nature" than did Cuvier. Cuvier, indeed, derided as "metaphysics" all speculation about the ultimate origins of life. Although a pious Protestant, he was not at all the obscurantist he is sometimes made out, his scrupulousness in excluding religion from the controversy making him all the more redoubtable an opponent.

Cuvier's strength was demonstrated when he joined battle with the scientist who had been first his patron, then his colleague, and finally his bitter antagonist, Geoffroy Saint-Hilaire. As Cuvier shared with Lamarck a passion for fossils, so he shared with Geoffroy an interest in comparative anatomy. Geoffroy's imagination was even more sweeping than Lamarck's. It was his theory that the entire animal kingdom exhibited the same basic anatomical type, the auditory bone of mammals deriving from the cranial bone in the fish, the invertebrates having identifiable vertebrate structures, and so on. The implications of this theory for the relation and evolution of species were obvious, for not only were all the higher orders derived from the lower, but all partook of the single essence of "animality."

Geoffroy had elaborated this theory as early as 1794, but it was not until 1830 that he and Cuvier confronted each other in public. In that year Geoffroy submitted to the French Academy of Science a paper written by two younger disciples purporting to establish the vertebrate structure of the ink fish and containing an explicit attack upon Cuvier. Cuvier's refutation of both the ink fish study in particular and the theory in general was as conclusive as it was courteous. Geoffroy himself then personally entered the arena, abandoning the cause of the ink fish and substituting a "theory of analogues" for the "unity of plan" to which Cuvier had objected. These concessions failed to establish his case, and the great majority of the Academy gave the decision to Cuvier—a decision not seriously challenged even by later evolutionists. Huxley himself, reviewing the controversy many years later, was obliged to admit that however right Geoffroy might have been in his general thesis, he could not have been more wrong in detail, not one of his examples having survived criticism. One can more readily appreciate the motives of the atheist chemist, in Hertzen's *Memoirs*, who advised the young revolutionary to abandon the ephemeral pursuits of literature and politics and to cultivate natural science—the science of Cuvier, not the mysticism of Geoffroy, he warned.

What emerged from this debate, however, was not only the victory of the immutabilists but also the vitality of the defeated evolutionists. The theory of evolution may not have been an acceptable doctrine, but it was a respectable subject for controversy, and every majority decision against it was accompanied by a minority one in its favor. It was in testimony to the importance of the debate, rather than to its particular outcome, that Goethe, an ardent supporter of Geoffroy, declared it to be an event far more memorable than the political revolution convulsing France at the time. Now that there was an open conflict between the old scientific order and the new, he exulted, discoveries would no longer be smothered behind closed doors, mind would rule over matter to fathom the "inherent law" of nature, and "there will ·be glances of the great maxims of creation, of the mysterious workshop of God!"[40] The debate, in fact, was not closed. In 1834 the University of Munich, itself a stronghold of Catholicism and orthodoxy, offered a prize for a thesis on "The Causes of the Mutability of Species." And other reputable scientists spoke up in favor of different aspects of the evolutionary theory: the geologist Von Buch, the botanist de Candolle, the anatomist Oken, more tentatively even Humboldt. If their contributions were less spectacular than those of Lamarck or Geoffroy, they were also often more sober and acceptable.

And there were others less well known and possibly less sober, whose writings, published in these first decades of the nineteenth century, might have influenced Darwin, had he chanced to read them. One was James Cowles Prichard, whose *Researches into the Physical History of Man* went through several editions between 1813 and 1847, but whose claim to be a serious predecessor of Darwin is confuted by the contradictory views expressed in the different editions; the edition Darwin read, which was the most popular one, also happened to be one of the least evolutionary, so that Darwin, like others, classed him among the immutabilists. Another was William Charles Wells, several of whose papers for the Royal Society attracted much attention (the famous one, on dew, being used by Herschel and by John Stuart Mill after him as a case study of experimental inquiry). His observation that nature could accomplish with equal efficacy, although more slowly, what domestic selection had been able to do in the promotion of varieties, was contained in the little known essay, "An Account of a Female of the White Race of Mankind, Part of Whose Skin Resembles that of a Negro." Read in 1813 and published after his death as an addendum to his other essays, it made little or no mark, perhaps because Wells modestly presented his theory—essentially a theory of natural selection—as an essay in "conjectual reasoning."[41] And there was a Patrick Matthew who, after the appearance of the *Origin*, took to celebrating himself as the "Discoverer of the Principle of Natural Selection," claiming the distinction on the basis of an appendix to a work on *Naval Timber and Arboriculture* published in 1831.

Darwin did not, as it happened, read Wells or Matthew until after the éclat of his own book had focused attention upon their obscure productions. What he was familiar with, in this formative period, was the more conventional evolutionary literature ranging from Buffon through Von Buch. And although he later tended to dismiss these predecessors rather cavalierly, he was at this time more appreciative, describing his own task as little more than an application of the principles they had laid down. In his early notebooks he experimented with different versions of a disclaimer of originality intended to preface his forthcoming work:

> State broadly scarcely any novelty in my theory, any slight differences the opinions of many people . . . The whole object of the book is its proof, its extension, its adaptation to classification and affinities.

> Seeing what Von Buch (Humboldt), G. H. Hilaire [*sic*] and Lamarck have written I pretend to no originality of idea (though I arrived at them quite independently and have read them since). The line of proof and reducing facts to law only merit, if merit there be, in following work.[42]

The parenthetical reservation that he had come to his ideas independently was not entirely candid, for he had, in fact, read the standard works of evolution before formulating his own theory. It is unfortunate that even in this emasculated form the disclaimer was not suffered to appear in the *Origin*.

An even more striking omission in Darwin's "Historical Sketch" was the name of Lyell. Here it was not Darwin's immodesty or ungraciousness that was at fault, but rather Lyell's vacillation and confusion. On other subjects Darwin never tired of acknowledging, privately and publicly, his debt to Lyell. And most contemporaries agreed that Lyell was the most important single influence preparing the way for Darwin. Yet, in spite of this testimony, Lyell now figures in the history of evolutionary thought not as one of Darwin's predecessors but as one of his followers. How it happened that the master should be so ignominiously demoted to the rank of disciple is one of the most tantalizing problems of Darwiniana.

In the first two volumes of his *Principles of Geology*, published in 1830 and 1832, Lyell dealt with the question of species at some length. That he should have felt called upon to do so in a treatise on geology testifies both to its importance for all naturalists and to its special significance for him. The uniformitarian, one would think—as many of his contemporaries did think—would naturally be an evolutionist, the theory of evolution doing for biology what the theory of uniformitarianism did for geology. Yet this is just what Lyell purported to deny. Uniformitarianism, he said, did not necessarily imply a theory of evolution. And most emphatically it did not imply the doctrine that man was the final stage in the "successive development of species."[43] Rationality, that sum of intellectual and moral attributes by which man is distinguished from the animals, was too unique a phenomenon to be regarded as just another step in the progressive order of nature. It was, Lyell protested, not so much a step as a leap.

To introduce the idea of a leap was a deliberate provocation, for it naturally recalled the basic tenet of uniformitarianism, the denial of leaps in nature. If man could be introduced by such a leap, what was there to prevent other such leaps, and what happened to the uniformity of nature? Lyell himself recognized the dilemma:

If such an innovation could take place after the earth had been exclusively inhabited for thousands of ages by inferior animals, why should not other changes as extraordinary and unprecedented happen from time to time? If one new cause was permitted to supervene, differing in kind and energy from any before in operation, why may not others have come into action at different epochs? Or what security have we that they may not arise hereafter? If such be the case, how can the experience of one period, even though we are acquainted with all the possible effects of the then existing causes, be a standard to which we can refer all natural phenomena of other periods?[44]

The case was so well put against himself that it admitted of no persuasive answer. And, indeed, Lyell's pretended answers were suspiciously half-hearted and contradictory. At one point he retreated slightly, intimating that the leap was less serious than it seemed, the introduction of "a savage horde" of men upon some small portion of the earth being of no greater consequence than the normal introduction of any new species into an area. But if this was so, if the "savage horde" was not so unique a creation, why had he insisted in the first place upon man's uniqueness, the uniqueness that required a leap? Several pages were given over to trying to extricate himself from this fresh difficulty. That he did not succeed is almost less important than the overwhelming impression of an argument devious and tortuous beyond the requirements of the problem and, probably, beyond the patience of even his more sophisticated readers.

The point of the argument became more obscure as it went on. In the first volume it was said that a leap was required in the case of man because of his uniqueness—the implication being that the other species, boasting no uniqueness, might well have evolved from each other. In the second volume, however, Lyell disavowed any such implication of mutability: "Species have a real existence in nature, and . . . each was endowed, at the time of its creation, with the attributes and organization by which it is now distinguished."[45] Lamarck was painstakingly criticized for assuming that the adaptation of an animal to its environment could result in the development of an entirely new faculty or organ, and for relying upon such verbal fictions as "the efforts of internal sentiment" and "the influence of subtle fluids." And all evolutionists were convicted with Lamarck of the same fallacies: the assumption that because species happen to be difficult to distinguish, they are not really distinct; or that because individuals can be modified, species too are transformable. The fact is, Lyell insisted, that there are no examples in nature of transformations of species.

As it is baldly stated here, the case against evolution seems clear and forthright; in its context, however, it was not so unequivocal. To the question of why the theory of mutability, in spite of its obvious faults, had been well received by so many naturalists, Lyell replied by adducing two circumstances in its favor: the practical difficulty of establishing the reality of species, and the theoretical objection to the repeated intervention of a First Cause. Once doubt had arisen regarding the reality and constancy of species, the amount of transformation of which they may be judged capable would seem to depend only on the quantity of time at their disposal. Given enough time, all becomes flux, the First Cause is dispensed with, and the naturalist finally "renounces his belief in the high genealogy of his species, and looks forward, as if in compensation, to the future perfectibility of man in his physical, intellectual, and moral attributes."[46] The explanation is so reasonable and amicable that one wonders where Lyell's own sympathies lay.

Again, suspicions are aroused by the unnatural emphasis given to some points. Thus the bulk of the discussion of variability was occupied not with the ostensible conclusion —that species were only variable up to a point, after which they died—but with numerous demonstrations of variability and the ability of species to accommodate themselves to changing circumstances. Similarly, the account of domestication, dwelling leisurely on the many remarkable changes induced in animals and plants, was hastily terminated by the observation that hybrids were generally sterile—a conclusion both anti-climactic and irrelevant, since the domesticated forms in question were admittedly not sterile. Sometimes the ambiguity was less a matter of what was said than what was left unsaid. Refuting the evolutionary thesis, Lyell referred to the mummies of animals removed from Egyptian tombs and found to be identical with species now existing—a discovery that had been widely publicized by Cuvier for the same polemical purpose. No less familiar, however, was Lamarck's rebuttal: that since the physical conditions of Egypt had not changed much in the past two or three thousand years, the species might be expected to remain the same; and that in any event a few thousand years was very brief "with reference to the time occupied by the great changes occurring on the surface of the earth."[47] This last point must have raised familiar echoes for Lyell, who had not only just finished reading Lamarck but had so often made the point himself. It would seem that when he did not actually go out of his way to provide ammunition for his ostensible enemies—as in the discussion of variability and domestication—he apprised the knowing reader, familiar with the literature of the controversy, with its exact location.

The general effect is of an argument pulling in two directions at once. Even while ostensibly arguing against mutability, Lyell was also insistent that if some species die out, as they have been known to do, others must be arising to take their place. And although he admitted that he had no definite evidence of the creation of new species, he did explain why such evidence would be hard to come by: The period of human observations has been minute compared with the history of the animate world, and species may have originated even in recent times only to escape detection; within living memory the numbers of known animals and plants in some classes have doubled and even quadrupled, and it is impossible to say which have only now been identified and which have just come into existence. The point about these arguments is that they serve equally well the purposes of the evolutionist, the evidence of transformation being exactly on a par with that of the creation of new species. When Lyell further confessed himself unable to explain how these new species came into existence—"if we could imagine the successive creation of species to constitute, like their gradual extinction, a regular part of the economy of nature"[48]—the balance was thrown in favor of the evolutionists, who had only to apply the principles of uniformitarianism to the problem of creation.

Lyell was too shrewd a polemicist not to be unaware of the ambiguous effect of his argument. And his letters written at the time confirm the suspicion that his basic sympathies were with the evolutionists. When he first read the work of Lamarck, he reported himself delighted with it as with a superior novel; the geologist, he explained, was particularly susceptible to its imaginative appeal, since he knew so well "the mighty inferences which would be deducible were they established by observations." Unfortunately, he found, they were not established by observations, and he reluctantly—reluctantly, because he said he had none of the common theological objections to Lamarck —declared him an "advocate on the wrong side." At the same time he said he agreed with Lamarck on such points as the age of the earth, and confessed himself pleased that Lamarck had had the courage and logic to pursue his argument to the point where man would have had to emerge from the orangutan. "After all, what changes species may really undergo! How impossible will it be to distinguish and lay down the line, beyond which some of the so-called extinct species have never passed into recent ones."[49]

In the case of Lamarck, Lyell's ambivalence apparently derived from a sympathy for his general aims combined with a genuine distrust of his facts and method. In the case of evolutionary theory in general, this ambivalence was more complicated. A letter to a disciple about to review the *Principles* explained why he had avoided the metaphysical questions of beginnings and ends, thus coming perilously close to the evolutionist position: "It is not the beginning I look for, but proofs of a *progressive* state of existence in the globe, the probability of which is *proved* by the analogy of changes in organic life." The idea that "probability" can be "proved"—and proved by analogy—

suggests that his scruples about evolution were often far removed from either logic or fact. That his equivocation was deliberate, being inspired largely by thoughts of social expediency, is borne out by the rest of the letter:

> I was afraid to point the moral, as much as you can do in Q.R. [*Quarterly Review*] . . . If we don't irritate, which I fear that we may . . . we shall carry all with us. If you don't triumph over them, but compliment the liberality and candour of the present age, the bishops and enlightened saints will join us in despising both the ancient and modern physicotheologians. It is just the time to strike, so rejoice that, sinner as you are, the Q.R. is open to you. If I have said more than some will like, yet I give you my word that full *half* of my history and comments was cut out, and even many facts; because either I, or Stokes, or Broderip, felt that it was anticipating twenty or thirty years of the march of honest feeling to declare it undisguisedly.[50]

As is often the case with subterfuges of this sort, Lyell's opponents were less taken in by it than his friends. Thus Sedgwick never doubted that as a uniformitarian, Lyell must also accept mutability, including the mutability of man. Seven years—and four editions of the *Principles*—later, he was strengthened in his opinion, going so far now as to include Lyell among those "infidel naturalists" who thought that the creation of new species was going on even at the present day as part of the regular order of nature. Lyell's reply was a masterpiece of vacillation. He protested against the association with infidels, but not against the views rightly imputed to him. He objected that he had never claimed the doctrine of the origin of new species to be capable of absolute proof, and that he had not presented it as his own opinion but had left it entirely to be inferred by the reader. At the same time he repeated how "unphilosophical" it was to suppose that no new species had been introduced to replace the dead; and he concluded that the real question was not whether species were still coming into existence but how they did so.[51]

While Sedgwick was taking Lyell to task for being a crypto-evolutionist, Herschel, who had himself been converted to evolution by uniformitarianism, was criticizing him for deserting his own principles. Lyell disarmingly replied that he was pleased to find Herschel agreeing with him that new species originated "through the intervention of intermediate causes." "I left this rather to be inferred," he candidly added, "not thinking it worth while to offend a certain class of persons by embodying in words what would only be a speculation."[52] As "speculation," he confessed himself well disposed to the theory of mutability. And as speculator, indeed, he proved himself in this letter more daring than most evolutionists, going further in anticipation of the theory of natural selection than had any of Darwin's other predecessors. It was in June 1836, before Darwin's return from the Beagle, that Lyell wrote:

> When I first came to the notion, which I never saw expressed elsewhere, though I have no doubt it had all been thought out before, of a succession of extinction of species, and creation of new ones, going on perpetually now, and through an indefinite period of the past, and to continue for ages to come, all in accommodation to the changes which must continue in the inanimate and habitable earth, the idea struck me as the grandest which I had ever conceived, so far as regards the attributes of the Presiding Mind. For one can in imagination summon before us a small part at least of the circumstances that must be contemplated and foreknown, before it can be decided what powers and qualities a new species must have in order to enable it to endure for a given time, and to play its part in due relation to all other beings destined to coexist with it, before it dies out . . .
>
> It may be seen that unless some slight additional precaution be taken, the species about to be born would at a certain era be reduced to too low a number. There may be a thousand modes of ensuring its duration beyond that time; one, for example, may be the rendering it more prolific, but this would perhaps

make it press too hard upon other species at other times. Now if it be an insect it may be made in one of its transformations to resemble a dead stick, or a leaf, or a lichen, or a stone, so as to be somewhat less easily found by its enemies; or if this would make it too strong, an occasional variety of the species may have this advantage conferred on it; or if this would still be too much, one sex of a certain variety. Probably there is scarcely a dash of colour on the wing or body of which the choice would be quite arbitrary or which might not affect its duration for thousands of years . . . But I cannot do justice to this train of speculation in a letter, and will only say that it seems to me to offer a more beautiful subject for reasoning and reflecting on, than the notion of great batches of new species all coming in, and afterwards going out at once.[53]

If Lyell is to be judged by his private rather than public sentiments, he must be accounted among Darwin's predecessors. His misfortune, it may be hazarded, was that if his views of species were in advance of his age, his faith in tradition and his notion of the philosopher as one who avoids unsettling established beliefs and institutions were behind the times. He must have been one of the last to carry over the aristocratic idea that a thinker has not only the right but also the obligation to conceal his true thoughts from the public. He was as consciously an esoteric philosopher as most of the sages of antiquity. As early as 1827, he could be found rebuking a geologist who had espoused the uniformitarian cause with more zeal than discretion:

I think he ran unnecessarily counter to the feelings and prejudices of the age. This is not courage or manliness in the cause of Truth, nor does it promote its progress. It is an unfeeling disregard for the weakness of human nature, for as it is our nature (for what reason Heaven knows), but as *it is* constitutional in our minds, to feel a morbid sensibility on matters of religious faith, I conceive that the same right feeling which guards us from outraging too violently the sentiments of our neighbours in the ordinary concerns of the world and its customs should direct us still more so in this. If I had been Sir A. Campbell, I would have punished those Christian soldiers who dug up the idols of the Burmese temples in the late campaign, and sent them home as trophies. To insult their idols was an act of Christian intolerance, and, until we can convert them, should be penal. If a philosopher commits a similar act of intolerance by insulting the idols of an European mob (the popular prejudices of the day), the vengeance of the more intolerant herd of the ignorant will overtake him, and he may have less reason to complain of his punishment than of its undue severity.[54]

Two years later he similarly condemned Milman's "crime" in publishing his controversial *History of the Jews* as a "popular book."[55] At the same time he hoped that the Milman scandal would create a diversion in his own favor, for it was a popular book he was himself preparing. "How much more difficult," he wrote, having in mind more than the obvious difficulty of exposition, "it is to write for general readers than for the scientific world." And he congratulated himself on having succeeded in expressing himself without exposing himself, two of his friends assuring him that no one save an ultramontane could be offended by his book.[56]

The modern reader will almost certainly be distressed by what strikes him as the low tone of these confessions. That Lyell was able to make them so freely suggests that they were not felt to be so reprehensible then as they would be today. By now democracy has succeeded in denigrating what was once a respectable, even sacred, literary convention, the convention of esoteric truth. Yet already in his time the convention was becoming obsolete, so that some of his own friends failed to recognize it—Darwin himself, like Herschel, taking the *Principles* at face value. In the 1840's and '50's the situation, as will be seen, changed, and if then, after Darwin had become Lyell's intimate, he still persisted in assigning Lyell to the ranks of the immutabilists, it was not so much a reflection of Darwin's obtuseness as of Lyell's withdrawal and retreat. But in 1837 and 1838, when Darwin was seeking his predecessors and authorities, Lyell properly belonged among them.[57]

CHAPTER 9

PROGRESS, DIGRESSION, AND COLLABORATION

A biography of Darwin inevitably becomes an exegesis on his autobiography, the opinions and illusions a man has about himself being the most revealing thing about him. Pages may be required to elucidate a single remark about the Baconian method, a simple testimony to the inspiration of Malthus, or the seemingly innocent chronology following his discovery of the principle of natural selection:

> But I was so anxious to avoid prejudice, that I determined not for some time to write even the briefest sketch of it. In June 1842 I first allowed myself the satisfaction of writing a very brief abstract of my theory in pencil in 35 pages; and this was enlarged during the summer of 1844 into one of 230 pages, which I had fairly copied out and still possess.[1]

The studious delay would seem to confirm his intention to "avoid prejudice," to collect more facts and examine the theory more carefully before finally committing himself to it. In fact, however, although three and a half years intervened between the reading of Malthus in October 1838 and the beginning of his first sketch in May 1842, the interval sounds longer than it was. For during those years the origin of species could occupy only leisure moments of his thought. He was more than usually ill at the time, and when he could, he had more pressing matters to attend to:

his book on *Coral Reefs*, papers for the Geological Society, and work connected with the *Zoology of the Voyage of the Beagle*. After correcting the last proofs of *Coral Reefs* in May 1842, he left Down for a month's vacation at Maer and Shrewsbury. It was then, before going on to a short geological expedition in Wales, that he wrote his first sketch of the species theory.[2]

Nor did the two years following the writing of this first sketch leave much more time for research on species. As soon as he returned from Wales in the summer of 1842, he started work on the *Geology of the Volcanic Islands*. After its publication in the spring of 1844, he complained that it had cost him eighteen months, which meant that it had occupied very nearly all his working time between the middle of 1842 and the beginning of 1844. Yet it was early in 1844, immediately after finishing the proofs of this book, that he started his second, much enlarged sketch of the species theory, completing its 189 pages in July. This second sketch—in length, detail of treatment, and formality of composition—was as much a finished work as many a first edition (notably the first edition of Malthus' *Essay*).

Darwin expressed his own high regard for it by having it copied, proofread, and bound, the pages of text (231 in this copy) being interleaved with blank sheets to permit later amplification. He also provided for its disposition in the event of his death. Attached to the manuscript was a letter addressed to his wife: "I have just finished my sketch of my species theory. If, as I believe, my theory in time be accepted even by one competent judge, it will be a considerable step in science."[3] There then followed his "most solemn and legal request" that if he should die before completing his work, four or five hundred pounds be assigned for the editing and publication of this sketch. He named several editors in the order of preference: Lyell, Forbes, Henslow, Hooker, Strickland, and Owen. (Ten years later, still anxious about the future of his work, he amended the list, putting Hooker first; it was probably then that he struck out Owen's name.)

The rest of Darwin's instructions cast some light upon the extent of his research at this time. He directed that those of his books that were scored or contained page references be turned over to the editor, together with quotations from other works collected in eight or ten portfolios. That he had no great respect for the state of his materials may be deduced from the wide latitude he not only permitted but enjoined upon his editor, the several hundred pounds being intended as remuneration for the long labor of examining the references and notes, and enlarging, altering, and, above all, correcting—Darwin underlined this last—the sketch. He left it to the editor's discretion whether the added facts should be included in the body of the text or introduced in the form of notes or appendices, cautioning him that "many of the scraps in the portfolios contain mere rude suggestions and early views, now useless, and many of the facts will probably turn out as having no bearing on my theory."[4]

It is not surprising that his research was not more advanced. The only period he had for sustained work on the subject was in the summer of 1837, after finishing his *Journal* and before setting seriously to work on the geology of the voyage; although even then some time was spent in reading proofs, negotiating for the publication of the zoology volume, and preparing the scheme of this cooperative work. And this was before his reading of Malthus and his formulation of the idea of natural selection. For the rest, his research was largely the by-product of his other work, supplemented by incidental reading, occasional inquiries of friends, or apologetic requests for specimens. This is not to depreciate the information accumulated in this way. But it may suggest reservations about Darwin's own enthusiastic and somewhat ingenuous report of his progress. While preparing his sketch in 1844, he wrote to Hooker:

> I have been now ever since my return engaged in a very presumptuous work, and I know no one individual who would not say a very foolish one. I was so struck with the distribution of the Galapagos organisms, etc. etc., and with the character of the American fossil mammifers, etc. etc., that I determined to collect blindly every sort of fact, which could

bear any way on what are species. I have read heaps
of agricultural and horticultural books, and have never
ceased collecting facts. At last gleams of light have
come, and I am almost convinced (quite contrary to
the opinion I started with) that species are not (it is
like confessing a murder) immutable.[5]

It was not often that Darwin dramatized himself—and the
facts—so blatantly. The "gleams of light" appearing only
after years of blind investigation, the implication that until
only recently he had been a firm immutabilist, the anguish
of conversion—none of this can be taken seriously. Darwin
was, in fact, as little disturbed by his conclusion as he was
surprised by it.

What he might more justly have prided himself on was
not the slowness but rather the speed with which his theory
matured. It was only because of his peculiar combination
of modesty and pride—modesty in regard to his present
accomplishments, pride in his ambitions for the future—
that he suffered the manuscript of 1844 to go by the name
of a "sketch." Reading it after Darwin's death, Huxley was
surprised to find how fully developed it was, resembling
the *Origin* not only in the main lines of argument but
even in the details.

The earlier sketch of 1842, which was genuinely a sketch
—elliptical, unsyntactical, with many private references—is
an even more startling anticipation of the *Origin*. Like the
Origin, it opened with a statement of the principle of variation, and included such subtleties of the principle as that
variation is greater when several generations of a species
are exposed to new conditions; that while some variations
originate as a result of exposure to new conditions, others
originate in the process of reproduction; and that the tendency to vary is counteracted by a tendency for variations
to be lost in crossbreeding unless preserved by isolation or
selection. Again, like the *Origin,* the discussion proceeded
from domestic selection to natural selection, with nature
found to be more skillful than man in producing organisms
that were well adapted to their environment—as if "a being
infinitely more sagacious than man (not an omniscient cre-

ator)"[6] had diligently pursued a policy of selection and perpetuation over many thousands of years. And there were other more obvious anticipations of the *Origin* in the discussion of the geographical distribution of species, their affinities and analogies, and the evidence of morphology, embryology, and abortive organs. Nor were the difficulties of the theory ignored: the supposed sterility of hybrids (countered by the idea of a "gradation" of sterility[7]); the intricacies of such organs as the eye and ear; and the absence of intermediate forms in the geological record. Lyell's simile, in which the geological record figures as a history book with several pages or chapters torn out, appeared here, as later in the *Origin*. And the conclusion, far more carefully written than the rest of the sketch, set the tone of the memorable closing passage of the *Origin*:

> It accords with what we know of the law impressed on matter by the Creator, that the creation and extinction of forms, like the birth and death of individuals, should be the effect of secondary means. It is derogatory that the Creator of countless systems of worlds should have created each of the myriads of creeping parasites and slimy worms which have swarmed each day of life on land and water on this one globe. We cease being astonished, however much we may deplore, that a group of animals should have been directly created to lay their eggs in bowels and flesh of others,—that some organisms should delight in cruelty,—that animals should be led away by false instincts,—that annually there should be an incalculable waste of eggs and pollen. From death, famine, rapine, and the concealed war of nature, we can see that the highest good, which we can conceive, the creation of the higher animals has directly come. Doubtless it at first transcends our humble powers, to conceive laws capable of creating individual organisms, each characterised by the most exquisite workmanship and widely-extended adaptations. It accords better with the lowness of our faculties to suppose that each must require the fiat of a creator, but in the

same proportion the existence of such laws should exalt our notion of the power of the omniscient Creator. There is a simple grandeur in the view of life with its powers of growth, assimilation and reproduction, being originally breathed into matter under one or a few forms, and that whilst this our planet has gone circling on according to fixed laws, and land and water, in a cycle of change, have gone on replacing each other, that from so simple an origin, through the process of gradual selection of infinitesimal changes, endless forms most beautiful and most wonderful have been evolved.[8]

It might be thought that the words "creation" and "creator" would constitute an important difference between this early sketch and the *Origin*, but, in fact, they occur in both. Another point was one that Darwin made much of but that others might be inclined to take more lightly. In his autobiography Darwin rebuked himself for having overlooked in his first two sketches "one problem of great importance":[9] why organic beings, descended from the same stock, diverge in character as they become modified. He could recall, he wrote, the very spot on the road where the solution came to him: the divergencies came about because the modified offspring became adapted to diversified places in the natural economy. It would be foolhardy to quarrel with so vivid a memory. But one may wonder, why it took so long for him to grasp both the problem and the solution and whether, in fact, he had not already implicitly understood it, without consciously formulating it. He certainly seemed to come very close to making the point in the first sketch; and it would appear to be implied in the very idea of his "theory of descent."[10]

Compared with these minor differences, the similarities were overwhelming. This was true not only of the details and structure of the argument but also of its strategy. Thus in the sketch, as in the *Origin*, he specifically excluded from his argument the development of man. Yet in both he could not resist the temptation to comment on it. In the sketch he noted as an aside: "Good place to introduce, saying reasons hereafter to be given, how far I extend theory, say to all mammalia—reasons growing weaker and weaker."[11] Considering the strength of his private conviction and the long temptation, dating back to 1837, under which he labored, it is remarkable how firmly he persisted in his resolution not to discuss the evolution of man.

It might have been expected that in 1844, after finishing the second sketch, Darwin would have set to work in earnest on his *chef-d'oeuvre*. Instead, the decade after 1844 passed, as had the years before, in work related to but not directly concerned with the *Origin*. For more than two years he worked on the *Geology of South America,* taking time out only to revise his *Journal* for a second edition.[12] In October 1846, with both the revised *Journal* and the *Geology* finally completed, he turned to what he debonairly called "a little zoology," and then, "hurrah for my species work."[13] This little zoology, a study of Cirripedia or barnacles, occupied him for all of eight years and resulted in four volumes: two (totaling over a thousand pages) on the existing forms, and two slight volumes on the fossil forms.

After Darwin's death, Huxley praised him for the "scientific insight" that made him see the need for training in biology, anatomy, and taxonomy, and for the "courage" that permitted him to acquire that training by years of patient toil.[14] But Huxley had not been personally acquainted with Darwin in the forties; later, learning of his deficiencies in these disciplines, he assumed that Darwin had sought to repair them. Hooker, who had known Darwin well at the time, had no such illusions about his friend's insight or courage, realizing that he had not contracted for an eight years' apprenticeship but only for a few months' casual work. The project had started as an investigation of a particular barnacle found on the coast of Chile; this led to the examination of other forms of barnacles and eventually to a study of the entire group. It was not to acquire training that he pursued the work but only, as his son later explained, because of his "dogged love of doing anything to the best of his power."[15]

Although the barnacles project was not undertaken as a deliberate course of apprenticeship, it probably had that effect. Hooker has said that only after the completion of the barnacles work did Darwin regard himself as a "trained naturalist."[16] It also had other results, such as providing an example of one of those "intermediate forms" for which Darwin was always seeking; this was a bisexual barnacle harboring a microscopic male parasite, which Darwin took to confirm his theory that "an hermaphrodite species must pass into a bisexual species by insensibly small stages."[17] More important was the overwhelming evidence it furnished of variability, for even the same organ of the same species of this primitive crustacean was found to vary, and Darwin was confronted with the problem that vexes all systematists—that of distinguishing between varieties and species: "After describing a set of forms as distinct species, tearing up my MS., and making them one species, tearing that up and making them separate, and then making them one again (which has happened to me), I have gnashed my teeth, cursed species, and asked what sin I had committed to be so punished."[18]

Nevertheless, it did not take eight years to find out what all naturalists knew: how difficult it is to distinguish between species and varieties or to create a natural classification of species. Darwin had known this long before. In fact, his species theory had little effect upon the barnacles work. His belief in mutability had not influenced his classifications, although he sometimes felt the folly of "discussing, and doubting, and examining over and over again, when in my own mind the only doubt has been whether the form varied today or yesterday."[19] And by the same token, as he himself admitted, the barnacles work had little effect on his species theory, except perhaps to make him more aware of some of its difficulties. Even the biological training he received came too late to influence the basic structure of his theory. To say that Darwin became a "trained naturalist" only after he had served his time with the barnacles is to refuse the title to him when he required it most—when the theory of the *Origin* was not only conceived but elaborated in detail and at length, which was long before the barnacles study. Darwin himself did not think the work was worth the time and energy that went into it; had he foreseen what a chore it would turn out to be, he later confessed, he would never have undertaken it —hardly the sentiment of the man of courage and insight depicted by Huxley. By 1854, when it was finally finished and the last of the ten thousand barnacles had been sent out of the house, he professed to hate them as not even the homesick sailor on a slow ship hated them. When Bulwer-Lytton introduced in one of his novels a Professor Long, the author of two huge tomes on limpets, Darwin recognized himself in the character.

It is the biographer, more than Darwin, who has cause for impatience, having to put up with a subject who seems almost perversely reluctant to consummate his work, fulfill his mission, and bring his biography to its inevitable climax. However tedious the barnacles were for Darwin, he at least did not labor under the lash of a foreordained future and could yield himself to the satisfactions of the moment: a work well done, the award of the coveted Royal Society medal for biology in 1853, and the knowledge that his name was already among those of the foremost scientists of England.

The species work was not entirely ignored during these years. When Hooker wondered what his inquiries about "ornamental poultry" had to do with barnacles, Darwin told him: "Do not flatter yourself that I shall not yet live to finish the Barnacles, and then make a fool of myself on the subject of species, under which head ornamental Poultry are very interesting."[20] As the years passed—he was forty-five before the barnacles work was finished—his friends became accustomed to the flow of questions which made of his correspondence a course in research, and of research an exalted correspondence course. To relations, old school friends, scientific acquaintances, and even strangers, the inquiries went out: "Whether offspring of male muscovy and female common duck resembles offspring of female muscovy and male common?"[21] "Can St. Helena be classed, though remotely, either with Africa or

South America?"22 "Have you any good evidence for absence of insects in small islands?"23 "At what age [do]
nestling pigeons have their tail feathers sufficiently developed to be counted?"24 "Did you ever see black grey-
hound (or any subbreed) with tan feet, and a tan coloured
spot over inner corner of each eye. I want such case, and
such *must* exist because theory tells me it ought."25 The
smallest contributions, he assured one friend, would be
thankfully accepted: the descriptions of offspring of
crosses of domestic birds and animals, the carcass of a half-
breed African cat, or specimens of lizard's and snake's eggs
(for which schoolboys were offered a reward of two shillings a dozen, to be increased at his discretion).

In the curious medley of amateurs and professionals that
made up the community of natural scientists, no one
seemed to be surprised by this leisurely and haphazard
mode of research. If they had doubts whether anything
would finally come of it, it was on account of Darwin's ill
health and his apparent dilatoriness. What they did not
question was that a systematic treatise could emerge from
so strangely assembled a body of information. In this they
perhaps displayed a wisdom superior to that of a later
generation of more highly professionalized scientists, who
cannot work except under laboratory conditions and who
would sooner abandon their problem than compromise
their methods. It is true that much of the information
Darwin received was unverified. But it was the only way
any one man could hope to amass so heterogeneous a mass
of information, particularly one as incapacitated as Darwin
(although he did breed pigeons himself and conducted
other not too strenuous experiments). In a sense, what
Darwin had improvised was the modern device of the
collective project—a more personal version of it, since those
who contributed data did not take responsibility for the
conclusions and were treated more as confidants than as
collaborators.

The first of Darwin's confidants, although the least satisfactory of them, was Lyell. He was also the least sympathetic of his friends. His "Lord Chancellor" manner and

social snobbery alternately amused and dismayed Darwin.[26] A more serious conflict of temperament and cause for dismay was Lyell's subtle and complex mind, which could not view the problem of mutability as a simple scientific one but saw it refracted through other, less transparent mediums—religious, social, esthetic, and emotional. His resistance to the theory expressed itself now in one, now in another of these terms, but always obliquely, so that Darwin never had the satisfaction of knowing when or why he would suddenly balk. In 1840 Darwin might be heartened by a passage in the sixth edition of the *Principles* which seemed to announce his conversion: "Animate creation is in a state of continual flux. . . . The extinction and creation of species has been and is the result of a slow and gradual change in the organic world."[27] But four years later Lyell had regressed to the point where Darwin felt bound to include him among the opposition.[28]

During the next fifteen years Darwin went on soliciting facts from Lyell and exchanging views with him, occasionally hopeful of his conversion, but more often despairing of his shifting grounds and conflicting loyalties. In July 1856 Darwin triumphantly announced to Hooker: "From Lyell's letter, he is coming round at a railway pace on the mutability of species, and authorises me to put some sentences on this head in my preface."[29] But a fortnight later Lyell was expressing himself to Hooker in the tone not of a humble convert but of a superior magistrate, suspending judgment in regard to the ultimate truth of Darwin's theory and ruling against it for the present on the grounds of expediency. And now it was the expediency of science itself, as well as religion and morality, that spoke against it. Turning his old argument on its head, Lyell now conceded that mutability was the wave of the future: "Whether Darwin persuades you and me to renounce our faith in species (when geological epochs are considered) or not, I foresee that many will go over to the indefinite modifiability doctrine." But this, he argued, was all the more reason for superior men to withstand the tide. A mass conversion to mutability would have the unfortunate effect of debasing scientific standards, the "species multipliers" being quick to take advantage of a theory that seemed to make the boundaries of species an artificial or human contrivance dependent upon their own idiosyncrasies. "So long as they feared that a species might turn out to be a separate and independent creation they might feel checked; but once abandon this article of faith, and every man becomes his own infallible Pope." Moreover, Lyell now discovered, Darwin's theory was of no immediate practical importance to science:

> In truth it is quite immaterial to you or me which creed proves true, for it is like the astronomical question still controverted, whether our sun and our whole system is on its way towards the constellation Hercules. If so, the place of all the stars and the form of many a constellation, will millions of ages hence be altered, but it is certain that we may ignore the movement *now*, and yet astronomy remains still a mathematically exact science for many a thousand year.[30]

To remove the issue to the stratosphere, where it would be of no practical concern, was the most desperate but least successful of Lyell's strategies. For what troubled him most, in fact, was precisely the practical, terrestrial consequence of the theory. He would have been willing to concede it in regard to species, had it not been necessary to implicate man as well. But the transmutation theory, so far from dwelling on the safe plane of astronomical time and space, was obviously a clear and present danger to man on earth. This was Lyell's fear, as he confessed after the publication of the *Origin*. In the meantime, however, it remained unsaid, perhaps because he hoped the theory would collapse before it reached that stage.

If Lyell did not dissociate himself from Darwin during this period, it was not only because he was still ambivalent in his own mind, being attracted to the theory at the same time that its implications repelled him, but also because he could not bring himself to cut his ties with what might become the ruling theory. He had even less desire to be behind the times than to be too far ahead of them. He may also have felt that his presence in the counsels of the invading army would have a salutary, cautionary effect. Or if it did not actually succeed in moderating the intemperates, it would at least make him privy to their intentions. Knowing the worst, he would be in a position to do the best—if not against them, then with them. His continued association with Darwin was enjoined upon him not only by his conscience as a friend and scientist, which obliged him to put his resources at Darwin's disposal, but also by the instinct of the diplomat, for whom the breaking off of relations is an evil almost as disastrous as war itself.

A less complicated and more satisfactory confidant was Sir Joseph Hooker. Twenty years younger than Lyell and eight years younger than Darwin, Hooker represented a generation of natural scientists more professional and specialized than the old. The son of the famous botanist Sir William Hooker, he attended the University of Glasgow where his father held the chair of botany; thus he had the advantage not only of the superior professional training of the Scottish university but also of knowing from the start exactly what career he was going to follow. He later accompanied his father to Kew when Sir William assumed the post of director of the botanic gardens, so that he continued to enjoy excellent facilities for research. Compared with him, Lyell and Darwin appear more than ever as autodidacts and dilettantes.

When Hooker, then twenty-two, applied for the post of naturalist on board the Erebus, he was told that the man appointed would have to be as well known as Mr. Darwin. (This was early in 1839, before Darwin's *Journal* had appeared but after he had circulated for some years in London scientific society.) Hooker sensibly retorted: "What was Mr. Darwin before he went out? He, I daresay, knew his subject better than I now do, but did the world know him? The voyage with FitzRoy was the making of him."[31] In fact, Hooker knew his subject far better than Darwin had before the voyage—Darwin would have been hard put to it to say just what his subject was—and he received the appointment. It was then, while working hard to get his degree before he sailed, that he first met Darwin in person;

a little later he borrowed the proof sheets of the *Journal* from Lyell's father, sleeping with them under his pillow so that he might read them the moment he awakened.

Upon his return from the four-year expedition to the Antarctic, where he repeated part of Darwin's itinerary, he resumed the acquaintance with Darwin, which rapidly developed into intimate friendship. After a time in London and a brief period at Edinburgh University, he took up residence near his father at Kew where he spent the rest of his life, married to Henslow's daughter and, after his father's death, succeeding to the directorship of the botanic gardens. With Darwin at Down and Hooker at Kew, most of their communications were by post; when their correspondence was transcribed after their deaths, it was found to come to over fifteen hundred pages in typescript.

Darwin once complained that there seemed to be something antipathetical in the mental attitudes conducive to systematizing and generalizing.[32] His, he confessed, was the generalizing mind; Hooker's, it is clear, was the systematizing. And since botany was Darwin's weakest field, he leaned heavily upon the master systematizer, Hooker. The constant flow of letters was interrupted only by occasional meetings. Hooker has described the gatherings at Down in Darwin's less ailing periods, when several scientist friends would assemble—each of them, during the course of the day, finding himself closeted alone with Darwin to talk about his particular specialty. When Darwin's health was poor, Hooker would visit alone for days or weeks at a time, bringing his work with him and enjoying Darwin's company at specified times. "It was an established rule," he recalled, "that he every day pumped me, as he called it, for half an hour or so after breakfast in his study, when he first brought out a heap of slips with questions botanical, geographical, etc., for me to answer, and concluded by telling me of the progress he had made in his own work, asking my opinion on various points."[33] After one of these visits, Darwin wrote: "I learn more in these discussions than in ten times over the number of hours reading."[34]

Darwin also profited from experiments carried out at Kew, often at his instigation, and specimens traveled from Kew to Down, while Hooker compiled lists and statistics relating to distribution, variation, transportation, cross-fertilization, and hybridization. Hooker also read and criticized Darwin's manuscripts—the 1844 sketch and the several versions of the *Origin* chapter by chapter. (Part of the manuscript of the *Origin* was lost when, by a typical Victorian economy, it found its way into a drawer full of papers, on the backs of which the Hooker children were permitted to draw. Fortunately, Darwin had kept a rough draft of it.) The chapter on geographical distribution, the first written by Darwin and one of the most crucial in his book, was so greatly indebted to Hooker that Darwin wrote: "I never did pick anybody's pocket, but whilst writing my present chapter I keep on feeling (even while differing most from you) just as if I were stealing from you, so much do I owe to your writings and conversation: so much more than mere acknowledgments show."[35] Examining the record of their collaboration many years later, Darwin's son agreed that "without Hooker's aid Darwin's great work would hardly have been carried out on the botanical side."[36]

In spite of this extensive collaboration, however, Hooker was not, during all these years, a disciple of Darwin. He accepted Darwin's theories up to a degree: the considerable variability of species, the means by which they were transported and distributed, the relationship of allied species, and the kinship with their fossil predecessors. But he stopped short at the crucial point of transmutation. His biographer says that when he went on an expedition to India in 1847, it was as a scientist "possessed but not converted" by Darwin's theories, and that he was disappointed when India failed to provide evidence of transmutation and thus assist his conversion.[37] Seven years later he dedicated his *Himalayan Journal* to Darwin, but although he was then more than willing to accept Darwin's theory if only he could believe it—he had none of Lyell's *arrière-pensées*—he felt obliged in his own work to act on the conventional theory of the fixity of species. Even while he was discussing with Darwin the suggestive example of a single plant containing on different branches flowers previously identified as three distinct genera, he was unconvinced of the theory of mutability; and Lyell was able to describe Hooker as one who, like himself, was clinging to the orthodox faith in spite of these "ugly facts"—apparently Hooker's term for the "abnormal vagaries" found in nature.[38] Each reading of Darwin's manuscript left him more shaken but still not entirely convinced. As late as January 1859, he had only reached the point of believing that although the doctrine was not certainly true, it was at least more useful than any other:

> What I shall try to do is, to harmonise the facts with the recent doctrines, not because they are the truest, but because they do give you room to reason and reflect at present, and hopes for the future, whereas the old stick-in-the-mud doctrines of absolute creations, multiple creations, and dispersion by actual causes under existing circumstances, are all used up, they are so many stops to further enquiry; if they are admitted as truths, why there is an end of the whole matter, and it is no use hoping ever to get to any rational explanation of origin or dispersion of species—so I hate them.[39]

With the best will in the world, he could go no further, and with this Darwin had to be content.

It is curious that Darwin should have had so little success in making converts out of his confidants, and most curious in the case of the third, Thomas Henry Huxley. For Huxley not only had the willingness to believe, which came from close association with Darwin and from the absence of mental reservations, but he had in addition what not even Hooker had: a positive, ideological incentive to believe in just such a theory as that of evolution. An agnostic in religion and a reformer by temperament, Huxley might have been expected to be an immediate and enthusiastic convert to Darwinism. If Hooker, on purely scientific grounds, was becoming impatient with theories of creation because they were "all used up," leaving no scope for reason or reflection,[40] Huxley had all the more reason for disaffection. That he did not, again like Hooker, immediately announce himself a convert testifies both to his own scientific scruples and to the unpersuasiveness of Darwin's theory.

As Hooker was eight years younger than Darwin, so Huxley was eight years younger than Hooker. The son of a schoolmaster turned clerk, he was almost completely self-educated—not only, as was Darwin, self-educated in science because it was not provided for in the formal curriculum, but self-educated in having had almost no formal education at all. Fortunately, his father had a small library and the boy a driving intellect, so that he read all the books he could find and discussed them with his large and intelligent family, particularly with his favorite brother-in-law, a doctor. By the time he was twelve, he was getting up at dawn to read Hutton's *Geology* or Hamilton's *Logic*. At thirteen or fourteen there set in the chronic, life-long indisposition which he characterized as "hypochondriacal dyspepsia,"[41] but which was not allowed to interfere with his study of German, physiology, chemistry, physics, and mathematics, his experiments in electricity, or his reading of Hume, Mackintosh, Guizot, Lessing, Goethe, and Carlyle. The scheme he then devised for the classification of all knowledge was also intended as a personal program of study. He was not yet sixteen when he began his apprenticeship with a doctor in the East End of London, matriculated at the University of London, and started to receive a variety of honors culminating several years later in a gold medal and his M.B. degree.

Only when, at the age of twenty-one, he joined H.M.S. Rattlesnake as assistant surgeon and part-time naturalist did his career momentarily resemble that of Darwin or Hooker. The ship took him as far as New Guinea, landed him in Australia long enough to become engaged to the young woman who eventually became his wife, and returned him to England after four years and much research in invertebrate zoology. But even after his return, as a Fellow of the Royal Society and the author of several distinguished scientific monographs, he had to struggle for a living. Not until 1854, when he was almost thirty, did he receive his first appointment, as lecturer at the Government School of Mines in London with a salary of two hundred pounds a year. Married the following year, after an engagement of seven years, he was favored with Darwin's monitory congratulations: "I hope your marriage will not make you idle; happiness, I fear, is not good for work."[42]

Darwin and Huxley had met in 1851, after Huxley's return from his voyage and while Darwin was still occupied with his barnacles. Darwin's first letter to him, written in October 1851, was a recommendation for the professorship of natural history at the University of Toronto, one of several such abortive schemes.[43] By 1854 they were on intimate terms, Huxley having been established as Darwin's expert in zoology. Finding his mind "as quick as a flash of lightning and as sharp as a razor,"[44] Darwin also turned to him for discussion of the more general questions of natural history, as well as for advice regarding the practical politics of science—who should be sponsored for what medal, or what action taken at the Royal Society Council meetings.

Yet, like Hooker, Huxley was not then a convert. He later described his attitude as "a plague on both your houses."[45] His investigation of molluscs, on board the Rattlesnake, had shown him what he took to be an archetypal structure shared by all the varieties, from which he deduced that the divergencies within any one group must have come about as a result of modification from an original type; he even went so far as to use the word "evolution"—not then used by Darwin—to describe the concept. This was the burden of a monograph published by him in 1853 and the main theme in his translation of selections from the work of Von Baer, from whom he had adopted the idea of the archetype.

So far Huxley was prepared to go toward a theory of evolution. But it was not evolution as Darwin understood it. The state of "active skepticism"—the term he had borrowed from Goethe to describe his own state of mind—that provoked him to be an evolutionist among the ortho-

dox also provoked him to be orthodox with the evolutionists. Huxley recalled with embarrassment that at his first
interview with Darwin he had expatiated, with all the confidence of youth, upon the sharp lines of demarcation separating species and the absence of transitional forms. At the
time he was not aware that Darwin held other views, but
even after he was enlightened, he persisted in his opinion.
Nor was he merely playing the part of devil's advocate. He
was sincerely persuaded that although within any one species there was evolution, there was none between species,
each species having its own unique archetype—a theory that
might be described as transmutation without transition. In
two lectures before the Royal Institution, in 1854 and
1855, he stated that although the embryos of the several
groups of animals revealed traces of a common plan, no
such plan was to be found among the adult forms, and he
explicitly denied the possibility of progressive development
from one group to another. As late as 1857, when he discovered that the peculiar anatomy of the brain which was
said to distinguish man from the apes was not, in fact,
peculiar to man, he saw no reason to abandon his theory
or to embrace evolution, since man and the apes, belonging
to the same major group, would naturally share the same
archetypal structure.

Although this theory of the archetype was satisfactory
so far as it went, it did not go far enough, since it left the
problem of the origin of the archetype unsolved. Yet, even
after years of familiarity with Darwin's views, Huxley remained unconverted, in a state of "critical expectancy."[46]
It was not until he read the Wallace-Darwin papers of
1858 and the complete text of the *Origin* the following
year that he announced himself a convert.

The situation was so odd that even those who were intimately involved were apt to be deceived. That the men
most closely associated with Darwin—those who fed him
material, discussed with him his theories, read his manuscripts, and were afterward prominent among his partisans
—were not, during this long tenure of collaboration, also his
converts and disciples seemed too singular to be believed,

even by those who might have been expected to know it best. The situation was made more anomalous by the circumstance that each of these confidants—Lyell, Hooker. and Huxley—suspected the others of believing more than he himself did and more than, in fact, they all did. While Lyell was firmly denying Sedgwick's charge that he was a crypto-evolutionist, he was warning others that Hooker and Huxley were getting more deeply implicated than they knew. Hooker, on the other hand, knowing his own reservations, assumed Huxley to be the true believer; while Huxley, dramatically conscious of his own skepticism, thought that Hooker was *capable de tout* in the way of evolution.[47]

The anomaly was not confined to these three, for they were only the most prominent of those who were privy to Darwin's views but resisted conversion until the publication of the *Origin*. The American botanist Asa Gray had been given a written abstract of his theory; Darwin's cousin, W. Darwin Fox, a clergyman and amateur naturalist who fed him a constant stream of information, was aware of the point of the inquiries; as was the geologist Leonard Horner, who, as early as 1842, was transmitting suggestions about "intermediate creatures."[48] The roster of those whose brains Darwin picked would have made a very respectable scientific society,[49] and he made it a point to acquaint almost every one of these with the general nature of his theory, so that no one could later complain of being used unfairly. Not until 1857, when he sent the manuscript of his abstract to America, did he acquire discretion:

> You will, perhaps, think it paltry in me, when I ask you not to mention my doctrine; the reason is, if anyone, like the Author of the *Vestiges*, were to hear of them [*sic*], he might easily work them in, and then I should have to quote from a work perhaps despised by naturalists and this would greatly injure any chance of my views being received by those alone whose opinions I value.[50]

Nor was it only his collaborators who were in his confidence. To the extent to which not only the raw material but even the very terms of his theory were common to the entire scientific community, a considerable part of it was in the public domain. The sense in which it was a collective enterprise is not exhausted by reference to the mountain of letters which Darwin sent and received (the five published volumes of his correspondence are only a small part of those extant in manuscript). There was also the large body of scientific literature bearing on the subject. The interesting thing is that the work of his opponents was often as helpful as that of his friends. For the entire community—mutabilists and immutabilists alike—shared the terms of reference and frame of intercourse within which the theory of natural selection developed. Except for the pure systematists, who were neutral in the controversy, the main subjects of their research were also the heads of his theory: variability, geographical distribution, the geological record, adaptation of organs, affinities of types, embryological development, comparative anatomy, the sterility of hybrids, and the effects of crossbreeding; almost the only unfamiliar term was natural selection itself. It is for this reason that Darwin could communicate with his colleagues so easily and could utilize their findings so readily. And it is for this reason too that questions of priority should inevitably arise. What is remarkable is that Darwin was not anticipated more often and more seriously.[51]

It is not surprising to find discredited the conventional, romantic view of Darwin as an invalid genius working out his revolutionary discovery in silence and isolation and finally exploding it before his unsuspecting colleagues. What is surprising is to find that his colleagues, so well prepared for the explosion, should nevertheless have been so taken aback when it came.

CHAPTER 10

CLIMATE OF OPINION ON
THE EVE OF THE *ORIGIN*

IN Disraeli's novel, *Tancred*, published in 1847, a fashion-able lady urges the hero to read the current best-seller, "Revelations of Chaos":

> You know, all is development. The principle is per-petually going on. First, there was nothing, then there was something; then—I forget the next—I think there were shells, then fishes; then we came—let me see—did we come next? Never mind that; we came at last. And at the next change there will be something very superior to us—something with wings. Ah! that's it: we were fishes, and I believe we shall be crows. But you must read it.

Waving aside Tancred's protest, "I do not believe I ever was a fish," she goes on with serene confidence:

> Oh! but it is all proved. . . . You understand, it is all science; it is not like those books in which one says one thing and another the contrary, and both may be wrong. Everything is proved—by geology, you know.[1]

"Revelations of Chaos" was a parody of *Vestiges of Cre-ation* published anonymously in 1844—the boldest and most comprehensive statement of the evolutionary doctrine since Lamarck. Something of its character may be deduced from the assortment of celebrities variously reputed to be its author: Thackeray, Lyell, Prince Albert (then twenty-five years old), Byron's daughter—and even Darwin, a rare case of self-anticipation. (Darwin himself identified the author correctly.) And its influence may be gauged by the fact that it went through four editions in its first seven months and ten editions in as many years. Later it was to be one of the chief priority claims brought against Darwin—although not by the author, who knew better. It was, properly, not a claim against Darwin, whose views by then were firmly settled, as against the public, whom it initiated into some of the mysteries of the doctrine. Like the other priority claims dating from this time, this one reflected not upon the inventiveness or derivativeness of Darwin's mind but only upon the receptivity of his readers—a distinction that is obvious now but was generally forgotten in the heat of the controversy.

The author of the *Vestiges* (as it emerged forty years later) was Robert Chambers, journalist and popular educator, editor of the *Chambers' Journal,* and author of some thirty books on history, literature, and biography. When he decided to invade the new and strategic field of science, he removed himself from the distractions of Edinburgh society to the quiet academic retreat of St. Andrews, where he spent two years mastering the several scientific disciplines and preparing his own treatise. (At the same time he continued his frequent contributions to the *Journal.*) The work was issued anonymously because he realized that his name would not lend authority to it, and because he feared that the abuse it received would only bring discredit upon the other enterprises with which he was associated.

What is remarkable is not that the book should have been so vulnerable to abuse and parody but that it should have been able to put up such a show of seriousness and erudition. The evidence for its thesis—the "vestiges" from which a true law of creation was to be deduced—were drawn from geology, anatomy, zoology, paleontology, and embryology. Its main inspiration, however, came from the ideas of natural law and the uniformity of nature:

If there is anything more than another impressed on our minds by the course of the geological history, it is, that the same laws and conditions of nature now apparent to us have existed throughout the whole time, though the operation of some of these laws may now be less conspicuous than in the early ages, from some of the conditions having come to a settlement and a close.[2]

It was this principle of uniformity that was invoked to discredit the primitive, anthropomorphic idea of a God who created the various species by His "personal or immediate exertion."[3] There was a God, the *Vestiges* insisted, but it was to His greater glory that He did not stoop at "one time to produce Zoophytes, another time to add a few marine mollusks, another to bring in one or two crustacea, again to produce crustaceous fishes, again perfect fishes and so on to the end."[4] Instead, He took the more dignified course of letting nature operate through the law of development:

> The idea, then, which I form of the progress of organic life upon our earth, is that the simplest and most primitive type, under a law to which that of like-production is subordinate, gave birth to the type next above it, that this again produced the next higher, and so on to the very highest, the stages of advance being in all cases very small—namely, from one species only to another; so that the phenomenon has always been of a simple and modest character.[5]

The orthodox party was not at all mollified by this promise of a more dignified divinity. One pamphleteer compared the *Vestiges* with an accomplished harlot who could sing as sweetly as a siren, but who was in her person "a foul and filthy thing, whose touch is taint, whose breath is contamination."[6] The *British Quarterly* denounced it as rank heresy; the *Athenaeum* included it among such kindred humbugs as alchemy, astrology, witchcraft, mesmerism, and phrenology; and the *Edinburgh Review* attacked it with all the rhetoric and learning of Sedgwick himself:

> All in the book is shallow; and all is at second-hand. The surface may be beautiful; but it is the glitter of gold-leaf without the solidity of the precious metal. . . . Sober truth and solemn nonsense, strangely blended, and offered to us in a new material jargon, break discordantly on our ears, and hurt our better feelings. . . . The world cannot bear to be turned upside down; and we are ready to wage an internecine war with any violation of our modest principles and social manners.[7]

The degree of Sedgwick's contempt may be measured in the suggestion that it might have been the work of a woman, partly for the charm of the writing but more for "its ready boundings over the fences of the tree of knowledge, and its utter neglect of the narrow and thorny entrance by which we may lawfully approach it."[8]

There was a small minority of critics who were more amiable, but Darwin and his friends were not among them. Strangely enough, it was the conception of natural law that irritated some of them. The *Vestiges* based its "law of development" on what it took to be the principle of uniformitarianism: the higher orders arising out of the lower because of their inherent proclivity to advance, an impulse implanted in them by God as part of his cosmic arrangement for the natural functioning of the universe. Huxley was goaded into writing a violent review, and although he later regretted its "needless savagery"[9]—one of the few such admissions in his long career as a savage polemicist— it prejudiced him against the theory of evolution for more than a decade. What particularly offended him was the "pseudo-scientific realism" of this philosophy of nature: the notion that science and theology could be reconciled by the simple expedient of having God create laws which the universe would then obey. Laws thus became the "angels or demiurgoi, who, being supplied with the Great Architect's plan, were permitted to settle the details among themselves."[10]

Darwin shared Huxley's low opinion of the *Vestiges*, although he commiserated with the victim of his attack. He was disturbed not only by its conception of law but by its assumption of necessary, progressive development. In his copy of the *Vestiges* was pinned a slip of paper with the memorandum: "Never use the word higher and lower."[11] If he sometimes forgot himself and did use these words, at other times he made it clear that in his evolutionary scheme, "higher" might also mean more degraded, as when the species Typhlops, a blind, degraded, wormlike snake, succeeded the true earthworm.

A more decisive objection for Darwin was the fact that the *Vestiges* had no clear theory of the mechanics of evolution, no idea of how this progression from lower to higher forms was effected. The first edition of the *Origin* dismissed the *Vestiges* briefly and contemptuously: "The author of the 'Vestiges of Creation' would I presume say, that after a certain number of unknown generations, some bird had given birth to a woodpecker, and some plant to a misseltoe [*sic*], and that these had been produced perfect as we now see them."[12] In later editions this sentence was replaced by a quotation from the *Vestiges* describing the two impulses to which organisms were reputedly subject: the impulse to advance and the impulse to be modified in accordance with their environment, neither of which, Darwin commented, accounted for the many and intricate adaptations in nature. He even preferred the Lamarckian theory to this vague talk about impulses; but the *Vestiges* had dismissed Lamarck curtly: "In the present day, we have superior light from geology and physiology."[13] Nor had it been more appreciative of Malthus, declaring him to be not only "disrespectful to Providence" but also, which was worse, wrong in his calculations.[14]

If these conceptions of natural law and evolution had not been enough to alienate Darwin and his friends, the errors of fact and flights of fancy would have done so. Although Darwin admitted that the book was admirably written and organized, he was not reconciled to it, warning Hooker that the author's "geology is bad and his zoology worse."[15] And his own copy of the book had such marginal comments as: "the idea of a fish passing into a reptile, monstrous."[16] There were enough such monstrosities to make Huxley complain of the "prodigious ignorance and thoroughly unscientific habit of mind" of the author.[17] Even one like Darwin, who had no *a priori* objection to the idea of spontaneous generation, for example, was apt to be put off by the assertion that when lime is laid on waste ground, white clover will spring up spontaneously; or by such examples of transmutation as oats cropped before maturity and allowed to remain in the ground over the winter coming up the following year in the form of rye. Nor was he likely to be persuaded by the argument that because a human fetus might retrogress, developing a three-chambered heart like that of the reptile or two-chambered heart like that of the fish, there might also be a comparable advance, with a fish developing a reptile heart or a reptile a mammal one:

> It is no great boldness to surmise that a super-adequacy in the measure of this under-adequacy (and the one thing seems as natural an occurrence as the other) would suffice in a goose to give its progeny the body of a rat, and produce the ornithorhynchus, or might give the progeny of an ornithorhynchus the mouth and feet of a true rodent, and thus complete at two stages the passage from the aves to the mammalia.[18]

Yet, in spite of these excellent reasons for dissatisfaction, Darwin was occasionally moved to sympathy with one who was, after all, on his side. By a temporary relaxation of the polemical rule that one's closest neighbor is one's worst enemy, he was sometimes willing to make common cause with the author of the *Vestiges*. Thus the "Historical Sketch" appended to later editions of the *Origin* concluded its criticism on a somewhat condescending but nevertheless friendly note: "In my opinion it has done excellent service in this country in calling attention to the subject, in removing prejudice, and in thus preparing the ground for the reception of analogous views."[19] He even commended it to Huxley, suggesting that if it did no other good, it might spread the taste for natural science.[20] And the caution to Hooker about the book's bad geology and worse zoology came as an afterthought to some more tolerant remarks: "I have been delighted with *Vestiges* for the multiplicity of parts he brings together though I do [not] agree with his conclusions at all. He must be a funny fellow: somehow the book looks more like a nine days wonder than a lasting work: it certainly is 'filling at the price.' "[21] All told, however, Darwin would probably have much preferred it not to have been written, for if it helped to cushion the shock of his book, it also helped to discredit it in advance. His nightmare was to be greeted as another "Mr. Vestiges," the author of a clever but absurd work.[22]

Where the *Vestiges* often read as a parody of science, the work of Herbert Spencer was more a parody of philosophy. Spencer was the Victorian *philosophe:* a civil engineer and journalist by profession, inventor by avocation, rationalist and atheist by conviction, and autodidact and popularizer by temperament. He pretended to no greater learning than he had, and no greater modesty, simply exploiting all his resources to the full. Charged with superficiality, he retorted that if his knowledge was quantitatively superficial, qualitatively it was the reverse, the important thing being not the number of facts in one's possession but the "cardinal truths" of which the facts were illustrative.[23] And it was in cardinal truths that he specialized. When a botanist friend once told him that had he known as much of plant structure as botanists do, he would never have developed his theories of morphology, he took it not as a criticism of his theories but as a compliment to his unorthodox way of arriving at the truth. And when Hooker patiently explained that one of his "illustrative" facts was only true as an exception, Spencer said it did not matter: "It would do just as well as an example."[24] Huxley's quip, that "Spencer's idea of a tragedy was a deduction killed by a fact,"[25] does not do justice to his extraordinary resiliency; there were, in fact, few tragedies in his universe because he rarely admitted the demise of either fact or deduction.

With the talent for generalizing upon whatever facts were available and speculating when the facts failed him, Spencer was the ideal propagandist and popularizer. Like Darwin, he suffered from sleeplessness and a variety of other nervous and physical ailments, but in spite of much changing of residence, experiments in distraction, and even periods of enforced idleness, he managed to produce a formidable body of literature, including comprehensive works on psychology, biology, sociology, ethics, and politics. He wrote about philosophy without having read any of the classical texts; he confessed that he had never been tempted to open his copy of Locke's *Essay* and had closed Kant's *Critique of Pure Reason* after reading and disagreeing with the first few pages. When he did condescend to read G. H. Lewes' *Biographical History of Philosophy,* perhaps because it was by a friend, it inspired him to undertake the writing of a psychology treatise, completed in little more than a year.

If his training in philosophy was slight, his training in science was only a little less so. He was a confirmed evolutionist at a time when his scientific knowledge consisted of little more than that of a schoolboy who collected stones and experimented with gases. He was converted to evolutionism by, perversely, Lyell's refutation of Lamarck. (Or not so perversely, if the ambiguity of the *Principles* is recalled.) When the *Vestiges* appeared, he had only to read the reviews to know that he would approve of it; and he was attracted to Humboldt's *Kosmos,* published the following year, because he had read quotations from it that seemed to suggest a theory of evolution. It did not much matter that when he actually read the *Vestiges,* he disagreed with its particular account of the evolutionary process; he frankly admitted that what interested him were not the details of the theory but its general import. He discussed the question with Huxley, and was amused by Huxley's quaint notion of "keeping judgment in suspense in the absence of adequate evidence." Huxley demolished every one of Spencer's arguments for evolution without in the least disconcerting him. There were, as Spencer saw it, only two possible alternatives, the theory of special creation and the theory of progressive development; and since he regarded the first as "intrinsically incredible," the second must be right. "Hence," he concluded, "fallacious as proved this or the other special reason assigned in support of it, my belief in it perpetually revived."[26]

It was in 1851, after having been an evolutionist for some time, that he came across the particular formula that was to govern most of his later work. He had been asked to write a notice of a new edition of Carpenter's *Principles of Physiology*. "In the course," he characteristically noted, "of such perusal as was needed to give an account of its contents,"[27] he came across the theory of Von Baer—that the development of all animals and plants was from homogeneity to heterogeneity. This idea of progressive differentiation was then added to the Lamarckian idea of adaptation to become his distinctive principle of evolution.

The theory was first advanced in an essay on the "Development Hypothesis" published in March 1852 in the positivist journal, the *Leader*. The argument was unashamedly abstract: If we do not actually know of any case of transmutation, neither do we know, by actual experience, of any case of special creation; "insensible gradations" plus an infinity of time can be supposed to have produced any changes; "surely if a single cell may, when subjected to certain influences, become a man in the space of twenty years, there is nothing absurd in the hypothesis that under certain other influences, a cell may, in the course of millions of years, give origin to the human race."[28]

A more memorable article, with the less provocative title of "Theory of Population," appeared the following month in the *Westminster Review*. There Spencer proposed that the "proximate cause of progress" was to be found in the pressure of population:

> For as those prematurely carried off must, in the average of cases, be those in whom the power of self-preservation is the least, it unavoidably follows, that those left behind to continue the race must be those in whom the power of self-preservation is the greatest —must be the select of their generation. So that,

whether the dangers to existence be of the kind produced by excess of fertility, or of any other kind, it
is clear, that by the ceaseless exercise of the faculties
needed to contend with them, and by the death of
all men who fail to contend with them successfully,
there is ensured a constant progress towards a higher
degree of skill, intelligence, and self-regulation—a better co-ordination of actions—a more complete life.[29]

Although the phrase "survival of the fittest" was used here
for the first time to elucidate one aspect of the evolutionary
process, Spencer did not realize, until it was too late, just
how much it might have elucidated. Nor was he any
wiser five years later, when he wrote an article with the
promising but unfulfilled title of "Progress: Its Law and
Cause." It was to be his lifelong regret that he had not
thought to make the simple connection of ideas which
would have produced the theory of natural selection.

If Spencer could not, to his chagrin, take credit for the
theory of natural selection, he could, and did, take credit
for coining the phrase "survival of the fittest" and popularizing both the word and the idea of "evolution." ("Evolution" had appeared in the first edition of Lyell's *Principles*,
but only in passing.) And not only popularizing but generalizing as well; for he prided himself on having fought
the battle of evolution on several fronts—biology, philosophy, psychology—before Darwin had so much as entered
the lists. (Later he took on logic as well in his essay, "On
an Evolutionist Theory of Axioms.") He complained that
his contributions were not appreciated; and it is true that
they were severely criticized for their eclecticism, pseudo-
science, speculativeness, and atheism. Taking notice only
of the last, he proudly assumed the mantel of the martyr
and heretic: "The days were days when the special-cre-
ation doctrine passed almost unquestioned."[30] In fact, at
about the same time his *Principles of Psychology* was published, there appeared another book to much the same
effect: Alexander Bain's *Senses and the Intellect*. Based on
the work of James Mill, it purported to show that the

mind of man had developed out of elements existing lower down in the evolutionary scale. Bain, a friend of John Stuart Mill and later of Spencer himself, pretended to no great originality of thought; he was merely setting down what had been familiar to two generations of philosophical radicals.

Spencer has been memorialized by William James as "the philosopher whom those who have no other philosopher can appreciate"[31]—a judgment with which it is all too easy to agree. The dilettante whose writing was as facile as his thinking, the inventor of mechanical gadgets and universal theories, the author of a dozen major works of which his autobiography alone has survived—the image is comic and pathetic. In fact, however, he was not quite the buffoon the professional would make of him. Even apart from his great merit as a popularizer, he was erudite enough, however amateurish and self-taught, to warrant the envy of most academically trained and specialized experts today. That a distinguished anatomist like Huxley should have been able to controvert his views on comparative anatomy is not surprising; what is surprising is that Huxley, who did not tolerate fools readily, found him sufficiently well-informed to discuss anatomy with him and even to read the proofs of some of his books.

Darwin's attitude toward Spencer was typically that of the specialist to the "universalist," admiring him in all fields save his own. As a philosopher, Darwin thought, he had no peer: "I suspect that hereafter he will be looked at as by far the greatest living philosopher in England, perhaps equal to any that have lived."[32] As a biologist, however, although wonderfully clever and a master in the "art of wriggling," he was unhappily not a master of the facts; Darwin confessed that he would much have preferred it had Spencer observed more and thought less.[33] And the essay on population, which Darwin read only after the publication of the *Origin*, he said was full of "dreadful hypothetical rubbish."[34] His final assessment of Spencer was so harsh in parts that it had to be deleted from the published version of the autobiography:

Herbert Spencer's conversation seemed to me very interesting, but I did not like him particularly and did not feel that I could easily have become intimate with him. I think that he was extremely egotistical. After reading one of his books I generally feel enthusiastic admiration of his transcendent talents, and have often wondered whether in the distant future he would rank with the great men, as Descartes, Leibnitz, etc., about whom, however, I know very little. Nevertheless, I am not conscious of having profited in my own work by Spencer's writings. His deductive manner of treating any subject is wholly opposed to my frame of mind. His conclusions never convince me; and over and over again I have said to myself, after reading one of his descriptions, "Here would be a fine subject for half a dozen years' work." His fundamental generalisations (which have been compared in importance by some persons with Newton's Laws!), which I daresay may be very valuable under a philosophical point of view, are of such a nature that they do not seem to me to be of any strictly scientific use. They partake more of the nature of definitions than of laws of nature. They do not aid me in predicting what will happen in any particular case. Anyhow they have not been of any use to me.[35]

With all the more popular manifestations of evolutionism, Darwin was out of sympathy, and the more popular they were, the less they engaged him. The *Vestiges* he read carefully if critically, Spencer more casually and impatiently, and Tennyson probably not at all. Tennyson exercised a royal suzerainty such as no poet is ever again likely to command, his domain stretching over science, philosophy, politics, and morality. Of these science was his particular province. He was an early convert to evolutionism, in spite of his study of Cuvier, and he read Lyell's *Principles,* the works of Herschel, the *Vestiges,* and other such books as they appeared. These studies, reflected in his poetry, earned him the title of the "poet of science." Science, not nature, for he was at the opposite pole from a

Wordsworth, who celebrated nature innocently and personally, who thought it "murder to dissect,"[36] and who despised the

> . . . fingering slave
> One that would peep and botanise
> Upon his mother's grave.[37]

Perhaps only Tennyson, in his time, would have instinctively represented a train, viewed late at night, as a great Ichthyosaurus.

In *In Memoriam* Tennyson staked his claim to both titles: "poet of science" and "poet laureate." It was written to commemorate the death of Arthur Henry Hallam, one of the most gifted of the Cambridge elite known as the "Apostles," who had died while touring Europe in 1833 at the age of twenty-three. The first of these verses was written soon after the event, and most of them had been completed long before the book was published in 1850. Although it was issued anonymously, both the publisher and the author saw to it that its authorship was not in doubt. A first printing of five thousand was sold out immediately, two more were issued the same year, and another two the following year. With Wordsworth's death, the poet laureateship was offered to Tennyson, largely, it is said, because of Prince Albert's admiration for anything savoring of science.

In Memoriam has been variously interpreted as a hymn to faith and a tribute to doubt. The interpretation currently in favor is that Tennyson was, by instinct and primitive conviction, a poet of doubt but that his desire, perhaps unconscious, to conciliate and soothe the orthodox made him pose as a poet of faith. The "philosopher-bard," the "State-mouthpiece"[38] had to stifle his doubts in a paean to faith:

> Strong Son of God, immortal Love,
> Whom we, that have not seen thy face,
> By faith, and faith alone, embrace,
> Believing where we cannot prove.[39]

In true poetic justice, this interpretation goes, the public face has obscured the private, so that today he is known not as the poet of doubt but as the poet of faith. "And having turned aside from the journey through the dark night, to become the surface flatterer of his own time, he has been rewarded with the despite of an age that succeeds his own in shallowness"; so T. S. Eliot has accounted for the decline of Tennyson's reputation.[40] A more sympathetic critic has suggested that this surface flattery was only the surface message of the poem, that in fact Tennyson, unable to repudiate orthodoxy openly, did so obliquely in an esoteric message of skepticism that can properly be understood only today.[41]

However Tennyson may have sought to soothe or flatter, his readers were not so easily deceived. Theologians, to be sure, were quick to seize upon his invocations to God, Bishop Westcott assuring Tennyson's son that "what impressed me most was your father's splendid faith (in the face of the frankest acknowledgment of every difficulty) in the growing purpose of the sum of life, and in the noble destiny of the individual man as he offers himself for the fulfilment of his little part."[42] Others were impressed not so much by the firmness of his faith as by his delicate balance between faith and doubt. The philosopher Henry Sidgwick said that what he found so moving in the poem was the conclusion that "humanity will not and cannot acquiesce in a godless world," that it must learn to live with doubt as well as faith, alternating between the two as night inexorably alternates with day.[43] If, for the majority of its readers *In Memoriam* acted as a "potent sedative" to ease their intellectual and moral anxieties, "a message of hope and reassurance to their rather fading Christian faith,"[44] to many others it raised more problems and anxieties than it removed. John Morley's impression was that it "lent the voice of pathetic music and exquisite human feeling to the widening doubts, misgivings, and flat incredulities of the time."[45] Huxley, as contemptuous of sedatives as he was of the easy consolations of religion, admired Tennyson so much that he was moved to write a poem commemorating his death. And Carlyle paid him the highest compliment by describing him in terms that would have been eminently suited to himself: "A man solitary and sad, as certain men are, dwelling in an element of gloom, carrying a bit of Chaos about him, in short, which he is manufacturing into a Cosmos."[46]

Whether the effect of *In Memoriam* was to stimulate or to subdue doubt, the interesting thing about it was the quality of the doubt it evoked. "Its faith is a poor thing," Eliot has remarked, "but its doubt is a very intense experience."[47] And doubt came from its vision of nature: blind, mechanical, ruthless, "red in tooth and claw." This was the nature that science, particularly geology, had recently brought to light:

> Are God and Nature then at strife,
> That Nature lends such evil dreams?
> So careful of the type she seems,
> So careless of the single life;
>
> That I, considering everywhere
> Her secret meaning in her deeds,
> And finding that of fifty seeds
> She often brings but one to bear,
>
> I falter where I firmly trod . . .
>
> "So careful of the type?" but no.
> From scarped cliff and quarried stone
> She cries, "A thousand types are gone:
> I care for nothing, all shall go.
>
> "Thou makest thine appeal to me:
> I bring to life, I bring to death:
> The spirit does but mean the breath:
> I know no more." . . .[48]

It is no wonder that the poet, witnessing the ravages of nature, should decide: "There lives more faith in honest doubt, / Believe me, than in half the creeds."[49]

Even in the final, conciliatory verses, where some semblance of faith is restored, it is a faith rising out of doubt, out of nature itself. It is evolution that becomes the final hope of man, the promise of salvation:

> . . . They say,
> The solid earth whereon we tread
>
> In tracts of fluent heat began,
> And grew to seeming-random forms,
> The seeming prey of cyclic storms,
> Till at last arose the man;
>
> Who throve and branch'd from clime to clime,
> The herald of a higher race
> And of himself in higher place . . .
>
> . . . Arise and fly
> The reeling Faun, the sensual feast;
> Move upward, working out the beast,
> And let the ape and tiger die.[50]

This is so much a scientist's image of nature that it almost recalls the verse of Erasmus Darwin. Huxley declared Tennyson to be "the first poet since Lucretius who has understood the drift of science."[51] And others went further, celebrating him as a predecessor of Darwin. For not only was "nature, red in tooth and claw" a far more dramatic and deliberate image than the prosaic formulations found elsewhere; it was also distinguished from them in being regarded as part of the evolutionary process itself, part of the development by which man arose out of the "seeming-random forms, the seeming prey of cyclic storms." If *In Memoriam* can be taken seriously as an integrated work, if its unity has not been impaired by its protracted composition or vitiated by a deliberate ambiguity, then the same nature which was at first represented as wantonly cruel and wasteful must also be credited with giving meaning and direction to life:

> . . . I see in part
> That all, as in some piece of art,
> Is toil cooperant to an end.[52]

It is probable that Tennyson later came to regret his youthful enthusiasm for evolution, although he did not entirely discourage the efforts of his admirers to establish him as a predecessor of Darwin; he was too vain to refuse any laurel, however unseemly it might be. When he met Darwin, he suspiciously asked him—or rather told him: "Your theory of evolution does not make against Christianity," to which Darwin affably agreed, "No, certainly not."[53] In spite of this assurance, however, Tennyson's misgivings about the "exaggerations" of the doctrine were not allayed.[54] It was pointedly in rebuke of Darwin that he once observed of the daisy he had just plucked, "Does not this look like a thinking Artificer, one who wishes to ornament?"[55] By the end of his life, in one of his last and least worthy poems, "By an Evolutionist," he had so simplified matters that the brute in man was conquered with age and the soul emerged triumphant:

> . . . I hear no yelp of the beast, and the Man is
> quiet at last
> As he stands on the heights of his life with a glimpse
> of a height that is higher.

Whatever Tennyson's early subterfuge or later retreat, however, many felt what one commentator said, that *In Memoriam* "noted the fact" for which Darwin, years later, was to provide the explanation.[56]

More influential than any one writer in preparing the way for the *Origin* was the mania for geologizing and fossil hunting that seized England in the forties and fifties—more influential because it gave each participant the sense that he himself was in possession of evidence bearing upon the new theories, that he himself was making significant discoveries. Spencer was initiated into the sport when he was working as a surveyor for the Birmingham and Gloucester Railway and competed with his co-workers for the fossils uncovered by the newly laid tracks. Tennyson used to go off on long geological expeditions with the local naturalist. When Henry Adams came to England, he found "gentlemanly geology" to be one of the standard features of the country-house week end.[57] Gentlemen geologist also had the disconcerting habit of turning professional. One famous doctor and amateur fossil collector, Gideon Mantell, became so absorbed by his hobby that he turned his home into a museum, drove out his wife and children, and finally abandoned his medical practice. Sir Roderick Murchison started on his career leading to the presidency of the Royal Society when his enterprising wife, herself an enthusiastic mineralogist and shell collector, determined to wean her husband away from fox hunting to more worthy pursuits; he was persuaded to take up geology by Sir Humphry Davy, who assured him that nothing more was required than the "quick and clear eye" of the sportsman.[58]

A good eye, a stout pair of legs, and a fondness for the outdoors were what induced so many men and even women to take to geology. It was the perfect hobby, suitable for any temperament or circumstance. It suggested the romance of nature, the poetry of mountains, cataracts, and gorges; yet one need venture no further than the neighboring countryside or nearby railway tracks. It was the rare occupation which ladies might follow without embarrassing themselves or their escorts, and which cut across class lines, appealing to the mechanic, the village chemist, the university student, and the squire alike. And it made no great demands upon the intellect. Murchison was assured that fox hunting and country living had already made him acquainted with "most mountain forms and features."[59] And the first of the British Association reports explained the popularity of geology by saying that since it dealt with "a lower order of facts," it could be readily understood.[60]

From the 1830's onward, and increasingly in the forties and fifties, geology was the most popular of the sciences. Harriet Martineau complained that five copies of an expensive work on geology were sold to one copy of the reigning popular novel. It was observed that the geology sections of the British Association were always by far the best attended; they were also the most picturesque, having a large contingent of ladies. At a meeting in Newcastle in 1838 over a thousand people sat through the regular meetings presided over by Lyell, while Sedgwick lectured to three thousand. In London young ladies deserted their embroidery and novels to attend Lyell's lectures at King's College, while their brothers at the university spent their long vacations scrambling up mountains in search of geological specimens to adorn the mantelpiece or library table. John Ruskin, himself an enthusiastic geologist, has described a conversation with one of these young men:

> "How beautiful," I said to my companion, "those peaks of rock rise into the heaven like promontories running out into the deep deep blue of some transparent ocean." "Ah—yes, brown, limestone—strata vertical, or nearly so, dip eighty-five and a half," replied the geologist.[61]

This popular zeal meant that a considerable body of opinion was being initiated into the mysteries of geology at a time when speculations about the origin of species were most rife, when even the orthodox doctrines were being modified and complicated until it was hardly possible to know where orthodoxy ended and heresy started—a state of affairs which, as both the orthodox and the heretics have always appreciated, is particularly conducive to the growth of heresy. Geology had started out bravely enough in the early 1820's, with Buckland holding firmly both to the deluge and to the Mosaic chronology. From the middle of the decade, however, almost every year brought some fresh item of evidence pointing to the co-existence of man with extinct animals. At first Buckland explained these away as the chance superposition of human bones on the ancient site of a fossil species. By 1836, however, in his *Bridgewater Treatise,* he was retreating from the orthodox theory at least to the extent of admitting a vast epoch of pre-Adamite time intervening between the creation of the world and the effective beginning of the Mosaic chronology. Thus antiquity, if it could not be allowed to man, could at least be allowed to inorganic matter. He also put less stress on the deluge; without denying it as a historical fact, he no longer put it forward as the primary and universal geological agency.

Buckland was the most cautious in retreat; with others the withdrawal was more precipitous. Whewell, Conybeare and Sedgwick publicly recanted the doctrine of the deluge, while reaffirming their faith in catastrophism and in the immutability of species. From a later perspective, the abandonment of the diluvial theory may be taken as the first step in a general rout. At the time, however, it seemed only that catastrophism, purged of its impurities, would be better able to withstand the excesses of uniformitarianism. When one group of clerics, led by Cockburn, the Dean of York, launched an attack upon these "defectors," the geologists protested their orthodoxy—a "higher," more subtle orthodoxy, they congratulated themselves. And they had the tacit endorsement not only of the British Association—Cockburn's attack had necessarily to be directed against the whole of the profession—but also of some of Cockburn's superiors in the Church. Buckland himself was shortly appointed Dean of Westminster, and geology continued to enjoy the favors of the most respectable and eminent. When the Archbishop of York came to dine with Buckland, Buckland's son was sent around the table with a silver box displaying the bones of a Siberian mammoth; and when Prince Albert visited Oxford in 1841, his itinerary inevitably included Buckland's geological museum.

This amity persisted as long as man was seen to come under the direct providence and special creation of God. When Lyell issued his proclamation of uniformitarianism, Sedgwick immediately recognized that whatever else catastrophism was obliged to concede—the deluge or the antiquity of the world—it could not compromise on the issue of man. Geology, he insisted, confirmed the tenets of natural religion:

> It tells us, out of its own records, that man has been but a few years a dweller on the earth . . . that man, with all his powers and appetencies, his marvellous structure and his fitness for the world around him, was called into being within a few thousand years of the days in which we live—not by transmutation of species, (a theory no better than a frensied dream), but by a provident contriving power.[62]

Gradually, however, even for the catastrophists, these few thousand years of man's existence on earth became prolonged into hundreds of millennia, as traces of paleolithic man, his flint axes and arrowheads began turning up in the 1840's and '50's, culminating with the discovery of the Neanderthal skull in 1856. The antiquity of the uni-

verse seemed to have as its corollary the antiquity of man.
And just as the antiquity of the universe worked in Lyell's
favor by giving nature all the time it needed in which to
develop itself out of its own resources and without divine
or catastrophic intervention, so the antiquity of man gave
an impetus to the "frensied dream" of man evolving by
means of a natural transmutation of species. By 1847 William Whewell, in the second edition of his *History of the
Inductive Sciences,* was giving fair warning that geology,
presenting the spectacle of one set of animals and plants
disappearing from the face of the earth and another appearing, confronted science with the alternative of either
boldly accepting the doctrine of the transmutation of species by natural agencies or of frankly recognizing the
miraculous nature of creation. He himself elected the second, in the full knowledge that it was neither a naturalistic
explanation nor the only reasonable one.

While the erosive movement went on, wearing away the
ground from beneath the orthodox without making it much
the firmer for the heretics, men clung to whatever theory
promised to reconcile the new facts of science with the
familiar concepts of religion. Hugh Miller was the most
popular of those who sought to mediate between religion
and geology. He was Chambers' opposite number, opposing evolution with much the same amateur enthusiasm and
skill that the author of the *Vestiges* brought to its support.
Both were hard-working, self-taught, didactic Scotsmen
with a passion for public enlightenment and a great gift
for popularization. Miller's *Footprints of the Creator and
Testimony of the Rocks*—the titles were patterned upon the
Vestiges of Creation—argued that geology revealed as often
a regression as a progression of life. In each of the geological epochs, he claimed, it was the higher form that preceded the lower. More important, and what particularly
impressed his readers, was his harrowing picture of the
animal world:

> The strong, armed with formidable weapons, ex-
> quisitely constructed to kill, preyed upon the weak,
> and . . . the weak, sheathed, many of them, in de-

fensive armour equally admirable in its mechanism, and ever increasing and multiplying upon the earth far beyond the requirements of the mere maintenance of their races, were enabled to escape, as species, the assaults of the tyrant tribes, and to exist unthinned for unreckoned ages.[63]

If Miller thus unwittingly helped prepare the way for the theory of evolution based on the survival of the fittest, his intention was the opposite: to warn his readers that God was the God of the Old as well as the New Testament and that it was weak and impious to see nature as a benevolent, progressively evolving system. This did not, however, mean that Miller read the Old Testament literally; like so many others, he interpreted the Mosaic "days" of creation as epochs, with the final three days corresponding to the three geological periods.

There were a dozen other schemes to reconcile the geological with the Mosaic record. One Presbyterian minister hastened to the defense of the New Testament God with a book on *The Primeval World,* in which he explained that the extermination and succession of species was the device by which a benevolent God sought to distribute the happiness of life as widely as possible, within the restrictions of a limited food supply: "When one race of animals have enjoyed their existence, they give place to another race: and thus the pleasure of existence is multiplied manifoldly." And as an afterthought: "True, this requires the introduction of death; but death is no great evil to the lower animals."[64] (He neglected to explain why the food supply should have been limited in the first place.) The author of the *Theology of Geologists* was less easily consoled. Refusing to believe that a merciful God could have created the hideous creatures of pre-historic times, he concluded that the fossils of these monsters were "the outer shells of devilish souls, diluted, so to speak, to the dozenth degree."[65]

Perhaps the most desperate of these schemes was that of Philip Gosse. Gosse was a hard-working, respected naturalist, a specialist in marine biology, the author of the de-

finitive *History of the British Sea-Anemones and Corals*
and of a dozen other more popular works on natural history
(as well as history proper), the inventor of the aquarium,
and the associate of the leading scientists in England. Darwin had met him for the first time in 1855 at a meeting of
the Linnean Society where Gosse had read a paper on a
new species of anemone, and he had been sufficiently impressed to enroll him in his corps of informants. After the
meeting of the Royal Society in 1857, first Hooker and
then Darwin himself had apprised him of the thesis of
Darwin's forthcoming work. According to Gosse's son and
biographer, this was part of the deliberate strategy devised by Lyell, "a great mover of men," to prepare a body-
guard of experts to shout down the "howl of execration"
which was sure to greet the *Origin*.[66] At the same time,
Lyell told him of his own plans to write a book on the
geological history of man. Gosse, a leading member of
that most fundamentalist of all sects, the Plymouth Brethren, was pleased with neither prospect; but respecting and
liking Darwin as he did not Lyell, he chose to direct his
advance rebuttal against the latter. The thesis of *Ompha-
los: An Attempt to Untie the Geological Knot* was that,
contrary to appearances, there had been no gradual change
of the earth's surface or gradual development of organic
life. The semblance of change and development, of a
world on which life had long existed, had been imprinted
on the earth at the very moment of the act of creation. Just
as Adam's navel (hence the title) had been intended to
stimulate a natural birth, so the fossils had been created
by God for the purpose of testing men's faith.

Lyell's strategy succeeded better than he could have
known. The howl of execration that awaited Darwin was
let loose on Gosse. Yet there was nothing really novel in
Gosse's argument; it was simply old-fashioned fundamentalism.[67] Gosse's misfortune was its double untimeliness:
geology had grown too big to be easily persuaded to abdicate; and good Victorians did not take kindly to the idea
of a God so unsporting and ungentlemanly as deliberately
to deceive the faithful. Charles Kingsley, a friend of Gosse,
said that it was not only his reason and his twenty-five
years' study of geology that rebelled against this theory,

but also his conscience. And Carlyle spoke the last word against it when he said that it violated the rules of the game: "Lying is not permitted in this universe."[68]

The attempts "to untie the geological knot" became more desperate as the knot was drawn tighter. For while it was as difficult for conscientious scientists as for conscientious clerics to accept the theory of transmutation, it was also becoming more difficult for both to accept the Mosaic account of creation. Moreover, the dilemma was no longer confined to professional scientists and clerics; the popularization of geology meant that more and more laymen were being caught up in it. Those barbarous young men who were too intent upon fossils and strata to see the beauties of nature, offended Ruskin's religious as much as his esthetic sensibilities: "If only the geologists would let me alone, I could do very well, but those dreadful hammers! I hear the clink of them at the end of every cadence of the Bible verses."[69]

Thus the 1850's, which have been apotheosized as the most tranquil, prosperous, and assured of all decades in English history, were, in fact, a period of intense spiritual anxiety and intellectual restlessness. It is not true, as too hasty contemporaries and too facile historians have suggested, that men were suddenly, at the end of the decade, deprived of a security of faith, a metaphysical consensus about beginnings and ends, and a certainty about the future which had made this the golden age of the modern world. All "golden ages" are suspect, the history of belief having always gone hand in hand with the history of doubt and dispute; and the golden age of Victorian England is as tarnished as most. In the very year of the Crystal Palace, that monument to Victorian complacency, as it is commonly thought, Thomas Carlyle was castigating his contemporaries for their "diseased self-listenings, self-questionings, impotently painful dubitations, all this fatal nosology of spiritual maladies so rife in our day."[70]

In this nosology of maladies, the most spiritually debilitating, Carlyle would have said, was the passion for science. And it was a passion which, by the beginning of the decade, had afflicted every class and institution in society. Mark Pattison observed the change from the Oxford of 1845, conservative, ecclesiastical, and theology-minded, to that of 1850, liberal, secular, and science-oriented. Elsewhere, where the tradition was less venerable and therefore more vulnerable, the disease made more progress. At University College, London, a seventeen-year-old student noted in his journal in 1852 a discussion "about the origin of species, or the manner in which the innumerable races of animals have become produced"; he himself did not doubt that "all animals have been transformed out of one primitive form by the continued influence, for thousands, and perhaps millions of years, of climate, geography, etc."[71]

Darwin once protested against the idea that the subject of the *Origin* had been "in the air" or that "men's minds were prepared for it."[72] Yet both were indubitably true. It was in the air and men were prepared for it—the public for evolution in general, and the scientific community for some special theory that Darwin was known to be working on. This is not, however, to say that because evolution was accessible it was necessarily accepted. The curious thing is that it was accepted most readily by those who knew it least; Tennyson probably made the most converts, while Darwin made none. This anomaly—that it was precisely Darwin, before 1859, who failed to convince his closest and most sympathetic colleagues, while lesser men were making converts by the score—suggests that the quarrel about evolution was not simply an incident in the supposedly chronic struggle between science and religion, truth and prejudice. Spencer was as much prejudiced in its favor as orthodox clerics were against it. And if Hooker or Huxley resisted conversion, it was not because of religious prejudice but only because of reservations about its scientific validity. Only amateurs who did not know enough to have such reservations were uninhibited in their enthusiasm. Such amateur enthusiasms, however, have a notoriously high mortality rate. It was by no means precluded, on the eve of the *Origin*, that evolution, like other "waves of the future," should have petered out before reaching the shore.

RECEPTION OF
THE *ORIGIN*

BOOK IV

CHAPTER 11

THE PUBLICATION OF THE
ORIGIN

In May 1856, on the urging of Lyell, Darwin started what was to be his magnum opus. By June 1858 he had completed about half of it, on a scale three or four times as extensive as the *Origin*. On June 18 the calamity that had been threatening for twenty years finally came to pass. The morning post brought a manuscript describing the theory of natural selection, not as a poetic device or a chance figure of speech but in Darwin's sense and to his very purpose. It had been submitted for his opinion by Alfred Russel Wallace, an English naturalist then exploring Malaya, who had earlier been in correspondence with him. Shattered, Darwin forwarded it to Lyell with a covering letter reminding him of his warning years earlier that he would be forestalled if he did not publish soon. "I never saw a more striking coincidence," he wrote. "If Wallace had my manuscript sketch written out in 1842, he could not have made a better short abstract! Even his terms now stand as heads of my chapters."[1] The only honorable thing to do, he thought (although Wallace himself had not proposed this), was to recommend the paper for publication in one of the scientific journals. Thus, he bemoaned, after all his labor, he was to be deprived of the title to originality. It was small consolation to know that his work would not be entirely lacking in value, since it would still have the merit of elaborating the details of the theory.

Lyell's genius for compromise intervened to salvage something for Darwin out of the wreck of his hopes. He proposed that both Wallace's paper and extracts from Darwin's sketch of 1844 be published together, establishing the rights of both to priority. Darwin, however, was dubious of the ethics of such a course. Would it be honorable to "take advantage" of Wallace's communication and rush into print when he had had no intention to do so? Would it not be "base and paltry" to publish merely in order to establish his priority? It was, in any case, a trumpery affair, and he was ashamed to have to confess to such feelings.[2] Lyell and Hooker, who had also been appealed to, finally convinced him that there was nothing dishonorable in the proposal, and after Wallace heartily agreed to it, Darwin consented to the joint publication of both papers.

At first sight, Wallace seems the least likely rival for Darwin's laurels. Although he was of Huxley's generation rather than Darwin's (he was fourteen years younger than Darwin), he was nevertheless more of an amateur, both in training and spirit, than even the young Darwin. Coming from an improverished, genteel, middle-class family, he attended the local grammar school, where he was given the conventional education that succeeded in teaching him, as he later thought, neither the classics nor anything else. Unlike Huxley, he showed no intellectual talent, and, unlike Darwin, he had not even the primitive interest in nature that the biographer can point to as a premonition of scientific distinction. His enthusiasms as a young man were rather political than scientific; he was a socialist of the Owenite variety, and a mild skeptic in religious matters. When he left school at the age of fourteen, he went to live with his brother to learn the trade of land surveying. It was then that he was introduced to the elementary facts of nature—learning, much to his surprise, that chalk was not the universal substratum of the land, that the pieces of stone he had known as "thunderbolts" were, in fact, fossils (like many surveyors, including Spencer, his brother was an ardent if uninformed collector of fossils), and that the different varieties of flowers, shrubs, and trees had distinctive names. It took another four years, however, for him to advance beyond the recognition of rose and buttercup, and to learn, from a shilling booklet published by the Society for the Diffusion of Useful Knowledge, the elementary classifications of botany.

A depression in the surveying trade, when he was twenty-one, induced him to apply for a teaching post. By private cramming at nights, he learned enough Latin to take the lowest class and, later, to make out the scientific names of birds and insects. He also discovered the town library, where he read, among other things, Humboldt's *Narrative,* Darwin's *Journal,* and Malthus' *Essay on Population,* and where he met the entomologist Henry Walter Bates, who initiated him into the delights of beetle collecting. It was also in this eventful year that he was introduced to those other subjects that were, later in life, to compete with natural history for his affections: spiritualism, psychical research, and mesmerism; he had earlier been converted to phrenology.

The following year he returned to surveying, but his friendship with Bates and his new interests occupied him more and more, until finally he was "bitten by the passion for species."[3] At first his passion was for collecting and describing, later for speculating about their origin. He read the *Vestiges,* Prichard's book, and other works on the subject, all of which he took to confirm the theory of evolution; years later he still thought that the *Vestiges* had been much undervalued and that it had been abused "for very much the same reasons" as the *Origin* later was.[4] About this time he also read a book describing a trip to the Amazon, and he was so taken with the delights of the country that he and Bates determined to make a similar expedition, paying their way by collecting specimens of insects, shells, birds, and animals, and selling their duplicates upon their return. They embarked in 1848 on the voyage that was to be Wallace's first real apprenticeship in natural science, as the Beagle had been Darwin's. He returned four years later in somewhat better financial circumstances and with a growing reputation in scientific circles as the author of *Travels on the Amazon and Rio Negro* (which did not, however, have a good sale).

In 1854 he left again on his more ambitious travels in Malaya. It was there that he wrote the article, "On the Law which has Regulated the Introduction of New Species," which appeared the following year in an English scientific journal. The law of the title was that "every species has come into existence coincident both in space and time with a pre-existing closely allied species"; and he suggested the analogy of the "branching tree" to represent the relations of species.[5] He did not, however, propose any mechanism for the succession of species. Darwin read the paper, having had his attention called to it by Lyell, and in response to a letter from Wallace he wrote that he agreed with "almost every word" of it. He also added that he had himself been working twenty years on the problem, had a "distinct and tangible idea" of its solution, and was preparing a work for publication which, however, was so extensive that although he had already written many chapters, he did not expect to go to press for another two years.[6] Wallace, duly warned off, nevertheless continued to work on the subject and even started to write a long book on it, although he had not yet any distinct and tangible idea such as Darwin's and expected to be spared the labor of finishing it by the appearance of Darwin's work.

It was a year later that the distinct and tangible idea—Darwin's own idea—occurred to him. He was resting at Ternate, an island near New Guinea, suffering from an attack of intermittent fever, when he happened to be reminded of Malthus' *Essay on Population* which he had read many years before.[7] He thought of Malthus' "positive checks to increase"—disease, accidents, war, and famine—which had the effect of keeping the population of savage races far below that of civilized peoples, when it suddenly occurred to him that similar causes might operate in the case of the animal population as well. (In fact, Malthus had himself derived his speculations about the human population from the more obvious situation among animals, but Wallace seems to have forgotten this.) Vaguely contemplating the enormous destruction that must take place every year to keep the faster-breeding species from over-running the earth, he put to himself the question: Why do some die while others live? The answer was obvious: The best fitted live.

> Then it suddenly flashed upon me that this self-action process would necessarily *improve the race,* because in every generation the inferior would inevitably be killed off and the superior would remain —that is, *the fittest would survive.* Then at once I seemed to see the whole effect of this, that when changes of land and sea, or of climate, or of food-supply, or of enemies occurred—and we know that such changes have always been taking place—and considering the amount of individual variation that my experience as a collector had shown me to exist, then it followed that all the changes necessary for the adaptation of the species to the changing conditions would be brought about; and as great changes in the environment are always slow, there would be ample time for the change to be effected by the survival of the best fitted in every generation. In this way every part of an animal's organization could be modified exactly as required, and in the very process of this modification the unmodified would die out, and thus the *definite* characters and the clear *isolation* of each new species would be explained. The more I thought over it the more I became convinced that I had at length found the long-sought-for law of nature that solved the problem of the origin of species. For the next hour I thought over the deficiencies in the theories of Lamarck and of the author of the 'Vestiges,' and I saw that my new theory supplemented these views and obviated every important difficulty.[8]

He waited impatiently until evening when his fever subsided and he was able to make notes of the theory, spent the next two evenings writing it out, and sent it off by the next post to Darwin. In the accompanying letter he expressed the hope that the theory would be as new to Darwin as it was to him, and requested that it be forwarded to Lyell if Darwin thought it of sufficient importance. Thus, in several literally feverish hours, was conceived and composed the four-thousand-word paper that was shortly presented, without revision, before the Linnean Society and formally gave Wallace the title of co-founder of the theory of natural selection.

Formally, but not in fact, Wallace acquired the title. Later those who wished to denigrate Darwin tended to exalt Wallace—not that there was any theoretical difference between the two at this time, but only because any means of denigration served the cause; to deny the originality of the *Origin* was generally the preliminary to a denial of its validity. Wallace himself was never a party to this stratagem. Like Darwin in so many ways—deficient in esthetic faculties, having no talent for languages, no wit in discourse, and little facility in writing, and suffering from a delicate nervous system and ailing constitution—Wallace also shared Darwin's modest temper, his want of assertiveness and egotism. There are few, if any, similar episodes of competing claims in which both parties acted with such generosity and good will. It is all the more unfortunate, therefore, that others should later have made claims on behalf of Wallace, which Wallace himself repudiated. At the fiftieth anniversary meeting commemorating the joint publication of their papers, he made it clear that although the idea of natural selection came to both of them independently, Darwin had the credit of twenty years of priority and work. Their relative contributions, he insisted, could be justly estimated as the proportion of twenty years to one week.

It is not the question of priority or of their relative contributions that is significant, but rather the coincidence of discovery. Wallace himself accounted for this coincidence. First, both were beetle collectors, and no other group of organisms is so impressive for "the almost infinite number of its specific forms, the endless modifications of structure, shape, colour and surface markings . . . and their innumerable adaptations to diverse environments." Second, both had what Darwin called "the mere passion for collecting" and what Wallace described as "an intense interest in the variety of living things . . . an almost childlike interest in the outward forms of living things." Third, both became travelers, collectors, and observers "in some of the richest and most interesting portions of the earth." And finally, both read and were provoked by Malthus. "The effect of that was analogous to that of friction upon the specially prepared match, producing that flash of insight which led us immediately to the simple but universal law of the 'survival of the fittest.' "[9] Of these, the last was clearly decisive, the first three qualifications being shared by a multitude of naturalists. And even the last, the reading of Malthus, was by no means a unique or even rare experience. It is apparent that it was not the coincidence of discovery that is surprising but rather the fact that the coincidence was so long delayed.

A less dramatic but no less significant feature of the Wallace episode was its epilogue: the presentation of their joint memoir to the Linnean Society. On July 1, 1858, the two papers were read under the title: "On the Tendency of Species to form Varieties; and on the Perpetuation of Varieties and Species by Natural Means of Selection." Wallace's contribution was the paper sent to Darwin, while Darwin's consisted of extracts from the 1844 sketch and part of a letter of September 5, 1857, explaining his theory to Asa Gray. The joint memoir was communicated to the Society by Lyell and Hooker, with a prefatory letter explaining the circumstances. Neither Wallace nor Darwin was present,[10] Wallace being thousands of miles away and Darwin as effectively removed by a condition of nervous prostration even more serious than usual. An epidemic of scarlet fever had just descended upon his family, carrying away his youngest child (his namesake, a feeble-minded two-year-old) and endangering the lives of the other children. Too distracted to compose a new abstract, he had contented himself with sending the Gray letter and the 1844 sketch to Lyell and Hooker to be used at their discretion.

Lyell and Hooker were both present at the meeting, and both made some remarks impressing the importance of the subject upon the audience. There was, however, no discussion. This was all the more remarkable, since a monograph designed to prove the immutability of species had been scheduled for the same meeting and withdrawn only at the last moment; yet even its author, the botanist George Bentham, did not offer dispute. Nor did the rest of the audience—some thirty fellows of the Society—seem unduly perturbed by the fact that they were attending to a thesis diametrically opposed to the one that had been scheduled. Nor were there any repercussions from the publication of the joint memoir in the *Journal of the Proceedings of the Linnean Society.* It was not mentioned in an article appearing in the *Edinburg Review* in January 1859, that denied the thesis of mutability and cited, among others, Cuvier, Geoffroy St. Hilaire, Owen, the *Vestiges,* and even Darwin's book on the barnacles. And when, a few months later, Thomas Bell, President of the Linnean Society and a firm immutabilist, reviewed the scientific progress of 1858, he found nothing memorable to record: "The year which has passed . . . has not, indeed, been marked by any of those striking discoveries which at once revolutionize, so to speak, the department of science on which they bear."[11]

Many years later Hooker tried to account for the silence which greeted the reading of the memoir by saying that the interest was so intense and the subject so ominously novel that the "old School" deliberately refrained from entering the lists before armoring. They did, he recalled, talk it over after the meeting with "bated breath," but those who might have spoken out were intimidated, he surmised, by its sponsors.[12] The bated breath must have been bated indeed to muffle the voices of the opposition not only during the meeting but also in their private correspondence following it and in their later recollections. For, apart from Darwin's intimates, there was little reference to it, either contemporaneously or retrospectively. Huxley, the only one to propose dating the public inaugural of Darwinism as July 1, 1858, was not himself present at the historic meeting, having been elected a Fellow only later that year.

At the time, the only public notice of it that came to Darwin's attention was the observation of a Reverend Haughton, who assured the Dublin Geological Society that were it not for the authority under whose auspices it appeared, it would not be worthy of remark. For science, as for the public at large, the effective inaugural of Darwinism had to await the publication of the *Origin* more than a year later. This was the most surprising feature of the Wallace episode: that the premature exposure of Darwin's theory should have provoked so little reaction, and that however much evolution and even natural selection should have been "in the air" and men's minds "prepared" for them, the real shock and awakening was still to come. Darwin himself, it appeared, proved incapable of anticipating the *Origin*.

In July Darwin removed his family from the disease-infested Down (six children in the village had died) to the bracing sea air of the Isle of Wight, where they all recovered, or at least settled down to their chronic ailments. There he began what he called his "abstract" of the species book—an abstract of the long work he had started two years earlier and of which several chapters had already been completed. Originally, this abstract was conceived as a paper or series of papers to be published by the Linnean Society. (The idea of a series was to evade the thirty-page maximum imposed on their monographs.) When it became clear that even a series would not accommodate it, Hooker suggested a grant to subsidize its independent publication as a volume. As the book grew, so did Darwin's ambitions, until he began to think that it might sell enough to pay expenses and warrrant commercial publication.

Lyell, in spite of his private reservations about the theory, was sufficiently impressed both by its importance and by its commercial prospects to recommend it to John Murray, who had brought out the revised edition of the *Voyage of the Beagle*. Darwin wrote to thank him and ask his advice. What terms of publication should he request? Did Murray know of the subject of the book, and, if not, how much should he tell him?

> Would you advise me to tell Murray that my book is not more *un*-orthodox than the subject makes inevitable. That I do not discuss the origin of man. That I do not bring in any discussion about Genesis, etc., etc., and only give facts, and such conclusions from them as seem to me fair.

> Or had I better say *nothing* to Murray, and assume that he cannot object to this much unorthodoxy, which in fact is not more than any Geological Treatise which runs slap counter to Genesis.[13]

Lyell replied that Murray, whatever he knew or thought of Darwin's views, was interested in publishing the book and only took exception to the terms "abstract" and "natural selection" in the title. Darwin reluctantly deferred to his—and Lyell's—objection to "abstract," although he would have liked to indicate that the work did not contain its full corpus of references and facts. "Natural selection," however, he could not dispense with; he proposed, however, to make it less cryptic by adding the explanatory phrase, "Through natural selection, or the preservation of favored races." Since the manuscript was still at the copyist's and would be there for another ten days, Darwin could only send Murray the headings of the chapters. It was, therefore, with great surprise and pleasure that he received, almost by return mail, Murray's offer to publish the book without waiting to see it, and upon the same generous terms Lyell commanded: two-thirds of the net profit. Darwin, always determined to do the honorable thing, accepted the offer on condition that Murrary could retract it after seeing the manuscript.

One reason for Darwin's scruples was his suspicion that Lyell might have induced Murray, against his better judgment, to accept the book. He was not, as it turned out, far wrong. Although Murray confirmed the agreement two weeks later after seeing the first three chapters, a reading of the entire manuscript left him dubious not only about its theoretical validity but also about its commercial promise. He himself, an amateur geologist, thought its theory "as absurd as though one should contemplate a fruitful union between a poker and a rabbit."[14] And his advisers were equally discouraging. The Rev. Whitwell Elwin, editor of the *Quarterly Review* and, like Murray, an amateur naturalist, thought it injudicious to publish an "abstract"—as he regarded it even after reading it—of views so controversial, and recommended that Darwin write instead a book based on his "observations on pigeons," which Lyell had assured him were "curious, ingenious and valuable in the highest degree." "Everybody is interested in pigeons. The book would be reviewed in every journal in the kingdom and would soon be on every library table."[15] Darwin, having spent eight years on barnacles, did not now propose to devote himself to pigeons and refused to be moved even by these seductive visions of fame. When he rejected the suggestion, Murray went ahead with plans to publish the manuscript as it stood. Murray was perhaps fortified by the recommendation of another reader to whom he had given the manuscript, George Pollock, a lawyer friend who found it to be "probably beyond the comprehension of any scientific man then living," but recommended its publication on the curious ground that "Mr. Darwin had so brilliantly surmounted the formidable obstacles which he was honest enough to put in his own path."[16] It was Pollock who induced Murray to publish one thousand copies of the book, instead of the five hundred he had originally intended. Had Darwin known of these behind-the-scenes decisions, he would not so blithely have written to Murray years later: "I think if you had sent the *Origin* to an unscientific man, he would have utterly condemned it."[17]

Darwin himself was more hopeful about the sale of his book than Murray. To Hooker he apologized for his presumptuousness in thinking that it might not incur too heavy a loss; he himself was surprised how much interest there was in the subject. Later, suspecting that he had been immodest, he begged Hooker not to spread it abroad that he expected the book to be "fairly popular" or have a "fairly remunerative sale," which was the most he hoped for it, since he would feel ridiculous if it should prove otherwise.[18] By the time the original printing of 500 copies was increased to 1250, he began to think that not only he but also Murray had been over-optimistic and that Murray might lose on it. It was probably the chore of proofreading that helped subdue his spirits, for he was shocked to discover that he had, as Murray pointed out, practically rewritten the book. Again his conscience suffered, and he offered to recompense Murray for the excessive corrections.

By the end of September the last proofs had been revised, and a month later copies of the book were being distributed to friends and critics. Publication date was November 24, 1859. A few days earlier the whole edition of 1250 copies, at 15 shillings each, had been disposed of at Murray's annual sale.[19] Neither before nor after this sale was the book given particular prominence on Murray's list or in the advertisements. In the first announcement of the publication, an advertisement in the *Saturday Review* of October 15, it was sixth in Murray's winter list of thirty-two works; it was preceded by a *Narrative of the Discovery of the Fate of Sir John Franklin and his Companions,* John Tyndall's *Glaciers of the Alps,* and books on the European Congresses, Wellington, and a Thomas Assheton Smith. And in later advertisements it appeared in the same modest type and inconspicuous position. The publication of a second, very slightly revised edition of three thousand copies on January 7, 1860, earned it the simple additional accolade, in the advertisements, of "fifth thousand."

The sale of the *Origin* was beyond everyone's expectations; yet it was not, by the literary standards of the time, particularly impressive; nor does it begin to measure the true influence of the book. By 1876, at which time the Darwinian revolution was virtually accomplished, it had sold sixteen thousand copies in England. *Adam Bede,* published in 1859 by the then little-known George Eliot and selling at thirty-one shillings and sixpence for the

three volumes, disposed of sixteen thousand copies in its
first year alone. Nevertheless, as works of science went—
and such sober works of science—the *Origin* was a popular
success. Mudie, whose chain of lending libraries made
him one of the chief forces in the dissemination (and
sometimes censorship) of culture, was moved to complain
that "the fairy tales of science, as narrated by a Huxley or
a Darwin, are beginning to be as eagerly demanded as the
latest productions of Miss Braddon or Mr. Wilkie Collins."[20]
Intellectual life in England had come a long way since
Samuel Johnson had confidently pronounced: "No man
ever read a book of science from pure inclination."

CHAPTER 12

THE DARWINIAN PARTY

BEFORE the publication of the *Origin*, Darwin had confided to Hooker his personal criteria of success:

> I remember thinking, above a year ago, that if ever I lived to see Lyell, yourself, and Huxley come round, partly by my book, and partly by their own reflections, I should feel that the subject is safe, and all the world might rail, but that ultimately the theory of Natural Selection . . . would prevail. Nothing will ever convince me that three such men, with so much diversified knowledge, and so well accustomed to search for truth, could err greatly.[1]

It may be that Darwin was subjecting himself to too easy a test when he took as his judges the three men with whom he was most in agreement; he could surely have found others with as much diversified knowledge and as zealous in the pursuit of truth who would have made more impartial examiners. But if his choice was deliberate in weighting the results in his favor, it was also deliberate in its order of priority, and within this order it was weighted against him. Lyell, Hooker, and Huxley—this proved to be, in fact, the inverse sequence of their conversions.

Lyell was the decisive test. Of the three, he offered most resistance to Darwin's theory, and, perhaps because of this, his opinion was most influential among scientists. To Darwin he was the "Lord High Chancellor" whose verdict would probably be more influential than the book itself in determining the success of his views.[2] As it happened, the fate of his book was decided before Lyell could bring himself to pronounce an unequivocal verdict.

Only a few years earlier, in revulsion against Darwin's "ugly facts," Lyell had taken refuge in "the orthodox faith."[3] By 1858 his personal distaste for these facts, and his sense of the impropriety of their public revelation, had so far abated that he could sponsor them before the Linnean Society and urge their publication. But it was one thing to think that they ought to have a hearing and another to subscribe to them. Even on the subject of their publication, he seems to have vacillated, agreeing at one point that Darwin would be better advised to publish a book on pigeons in place of the *Origin*. And when he read the proof sheets of the book, he characteristically remarked that while he was astonished at the cogency of the argument, he reserved judgment on its validity. In his address in September 1859 before the Geological Section of the British Association, he went as far as he could in praise of Darwin and his forthcoming book: "He appears to me to have succeeded by his investigations and reasonings in throwing a flood of light on many classes of phenomena . . . for which no other hypothesis has been able, or has even attempted to account."[4] He also gave Darwin permission to speak, in the *Origin,* of his former adherence to the theory of immutability and his present "grave doubts" on that score.[5] At the same time, it appeared that his grave doubts about immutability were matched by equally grave ones about mutability. Thus a long and detailed critique of the proofs of the *Origin* culminated in his opinion that the "continued intervention of creative power" was necessary for the origin of species.[6]

Lyell would clearly have liked to rest in some halfway house between immutability and mutability. But neither Darwin nor reason allowed him any repose. Again and again, Darwin welcomed him as a half-convert, only to follow with the warning that there was no compromise, no stopping point short of total conversion: "Do not, I beg, be in a hurry in committing yourself (like so many naturalists) to go a certain length and no further; for I am deeply convinced that it is absolutely necessary to go the whole vast length, or stick to the creation of each separate species."[7] If at any point the necessity for the intervention of a creator was conceded, then the theory of natural selection was utterly nullified. Indeed, it was in testimony to this that Lyell refrained from committing himself, for he himself had always maintained that what was true of animals was necessarily true of man and that if special creation was discarded for the one, it must also be for the other.

Back and forth went the dialogue between Lyell and Darwin, which was also a dialogue between Lyell and his alter ego. Alternately, Lyell might be found criticizing Darwin for not making his case strong enough, or advising him on how to make it seem stronger than, in fact, it was. "You are a pretty Lord Chancellor," Darwin thanked him, "to tell the barrister on one side how best to win the cause!"[8] Periodically, he would abdicate the Lord Chancellorship and assure Darwin that he intended publicly to announce himself the attorney for the defense. Darwin would then proclaim him "an entire convert"[9] and praise him for his courage in giving up a position he had held for thirty years: "Considering his age, his former views and position in society, I think his conduct has been heroic on this subject."[10] But no sooner had these congratulations been exchanged than Lyell would revert to type with such double-edged compliments as his praise of Hooker for raising "the variety-making hypothesis to the rank of a theory" and for showing "what grand speculations and results 'the creation by variation' is capable of suggesting, and one day of establishing"[11]—thus, with one stroke, demoting both Darwin and his theory to a rank inferior to Hooker and his systematic work. It was the same strategy of depreciation that prompted Lyell to subsume Darwin's theory under Lamarck's as if it was only a subcategory, a minor variation of the original classical doctrine. This was what vexed Darwin most: that Lyell should have made the vulgar mistake of confusing his "principle of improvement" with Lamarck's "power of adaptation."[12] If the two were basically the same, he asked Lyell, why had he been so adamant against Lamarck's theory in the *Principles,* and why had his conversion to mutability come only with the *Origin?* The answer, of course, which Darwin was too polite to suggest, was that Lyell cared no more for Lamarck now than then and that he invoked his name only because it was easier to make his obeisance to a senior and long-deceased master than to a junior and all-too-active competitor.

Originally, Lyell had proposed making his submission in a new edition of his *Geology.* By the beginning of 1860, inspired perhaps by the success of the *Origin,* he had decided to put it in the form of a book on the "geological history of man," to appear later that year.[13] For three years Darwin waited for the pronouncement that would put Lyell unequivocally on his side, venturing the jest that as Lyell used to caution him about man, so he now had to return the caution a hundredfold. When the long-awaited *Antiquity of Man* was finally published in 1863, it became apparent that neither cautions nor hopes had been warranted. For although Lyell conscientiously summarized the evidence for mutability as it applied to species in general and to man in particular, he did not commit himself to it. From such expressions as "should it ever become highly probable that the past changes of the organic world have been brought about by the subordinate agency of such causes as 'Variation' and 'Natural Selection' . . . ," or "Mr. Darwin labours to show . . . ,"[14] he intimated that Darwinism remained what it had always been: an interesting and respectable theory, but one that was far from proved. Again he featured Darwinism as a variation of Lamarckism and Hooker as a peer of Darwin. Again he suggested that man was the result of a leap of nature separating him at one bound from the species below him. And he concluded with the observation that transmutation in no way invalidated the idea of design or of a designer, the whole course of nature being "the material embodiment of a preconcerted arrangement."[15]

Darwin's disappointment amounted almost to a sense of betrayal. He could not see how this book could be reconciled with Lyell's private assurances on the subject, or how it fulfilled his promise to "go the whole orang."[16] "The best of the joke," he wryly commented to Hooker, "is that he thinks he has acted with the courage of a martyr of old."[17] On the theory that to be neutral at this stage of the controversy was, in effect, to be hostile, Darwin wished that Lyell had kept silent rather than declare himself so equivocally. Lyell, for his part, refused to take affront at Darwin's outspoken criticisms, justifying himself by arguing, first, that his book honestly represented his opinions as best he could make them out, and, second, that strategically it was better calculated to win supporters to evolution than any frontal attack. As usual, his rationale was a nice blend of reason, sentiment, and calculation:

> I find myself after reasoning through a whole chapter in favor of man's coming from the animals, relapsing to my old views whenever I read again a few pages of the "Principles" or yearn for fossil types of intermediate grade . . . Hundreds who have bought my book in the hope that I should demolish heresy, will be awfully confounded and disappointed. As it is, they will at best say with Crawford, who still stands out, "You have put the case with such moderation that one cannot complain." But when he read Huxley, he was up in arms again.
>
> My feelings, however, more than any thought about policy or expediency, prevent me from dogmatising as to the descent of man from the brutes, which, though I am prepared to accept it, takes away much of the charm from my speculations on the past relating to such matters. . . .
>
> I cannot go Huxley's length in thinking that natural selection and variation account for so much, and not so far as you, if I take some passages of your book separately.

I think the old "creation" is almost as much required as ever, but of course it takes a new form if
Lamarck's view improved by yours are adopted.

What I am anxious to effect is to avoid positive in-
consistencies in different parts of my book, owing
probably to the old trains of thought, the old ruts,
interfering with the new course.

But you ought to be satisfied, as I shall bring hundreds towards you, who if I treated the matter more
dogmatically would have rebelled.

I have spoken out to the utmost extent of my
tether, so far as my reason goes, and farther than my
imagination and sentiment can follow, which I suppose has caused occasional incongruities.[18]

As Hooker simplified it: Lyell was "half-hearted and
whole-headed."[19]

This tenuous balance was beyond Darwin's understanding. It was even difficult for Lyell to sustain, and in the
second edition of the *Antiquity,* published only a few
months after the first, he inserted a parenthetical phrase
that went further to commit him than almost anything
else in the book, although it was as deliberately awkward
and evasive as only Lyell could have put it: "Yet we
ought by no means to undervalue the importance of the
step which will have been made, should it hereafter become the generally received opinion of men of science (as
I fully expect it will), that the past changes of the organic
world have been brought about by the subordinate agency
of such causes as 'Variation' and 'Natural Selection.' "[20]
Darwin was somewhat mollified by this and other small
emendations, such as the alteration of "Mr. Darwin labours
to show" to "Mr. Darwin argues, and with no small success";[21] and even more when Lyell finally and unequivocally, in the tenth edition of the *Principles* published in
1867, announced his conversion to mutability (although
still with the tantalizing allusions to Lamarck).

Lyell may have given himself away in an ambiguous
remark concluding one of his apologias at the time of the
Antiquity: "I see too many difficulties to be in the danger

of many new converts who outrun their teacher in faith."[22] It is not clear whether the "teacher" was intended to refer to Darwin or to himself, whether he was cautioning against Darwin's impetuous converts or against his own—that is, Darwin. There are intimations that he resented the subversion of their natural relationship, the elevation of Darwin to the status of master and his own debasement to that of convert. Even while he was opposing the *Origin,* he took a perverse pleasure in criticisms directed against "Lyell and his friend."[23] And later, when Darwin's critics as well as his converts neglected to mention him, he felt slighted and resentful. The German biologist, Ernst Haeckel, in his *History of Creation,* was one of the few to pay homage to him as one of the founders of the theory of evolution, whereupon Lyell wrote to thank him:

> Most of the zoologists forget that anything was written between the time of Lamarck and the publication of our friend's *Origin of Species.* . . . I had certainly prepared the way in this country, in six editions of my work before the *Vestiges of Creation* appeared in 1842, for the reception of Darwin's gradual and insensible evolution of species, and I am very glad that you noticed this.[24]

Hooker was the second in Darwin's triumvirate of judges, outranked by Lyell in seniority and power but not in ability; in ability, Darwin pronounced him—although only after Hooker had announced his conversion—"by far the most capable judge in Europe."[25] It was with great satisfaction, therefore, that Darwin heard from him: "I expect to think that I would rather be author of your book than of any other on Natural History Science."[26]

Hooker was not only one of the first and most distinguished of Darwin's converts; he was also the first to apply the theory to a particular scientific problem. His introductory essay to the *Flora Tasmaniae,* completed and published only a few weeks before the *Origin* appeared, was a conscious attempt to apply the hypothesis of natural selection to the case of the Australian flora. While admitting that he had not proved the truth of the hypothesis, he insisted that it explained more of the peculiarities of that flora than any alternative theory and that it promoted inquiry where others hindered it. It was as a minor exercise in humdrum science, a ragged handkerchief flying beside the royal standard, that he compared his own essay with Darwin's, so that he was all the more astonished and distressed to find Lyell praising it so fulsomely, and to Darwin's obvious derogation. Had he known, he protested, that the *Origin* would appear so soon after his essay, he would have delayed its publication rather than seem to be antedating or competing with Darwin.[27]

Unlike Lyell, Hooker was not in the least abashed by his position as convert. He had, to be sure, some minor disagreements with Darwin; he thought, for example, that Darwin had attached too much, perhaps too exclusive an importance to natural selection. But these, he felt, were matters to be arbitrated by scientists working on the lines laid down in the *Origin*. Of the importance of the theory in general, he had no doubts. And he was the more firmly convinced of this, knowing how long he himself had resisted it. When he publicly announced his adherence, at the meeting of the British Association in 1860, he described himself as one who had been apprised of the theory fifteen years earlier, had vigorously argued against it, even while engaging in the scientific researches which had carried him around the world, and had only been persuaded of its truth when facts otherwise inexplicable became intelligible as a result of it. Thus, as he described it, conviction was "forced upon an unwilling convert."[28]

It is the unwilling convert who makes the most effective witness in a cause and is likely to be its most enthusiastic communicant. The eyes of the world and of posterity were upon science, Hooker warned one lagging colleague, and for the "credit of the age we live in," it was important that naturalists should have something better to show than was current a quarter of a century earlier. "Above all things," he counseled, "remember that this reception of Darwin's book is the exact parallel of the reception that every great progressive move in science has met with in all ages." The difficulties of the theory might be, as he confessed, appalling, but as long as the alternative was worse and as long as history gave promise of its ultimate vindication, progressive scientists had the duty of supporting it.[29]

If Darwin's opponents are sometimes charged with a partiality for the past, his friends may as justly be charged with a partiality for the future. It cannot be said that this partiality actually determined their conversion; for some time, indeed, it was in doubt which party the future would favor. But, having been converted, they were not long in arrogating the future to themselves. As Hooker was warning his friends not to offend posterity, so Huxley was predicting the inauguration of a new Augustan age in English science.

Huxley, rather than either Hooker or Darwin, was pre-eminently the modern intellectual. Regarding heresy rather than orthodoxy as the hallmark of truth, he deliberately flaunted the novelty of his views instead of smuggling them in, as was the older fashion, in the guise of tradition. Every great physical truth, he declared, had come into the world under the onus of blasphemy, and Darwinism was no exception. "In this nineteenth century, as at the dawn of modern physical science, the cosmogony of the semi-barbarous Hebrew is the incubus of the philosopher and the opprobrium of the orthodox." But the patient seekers after truth had learned to avenge themselves. "Extinguished theologians lie about the cradle of every science as the strangled snakes beside that of Hercules; and history records that whenever science and orthodoxy have been fairly opposed, the latter has been forced to retire from the lists, bleeding and crushed if not annihilated; scotched, if not slain."[30]

Huxley was the great avenger. Raging against the inferior status of scientists compared with clergymen, he looked forward to the time when he could get his heel "into their mouths and scr-r-unch it round."[31] The *Origin* gave him the opportunity. As self-designated bulldog, he announced that he was sharpening his claws to do battle with the pack of yelping curs at Darwin's heels. His first engagement was in the December issue of *Macmillan's Magazine* in an article called "Time and Life," based on a lecture given earlier that year before the Royal Institution and expanded to include an analysis of the *Origin*. By a stroke of luck and quick thinking, he was also able to capture the coveted review in the *Times*. As a matter of routine, the book had been handed for review to a regular staff writer, who, bemoaning his ignorance of science, was by chance referred to Huxley. Huxley promptly relieved him of the chore by writing the review himself, which appeared in the *Times* of December 26, prefaced only by a few paragraphs by the staff writer. He carried the battle into the next year with a lecture before the Royal Institution in February and a long article in the *Westminster Review* in April.

Years later, reviewing his pleadings on behalf of Darwin, Huxley insisted that his zeal never reduced him to the role of a mere advocate or blind partisan. It had been, he claimed, a matter of indifference to him whether Darwin's theory was ultimately proved true, so long as it was fully and fairly examined. And he did, in fact, often take the precaution of introducing it as a working hypothesis rather than an established fact. "Either it would prove its capacity to elucidate the facts of organic life, or it would break down under the stress," he reasonably proposed,[32] thus disarming the opposition and insuring against the future. But he made it clear that the hypothesis agreed with all the available facts—the "many apparent anomalies in the distribution of living beings" and the "main phenomena of life and organization"—and that it was utterly different from such speculative hypotheses as that of the *Vestiges*. Darwin, he insisted, "abhors mere speculation as nature abhors a vacuum"; "he is as greedy of cases and precedents as any constitutional lawyer, and all the principles he lays down are capable of being brought to the test of observation and experiment."[33] Admitting that Darwin had not proved that natural selection *did* actually operate to produce new species, he went on to say that such proof was unobtainable, that Darwin had proved all that was provable, which was that natural selection *must* so operate. And he concluded by saying that even if natural selection should turn out to be an inadequate explanation for the origin of species, Darwin would still have superseded all previous thinkers, much as Copernicus superseded Ptolemy; a Kepler or Newton might be required to correct his details, but they would not vitiate his great achievement.[34]

Thus Huxley managed to recover at the end what he had pretended to forfeit in the beginning. The working hypothesis turned out to be indistinguishable from an established theory, the "test of observation and experiment" was discounted in advance, and the verdict of posterity was assured from the start.

Even these small concessions to neutrality irked Darwin, who was piqued to have his theory passed off as a mere hypothesis. And other of Huxley's criticisms were serious enough to disturb him, such as the remark that the theory was faulty in not providing a *vera causa* of variation, or that species ought to be defined primarily in terms of sterility. What Darwin did not comment on was Huxley's strongest and most imaginative point of criticism. While Hooker was rejoicing because the *Origin* had revealed to him once again the truth of the dictum, *"Natura non facit saltum,"*[35] Huxley was deploring the fact that Darwin had unnecessarily burdened himself with so specious a principle. The strength of natural selection, he believed, was precisely the fact that it allowed for such jumps in nature, that, in his favorite formula, it accounted for transmutation without transition.[36]

Huxley's occasional reservations, however, were more than made up for by his enthusiasms in other directions. He gave Darwin fair warning that after a slow start he was gravitating toward his theories at such an accelerated pace that he would soon pass him. While Lyell was slowly and painfully feeling his way to the *Antiquity of Man*, Huxley was boldly lecturing to working men on "The Relation of Man to the rest of the Animal Kingdom" and publishing monographs demonstrating the humanlike structure of the brain of the higher ape. His professional colleagues remained unconverted, but his working men were so loyal that, as he reported to his wife, "by next Friday evening they will all be convinced that they are monkeys."[37]

In 1863 he published the substance of these lectures as *Man's Place in Nature*, and, by contrast to Lyell's book appearing at the same time, Huxley's seemed to Darwin all the more admirable. The *Daily Telegraph*, not discriminating between them, was only disturbed that the works of these heretical scientists should be torn from the hands of Mudie's salesmen as if they were novels.

Huxley and Hooker were Darwin's advance guard, with Lyell bringing up the rear. But it was not long before others joined the ranks. By March 1863 enough had committed themselves, to one degree or another, for Darwin to draw up a table of organization of the names and professions of fifteen of his more prominent adherents.[38] One of these, reviewing the *Origin* in the *National Review*, observed hopefully that as species emerged from a war of nature, so truth would emerge from the intellectual collision precipitated by this work.[39] Yet even Darwin's partisans could not agree as to the exact nature of the new truth.

The American botanist Asa Gray had so special an interpretation of it that Darwin represented him on his table as only a partial adherent. Yet Gray was Darwin's chief agent in America as Huxley was in England. He was one of those who had first read an abstract of the *Origin* and then the proof sheets of the individual chapters as they came from the printer, but who was not converted until he had experienced the dramatic effect of the printed, bound book. Once converted, however, he worked tirelessly in Darwin's interest, arranging for an American edition and extracting a token payment from the publishers of the pirated edition, reviewing the book and defending it against its detractors. He opened his campaign in March 1860 with an article in the *American Journal of Science and Arts* (more commonly known as "Silliman's Journal"), continued it with a discussion in the *Proceedings of the American Academy* in April, and brought it to a climax with a series of three articles in the *Atlantic Monthly* which were reprinted as a pamphlet and eventually incorporated in his book *Darwiniana*.

Gray's first reaction to the theory was typically American, although surprising in so ardent an abolitionist: "The prospect of the future is encouraging. It is only the backward glance that reveals anything alarming. . . . The very first step backward makes the Negro and the Hottentot blood relations. . . . Not that reason or Scripture objects to that, though pride may."[40] Reason and Scripture triumphed, and Darwin's prediction that Gray would come around was fulfilled: "For it is futile to give up very many species, and stop at an arbitrary line at others. It is what my grandfather called Unitarianism, 'a feather bed to catch a falling Christian.' "[41]

Unitarianism, indeed, proved to be not only capacious enough to accommodate fallen Christians but also fallen scientists, including Darwinists of several varieties. It is not surprising that the first sustained and sympathetic religious reading of the *Origin* should have come from America, where scientists were as jealous of their religious and moral reputations as theologians, where the leading naturalist, Louis Agassiz, might be described as "a sort of demagogue . . . [who] always talks to the rabble"[42] and the geologist James Dwight Dana as too "idealistic" to appreciate Darwin's naturalism.[43] Gray himself, whose descriptions these were, was none the less idealistic, or theistic, for all of his Darwinist naturalism. It was his thesis, that natural selection was "not inconsistent with natural theology,"[44] which later attracted so many converts and distressed so many others, leaving Darwin himself in a "hopeless muddle" about the whole affair.[45] His friends, Darwin was to discover, could be almost as quarrelsome as his enemies.

CHAPTER 13

THE ANTI-DARWINIAN PARTY

THERE were as many conflicting interpretations and private constructions among the entrenched opposition as among Darwin's attacking forces. Yet the defenders acted in concert more successfully than the invaders. Perhaps this was because the latter lost their natural leader when Darwin retired from the struggle, and because even the belligerent Huxley was too subtle a tactician, too readily carried away by his own wit and rhetoric, to be an effective commander. Thus it was the opposition that enjoyed the more militant and aggressive spirit, having something resembling a chain of command and a coordinated strategy of action.

One of the inspiring minds behind the opposition was Professor Sedgwick, who had once predicted great fame for his pupil, and who, as late as 1848, had proposed him in his own stead to write an important article on geology. But there were now serious principles at stake, religious and scientific, and on both counts Sedgwick was alienated. His biographer has said that next to scientific truth, his main concern was to demonstrate the teleological value of such truths[1]; and where teleology had made a partisan of the Unitarian Gray, it made an uncompromising enemy of the Anglican Sedgwick. Thus Sedgwick had to confess that although some parts of the *Origin* he admired greatly, other parts he found laughable, and still others he read with sorrow, "because I think them utterly false and grievously mischievous." That there was development in history he did not doubt; what was false and mischievous was Darwin's explanation of that development:

> There is a moral or metaphysical part of nature as well as a physical. A man who denies this is deep in the mire of folly. 'Tis the crown and glory of organic science that it *does* through *final cause*, link material and moral; and yet *does not* allow us to mingle them in our first conception of laws, and our classification of such laws, whether we consider one side of nature or the other. You have ignored this link; and, if I do not mistake your meaning, you have done your best in one or two pregnant cases to break it. Were it possible (which, thank God, it is not) to break it, humanity, in my mind, would suffer a damage that might brutalize it, and sink the human race into a lower grade of degradation than any into which it has fallen since its written records tell us of its history.

As if this were not enough—blasphemy, brutalization, and degradation—Sedgwick also accused Darwin of the sin of pride, citing the triumphant tone of the final chapter in which Darwin had appealed to the rising generation, a tone prophesying "of things not yet in the womb of time, nor . . . ever likely to be found anywhere but in the fertile womb of man's imagination." Such an appropriation of the future, Sedgwick objected, was very different from his own humble submission to God's revelation.[2]

It is strange that Sedgwick's letter should have been published among Darwin's correspondence, but not Darwin's reply. The reply was pitched on a lower tone, Darwin persisting in treating Sedgwick's rhetoric as if it called for concrete answers. Of the complaint that the *Origin* had shocked his "moral taste," Darwin could only protest that it was twenty years' labor and not bad instincts or taste that had gone into it. And when Sedgwick deplored it as grievously mischievous, he could only console him that if it was wrong, the mischief would soon be undone, since from so spirited a battle truth was bound to emerge. As for the appeal to the future which Sedgwick had found so offensive, Darwin cited others who had complimented him on his modesty; and his faith in the young as the best judges of a new doctrine was not, he assured him, invented for the occasion but was one of his long-held beliefs. The only moral question he would allow concerned the integrity and conscientiousness of his work. Would Sedgwick, he asked, have preferred him to conceal the fruits of his honest labor? For it was by these fruits that his work must be judged: "I cannot think a false theory would explain so many types of facts, as the theory seems to me to do. But magna est veritas and thank God, praevalebit."[3]

While Darwin was paying his respects to the "kind and noble heart" that had done him the "honourable compliment" of criticizing him,[4] he was elsewhere decrying the abuse of "Sedgwick and Co."[5] He discerned the hand of Sedgwick in what he regarded as the savage and unfair notice in the *Spectator*, but he did not guess the large part played in it by "and Co." nor the nature of the collaboration. Had he known of Sedgwick's lay counselors, he would have felt even more sorely treated. For the initiative for the *Spectator* review had come from the Archbishop of Dublin, who urged Sedgwick to make public what he and other like-minded men felt about the Darwin menace:

> I felt alarm at the apparent high favour and wide celebrity of Darwin's theory (which I suppose is Lamarck's cooked up afresh) because it was likely to establish *our* descent from Molluscs or Insects. I have very slightly alluded to it in my edition of Paley. . . . I have touched on it rather more, though still very slightly, in the Lecture on Civilization, which you probably are acquainted with.
>
> But my own paper emphasizes the improbability of the *last* step of all,—the advance of the *savage-man* into the civilized, without external help. I *doubt* the conversion of oats into rye: their conversion into appletrees, I *disbelieve:* but what I have undertaken to *disprove* is the conversion of the unaided savage into the civilized man.[6]

It may be doubted whether Sedgwick subscribed to this curious hierarchy of disbelief, in which the transformation of the savage into the civilized man was more strenuously to be disputed than the transformation of oats into rye and apple trees. Nevertheless, his reply must have been satisfactory, for their next several letters were devoted to the details of its publication. They decided that the substance of Sedgwick's letter would be communicated to a magazine by the Archbishop, "as having received it from one whom I consider a competent judge"—the advantage being, as the Archbishop assured him, that since Sedgwick would not enter into it at all, and the Archbishop only as the transmitter of the communication, neither of them could be "dragged into controversy." A week later the Archbishop happily reported that he had placed the piece with the *Spectator*.[7] After it appeared, Sedgwick blandly wrote to Owen about a "published letter . . . written without a shadow of a thought that the editor would send it to the Press."[8]

It was the high moral and religious tone of the *Spectator* notice that made Darwin suspect Sedgwick. What disturbed him, however, was not the reference to his "demoralized understanding" but rather the charge that he had laced together a series of facts on a single false principle. "You cannot," the review objected, "make a good rope out of air bubbles."[9] Sedgwick reverted to these themes in his address to the Cambridge Philosophical Society on May 7, 1860. Engaging the enemy on home territory, and with weapons more congenial than the formal and binding printed word, he felt safe in emerging from the shelter of anonymity. At first he seemed to be winning the battle. Having discharged his volley against Darwin's morals and methods, he was gratified to find himself supported by the next speaker, the Cambridge Professor of Anatomy, William Clark.[10] The attack boomeranged, however, perhaps because of the very vigor with which it had been executed, and Sedgwick found that he had antagonized at least one important person who would otherwise have been sympathetic. This was John Henslow, Darwin's former teacher (and now Hooker's father-in-law).

It was the gentleman in Henslow, rather than the scientist, who was aroused, so that where Darwin was offended by the slur on his scientific ability, Henslow resented the slur on his friend's character. As he later confided to Hooker, he could not sit by while Sedgwick dilated upon how "revolting to his own sense of right and wrong" were Darwin's notions. And since Sedgwick had made the mistake of alluding to him, he felt bound to declare himself publicly in Darwin's favor, refusing to allow that Darwin had been guided by any but the highest motives, and declaring that he himself believed him to have exalted rather than debased the Creator.[11] After Henslow had finished, Sedgwick, sensitive as ever to the temper of his audience and the esteem of his peers, hastened to aver his good opinion of Darwin and to explain that his "chief attacks" (presumably the moral and metaphysical ones) were intended not against Darwin but against the Oxford divine who, in a recent essay, had adopted Darwin's theory for his own heterodox purposes.[12] The retreat was skillful enough, considering how unexpected, to Sedgwick and Darwin alike, was Henslow's intervention on Darwin's behalf. For Henslow was even more orthodoxly religious than Sedgwick. And, indeed, he never did become a convert to evolution. He might be reported "shaken" by the evidence[13] and he might be found to concede the legitimacy of natural selection as a scientific hypothesis, but he was convinced that Darwin had pressed it too far, carrying it not only into the realm of the unproved but into that of the unprovable.[14] "An inquiry into the origin of species," he pessimistically (or perhaps, in view of his religious convictions, optimistically) concluded, was "about as hopeless as an inquiry into the origin of evil."[15]

There were hazards as well as advantages in the collaboration between layman and scientist, particularly where the layman was not satisfied with initiating and placing the article but insisted upon writing it as well. Samuel Wilberforce, Bishop of Oxford, whose acquisition of a First in mathematics many years earlier and long association with Oxford's scientific worthies had induced in him an allusion of scientific competence, could not be content, like the Archbishop of Dublin, with the passive role of literary agent. He humbled himself only to the extent of submitting to coaching by a professional scientist—in this case, Richard Owen. But where Sedgwick was entirely responsible for the *Spectator* notice, Owen was only partly responsible for that in the *Quarterly Review.* Observant readers of the July *Quarterly,* knowing of this collaboration, were amused to notice that a page of the article had been apparently substituted at the last moment, and were intrigued by the thought of some egregious error in which Owen had caught the bishop.[16]

"Soapy Sam," as the bishop was widely known (or "Samivel," as Huxley preferred it), was a ready, fluent orator, never at a loss for words or put off by a loss of facts. He explained away his nickname as glibly and good-humoredly as he explained away other embarrassing situations, saying that "though often in hot water, he always came out with clean hands."[17] Whether he came out with clean hands from his engagement with Darwin was a matter of opinion. It has been suggested that he entered into the controversy in an attempt to curry favor with the court and the low-church party, after having suffered from a too close identification with the high church and from the secession of his brother and brother-in-law to Rome. And it is true that he emerged from his campaigns against the *Origin* and *Essays and Reviews* with restored reputation. Yet, even without these incentives, Wilberforce would have been a natural, although perhaps less vociferous, enemy of Darwin.

After the conventional introductory sentiments of respect and good will that barely survived the professing of them, Wilberforce turned to what he described as the main argument of the *Origin:* "our unsuspected cousinship with the mushrooms."[18] "Is it credible," he asked, "that all favourable varieties of turnips are tending to become men, and yet that the closest microscopic observation has never detected the faintest tendency in the highest of the Algae to improve into the very lowest Zoophyte?"[19] Judged by this stern Baconian law of observation, he declared, Darwin's theory was wantonly hypothetical and conjectural, without the smallest basis in fact. And the vocabulary of hypothesis which Darwin employed so freely—"I can conceive," "it is not incredible," "I do not doubt," "it is conceivable"—Wilberforce objected to as dishonorable to science, having the effect of opening the doors of the temple of truth to the genii of romance:

> The whole world of nature is laid for such a man under a fantastic law of glamour; and he becomes capable of believing anything: to him it is just as probable that Dr. Livingstone will find the next tribe of negroes with their heads growing under their arms as fixed on the summit of the cervical vertebrae; and he is able, with a continually growing neglect of all the facts around him, with equal confidence and equal delusion, to look back to any past and to look on to any future.[20]

To blacken him still more, Darwin was made responsible for all the excesses of his grandfather as well. The kinship of blood, plus an intellectual kinship triumphantly demonstrated in their use of the same phrase ("it is not impossible"), gave Wilberforce the opportunity to entertain his readers with the more fanciful theories of Erasmus Darwin and with even lengthier and less relevant quotations from a parody of Erasmus composed by some of his contemporaries.

Declaring his firm resolve to confine himself to the scientific issue—"We cannot consent to test the truth of natural science by the Word of Revelation"[21]—Wilberforce proceeded to devote seven pages to a theological analysis of the *Origin*. He protested that he did not for a moment doubt the sincerity of Darwin's Christian faith, nor suspect him of being "one of those who retain in some corner of their hearts a secret unbelief which they dare not vent."[22] Yet he could not fail to point out that his theory was absolutely incompatible not only with the revealed word of God but also with His works and spirit: man's supremacy on earth, his gift of speech and reason, his free will and responsibility, his fall and redemption, the incarnation of the

Eternal Son, and the indwelling of the Eternal Spirit. When Biblical admonitions failed him, he had recourse to theologically minded quotations from Owen or from Owen's citation of Henry More: "And of a truth, vile epicurism and sensuality will make the soul of man so degenerate and blind, that he will not only be content to slide into brutish immorality, but please himself in this very opinion that he is a real brute already, an ape, satyr, or baboon."[23]

On matters scientific, moral, and theological, Owen was the authority most frequently and fulsomely invoked by Wilberforce, although occasionally Sedgwick and even Lyell were brought in for reinforcement, the latter being urged to abide by his *Principles* and give the lie to the insinuations that he had become a convert to Darwinism. Forgetting for the moment such domestic deviations as Erasmus or Charles Darwin or the author of the *Vestiges*, Wilberforce declared himself happy to take his stand with "the sober, patient, philosophical courage of our home philosophy" against those wild foreign theorists, Lamarck and Oken.[24]

Wilberforce's attack infuriated Darwin's friends as did no other single episode in the controversy. A quarter of a century later Huxley's indignation had still not abated. Looking back upon the reception of the *Origin*, he could find nothing more dishonorable, no better *pièce justificative* illustrating the vilification and misrepresentation to which Darwin had been subject, than the *Quarterly Review* article. Nor would he moderate this judgment when it was pointed out to him that the article was no longer anonymous but had been openly acknowledged by the Bishop; confession unaccompanied by penitence, he insisted, gave no absolution. Even the softer-spoken, milder-tempered Hooker was aroused by it. When Francis Darwin, reluctant to print Huxley's harsh judgment of it in the *Life and Letters* of his father, appealed to Hooker, he was surprised to find Hooker enthusiastically endorsing Huxley's views. As the "head and front" of the offending, Hooker insisted, the *Quarterly Review* deserved to be publicly pilloried: "It is not for us, who repeat *ad nauseam* our contempt for the persecutors of Galileo and the sneerers at Franklin, to conceal the fact that our own great discoverers met the same fate at the hands of the highest in the land of history and Science, as represented by its most exalted organ, the *Quarterly Review.*"[25]

Darwin himself was more tolerant. He pronounced Wilberforce to be "uncommonly clever" in finding so many conjectural expressions in the *Origin* and in making such "capital fun" out of him and his grandfather.[26] This good humor, compared with his bitterness about the *Spectator* notice, may be explained by the fact that he was less mindful and therefore more indulgent of the opinions of a bishop than of a scientist. He may also have been pleased with the unwitting evidence of the permeation of evolutionism provided by the same issue of the *Quarterly Review.* An essay entitled "The Missing Link" discussed the absence in the Anglican hierarchy of a deaconess to mediate between the clergyman and the district visitor!

When Huxley wanted to illustrate the errors in the *Quarterly Review* article, he specified two—one in comparative anatomy, and the other in paleontology. It was not without malice that he chose just those fields where Owen was deemed to be expert. (He had been much offended when Owen assumed the title of Professor of Paleontology in the course of delivering some lectures at Huxley's own school of Mines.) Whatever Owen's mistakes, however, he could not be said to share Wilberforce's views about creation. Indeed, so far was he from orthodoxy that he even qualified, for a time, as one of Darwin's predecessors or, at the least, sympathizers. Strongly influenced by Geoffroy St. Hilaire and Lorenz Oken, he developed a theory of homology that had distinct evolutionary implications. His first work in paleontology was a memoir on the toxodon discovered by Darwin in South America, in which he located the fossil as intermediate between two now widely separated groups. And when the *Vestiges* appeared a few years later, he wrote an appreciative note to the author, pointing out that he himself had taken much the same stand in his recent *Lectures on the Invertebrata.* To be sure, he never made his opinion of the book known publicly, and when Whewell, a few months later, asked him about it, intimating that he could not imagine Owen assenting to a scheme that ordered animals in a series ranging from the less to the more perfect, Owen hastened to assure him that he did not agree with it. He could not, however, resist playing with these heretical ideas, and, as late as 1857, he wrote of the anatomical resemblances between man and the higher apes:

> Not being able to appreciate or conceive of the distinction between the psychical phenomena of a chimpanzee and of a Boschisman [Bushman], or of an Aztec with arrested brain growth, as being of a nature so essential as to preclude a comparison between them, or as being other than a difference of degree, I cannot shut my eyes to the significance of that all-pervading similitude of structure—every tooth, every bone, strictly homologous—which makes the determination of the difference between Homo and Pithecus the anatomist's difficulty.[27]

When the *Origin* appeared, Owen, perhaps in pique at having been forestalled, perhaps in subservience to the powers-that-be, took his stand with the opposition, although with his usual secretiveness and equivocalness. Even before his collaboration with Wilberforce in the *Quarterly Review*, he had himself reviewed the *Origin* in the April *Edinburgh Review*. And although he was here the sole author, without the excuse of dissociating himself from another's views, he vigilantly maintained his anonymity, and long after the decent interval which satisfied the other participants in the controversy.[28] Yet few people acquainted with the affair doubted his authorship. Those who knew him best saw his handiwork in the repeated laudatory references to "Professor Owen" and in the fact that the reviewer took the occasion to notice several other books on the subject, including three by Owen himself (one of which dated back to 1855). Those who knew him less well and found it hard to credit such blatant self-advertisement carried on under the protection of anonymity, surmised that the article had been written by a disciple, possibly under Owen's direction. When Asa Gray suggested this, Darwin retorted that the most notable fact about Owen was that he had no disciples. Anonymity served a double purpose, permitting him not only to praise himself but also to criticize Darwin with an abandon he could not otherwise have enjoyed, for in private correspondence and conversation he had been surprisingly amicable, leaving Darwin with the impression that "at the bottom of his hidden soul" he went as far as he did himself.[29] Owen's criticisms included the familiar ones of faulty logic, inadequate facts, and an unproved theory, plus some distinctive ones of his own, such as his caviling at Darwin's use of the word "inhabitants" in the first sentence of the *Origin* to describe the flora and fauna of a region. Surely, he protested, "inhabitants" commonly referred only to the human species, in which case what could Darwin conceivably have meant in this crucial sentence? (In fact, Lyell had earlier established Darwin's usage of the word, and Owen, with his addiction to metaphysical-scientific argot, was hardly in a position to dispute so innocent an exercise of scientific license.) Sometimes this querulousness took a personal form, the more vulnerable Huxley often being made the victim in place of Darwin.

More typical of Owen's polemical strategy was the simultaneous attack from front and rear: not only were his opponents mischievously deluded in their beliefs, but they were also deluded in thinking them original, since they had been anticipated by Owen himself. Natural selection, he said, "is just one of those obvious possibilities that might float through the imagination of any speculative naturalist; only the sober searcher after truth would prefer a blameless silence to sending the proposition forth as explanatory of the origin of species, without its inductive foundations."[30] As Pascal had wagered with God, so Owen did with history. Either history would prove his enemies wrong, or, if right, Owen could claim priority. Later, when Darwinism seemed to be winning out, Owen was to become more extortionist in his claims; at this time he satisfied himself with laying the groundwork for them. Quoting from his own works, he enunciated such propositions as:

"the continuous operation of the ordained becoming of living things"; the "law of vegetative or irrelative repetition and of homologies"; and "the platonic *idea* [in Greek] or specific organizing principle [which] would seem to be in antagonism with the general polarizing force, and to subdue and mould it in subserviency to the exigencies of the resulting specific forms"—all of which, presumably, were more acceptable as scientific hypotheses than natural selection. So eager was he to assert his priority in deriving the "production of the succession of organisms" from a secondary cause or law[31] that he was even emboldened to criticize "literal Scripturalism": "We have no sympathy whatever with Biblical objectors to creation by law, or with the sacerdotal revilers of those who would explain such law."[32] Before the Scriptualists could pounce, however, he executed another about-face, making his peace with the orthodox and concluding with an axiom of Linnaeus: "Classification is the task of science, but species the work of nature"[33]—an axiom which no one, least of all Darwin, would have disputed, but which had a richly pious ring.

Just what Owen did believe in regard to the origin of species cannot easily be deduced, whether from this review or from his other works, and in the "Historical Sketch" prefacing each of the several editions of the *Origin*, Darwin was hard put to it to keep up with his claims. When he interpreted Owen's statement about the "continuous operation of creative power" as signifying a belief in the immutability of species, he was roundly abused; when he read a later passage as suggesting that natural selection may have had some part in the creation of species, this was as hotly denied; and when he quoted from a correspondence between Owen and the editor of the *London Review* in which the editor, like Darwin, took Owen to be claiming to have promulgated natural selection before Darwin, he was again charged with error. In the last edition he recounted this comedy of errors, consoling himself with the thought that no one else seemed to find Owen's writings any easier to penetrate and reconcile. Perhaps, he proposed, they ought both to abdicate, leaving priority to the author of "An Account of a White Female, part of whose skin resembles that of a Negro" or to the author of the appendix to "Naval Timber and Arboriculture."

Yet Owen was, even by Huxley's admission, England's leading anatomist. It is unfortunate that his weakness of character should have distorted and concealed his genuine abilities, that he should have been more vain of his social and court connections (he was a protégé of Prince Albert) than of his scientific standing, and more concerned with his scientific standing than with his achievements. Priority disputes, the competition for honors (and sometimes their usurpation), and an avidity for fame not only isolated him from his colleagues but also perverted his mind and spirit. When a Parliamentary committee rejected his scheme for a museum, his first thought was how Hansard would report his speech; later he declared himself satisfied to have his report immortalized "in the pages of a work which will last as long as, and may possibly outlast, the great legislative organization whose debates and determinations are therein authoritatively recorded."[34] Much of his vagueness and contradictoriness came from his desire to be all things to all men. Darwin expressed the general opinion of his colleagues "that Owen truckles to the approbation of those high in church and state."[35] What rankled in Darwin's mind, as it weighed heavily upon Owen's, was the prophesy that the *Origin* would be forgotten in ten years.[36] A shrewd speculator in reputations, Owen was willing to "hedge" with history to the extent of placing side bets on Darwin in the form of priority claims. But as long as the present and the future seemed to be with those "high in church and state" who disapproved of Darwin, Owen could be counted among them.

When Darwin predicted that his book would find more favor with intelligent laymen than with professional scientists, it was because he thought the scientists were too committed to the old conception of species to admit new ideas on the subject. What he did not expect was that their objection would be not only professional but also religious.

The religious issue having played no part in his own thinking, he was entirely unprepared for its prominence in the judgments of even the most professional and reputable scientists. And since he selected for rebuttal only points of scientific dispute, his letters do not suggest the gravity or pervasiveness of the religious question. Only from the reviews themselves can its full weight be felt.

It was the science contributor to the *Athenaeum* who left Darwin to the mercy of the divinity hall;[37] the zoologist Thomas Wollaston, in the *Annals and Magazine of Natural History*, whose main objection was that the theory was too materialistic and utilitarian, making of nature "a pestilent abstraction like dust cast in our eyes to obscure the workings of an Intelligent First Cause";[38] the botanist William Harvey, whose notice in the *Gardeners' Chronicle* concentrated on the case of a freak begonia which seemed to belie the principle of natural selection, but whose real concern, as it emerged in a "serio-comic squib" delivered at Dublin University, was "the cool manner in which the personal work and oversight of the Creator is reduced to a minimum, and the Creator Himself to the condition of a King Log, the nominal head over an irresponsible ministry";[39] and the zoologist-paleontologist Agassiz, who launched his most bitter attack on Darwin in the course of a series of lectures on "The Structure of Animal Life, being six lectures on the Power, Wisdom, and Goodness of God, as manifested in His Works."

Curiously enough, the reviews written by laymen and clerics, while also preoccupied with the religious implications of the doctrine, tended to be more tolerant and amiable than those of the professional scientists. Although the Reverend Mr. Dunns, a Presbyterian minister and amateur naturalist, writing in the *North British Review*, was not much taken with this "dreary discourse . . . so full of morbid views of creation"[40] and found it as ridiculous to suppose that natural selection was responsible for the origin of species as that Darwin's book might be held responsible for the zoological arrangements in the British Museum, he did not pretend that the *Origin* was a passing fancy. Seldom, he said, had an avowedly scientific work so speedily commanded the attention of both the general public and scientists. And while he heartily dissented from its findings, he as heartily acknowledged the great ability, the varied information and suggestiveness, and, astonishingly, the "classic beauty of style"[41] which it displayed.

The *Saturday Review*, as orthodox in religion as it was conservative in politics, was also noticeably respectful in its dissent. Like the *North British Review*, it acknowledged the *Origin* to be one of the most important works of the time, one that did not bear to be dealt with lightly. The reviewer praised Darwin for his remarkable ability and attainments, the long thought and labor that had gone into his work, and the candor and moderation with which he anticipated objections. And unlike most of the other critics, when he declined to enter into discussion of the religious issue he was as good as his word. Believing that no scientific theory could affect the conviction of man's moral and spiritual faculties, he felt that the critic could afford to be neutral, judging a scientific work solely on scientific grounds. To be sure, having reviewed the scientific evidence, he hastened to "relieve the anxiety," as he frankly put it, of some of his readers by assuring them that he was unconverted, that this latest and ablest attempt to penetrate the mystery of the origin of species was supported only by a mass of conjectures, and that although natural selection must be admitted as the chief means by which organic beings were modified, there were, nevertheless, limits to this modifying power, even if they were not discoverable by man. "For as explorers may travel along the shores of the Ocean of Truth, the horizon does but stretch the further before them, illimitable in its vastness."[42]

The *Saturday Review* had hit upon a casuistic device that the Catholics exploited more often than the Protestants, although it was available to both. By taking a lofty tone toward religion, making the truths of revelation absolute and independent, they also conceded to science an absolute and independent status. Thus Catholic journalists found fewer religious objections to the *Origin* than did the professors of science. The liberal Catholic *Rambler* took the occasion to point its customary moral: the near-sightedness of Protestant bibliolators (and Catholic zealots) who tried to fix the truth for all time and judge all learning according to a literal reading of the Biblical texts; and the far-sightedness of liberal Catholics who chose to leave science and other intellectual disciplines to their own devices, confident that ultimately religion and science would prove compatible. In this case, as in so many others, it maintained, the supposed religious issue at stake was a spurious one, the religious idea of creation meaning not the miraculous interference with the laws of nature but rather the miraculous institution of those laws.

The orthodox Catholic *Dublin Review* found an even more ingenious cause for satisfaction with the *Origin*, taking as its text not the vanity of bibliolatry but the vanity of all earlier science. Until Darwin, the *Dublin Review* said, it had been the prevailing theory among scientists that "the variations of the human frame amongst the different members of our race are so important and radical that it is impossible that the Mosaic narrative can be true, or that we can all be descended from a single pair."[43] Now Darwin came along to suggest that not only did all races have a single parent, but all animals came from some four or five or possibly even one prototype. In thus confounding itself, science had shown the frailty of those who were at the mercy of conflicting theories and the strength of those resting securely upon the mysteries of revelation. The *Dublin Review* was so delighted with the irony that Darwin should have been found to reaffirm the conventional Biblical doctrine that it was also inspired to praise him for his scientific bearing, his dispassionate reasoning, and his keen-sightedness. (But not so delighted as to condone the extension of his theory to "such unreasonable lengths" as to include man.)

It would not do to make too much of the paradox of scientists exhorting the public to greater theological awareness, while theologians exhorted them to a greater respect for the integrity of science. For scientists had their scientific objections to the *Origin*, as theologians, however so-

phisticated their casuistry, had their theological objections. If these occupy less space here than they might, it is
because the commonplace commands less attention than
the uncommon and also because a more extended account
will appear in later chapters. Yet it is interesting that religious motives should have played so large a part in the
responses of scientists—and not of one or two eccentrics but
of almost all. There may have been some justice in the
claim of the Catholics that while they themselves were too
firm in their faith to be moved by the vacillations of
science, the scientists, buffeted about by conflicting facts
and theories, had to be especially vigilant in the defense of
their religious faith. Theologians had not to reconcile science and religion at every stage in the here and now, because they were confident that ultimately science must
bear out the indubitable truths of revelation. Scientists, on
the other hand, could not take so long and sanguine a view.
The truths with which they were professionally concerned
were located in the here and now, and they could not
so easily continue in a state of suspended or divided
judgment.

What made matters worse was the fact that before the
appearance of the *Origin* science was satisfied that it had
come to an understanding with religion, a *modus vivendi*
consisting not in a positive statement on the origin of
species but in the negative assurance that the evolutionary
theory, at least, had been discredited. Speculations not so
very different from those of Darwin, at least in their bear-
ings on metaphysics and religion, had often been put forward and as often disproved, thus lulling the religious-
minded into a false security. This security was all the more
cherished because science had memories of a less halcyon
epoch when it had labored under suspicions of heresy and
had had to resort to concealment or evasion. Now that religion and science were finally reconciled, when men of
science were favorites of the court and church, it seemed
folly to give up this hard-won peace for such insubstantial
returns—indeed, for just another hypothesis.

Only a few weeks before the appearance of the *Origin*,
in a notice of two other works on natural history, the

Saturday Review had remarked on the happy state of science: "Writers on natural history have no longer any occasion for apology or vindication of their favorite science. It has had in bygone times its martyrs and its confessors, but those days are passed away, and its professors are now the priests of a dominant sect whose altars and oratories are erected in nearly every British home."[44] After the publication of the *Origin,* the *North British Review* regretted that this peaceful interlude had been shattered. The *Origin,* it said, might more properly have appeared under the title, "A Contribution to Scientific Speculation in 1720," since it was a retrogression to the pre-Linnaeus, pre-Cuvier epoch: "Thrust upon us at this time of day, when science has walked in calm majesty out from the mists of prejudice, and been accepted as a sister by a sound theology, it has reminded us of a word in the oldest and best of books, which we commend to Mr. Darwin and his followers: 'Shadows as the night in the midst of the noon-day.' "[45]

The notion that religion and science had been members of one happy family before the *Origin* came along to foster dissension and discord was not entirely illusory—and this in spite of the experiences of the 1840's and '50's. The *Vestiges,* the writings of Herbert Spencer or Lord Tennyson, the activities of a multitude of amateur geologists, even the monographs of scientists, all of which later appeared to have had the sole purpose of preparing the way for Darwin and widening the rift between science and religion, had at the time the opposite effect as well, hindering Darwin's advance and closing the gap between science and religion. While old-fashioned clerics like the Dean of York might deprecate the insidious influence of science, or sensitive souls like Ruskin might complain that the geologists' hammers were sounding the death knell of belief, the vast body of scientists, clergymen, and laymen were agreed that the community was admirably withstanding all temptations to subversion. Scientists were not being converted to mutability, however much their monographs seemed to verge on it; philosophers and poets were not embracing naturalism or atheism however they toyed with the ideas;

and the young men and women who climbed mountains and dug for fossils did not seem any the worse, metaphysically or religiously, for the experience. Indeed, the very presence of the toxic doctrines, the fact that they were so persistently in the air, seemed to have had an immunizing effect on the population. Sedgwick and his colleagues could congratulate themselves that they had been exposed to the worst and had survived without blemish. When the *Origin* came along, with a new and more virulent strain that defied their hard-won immunity, they were properly bewildered and indignant.

CHAPTER 14

PROGRESS OF THE CONTROVERSY

HUXLEY described the status of Darwinism in 1860: "There is not the slightest doubt that, if a general council of the Church scientific had been held at that time, we should have been condemned by an overwhelming majority."[1] Such a council was, in fact, held, but the majority against Darwin was not overwhelming, and the effect was less condemnation than confrontation.

The annual meeting of the British Association was held at Oxford at the end of June 1860. Darwin himself was absent, as he had been absent from the Linnean Society meeting two years earlier, from London when his book was published, and from every other controversial occasion. Two days before the opening of the meeting, he wrote Lyell that his stomach had "utterly broken down" and that he was obliged to go away for a water cure; for good measure, he added that his daughter Henrietta was still very ill.[2] After it was all over, he confessed: "I would as soon have died as tried to answer the Bishop in such an assembly."[3] A temperamental aversion to controversy had become a principled one. Like Newton, who found that "a man must either resolve to put out nothing new, or become a slave to defend it,"[4] Darwin soon discovered the advantages of retirement: "On principle I have resolved to avoid answering anything, as it consumes much time, often temper, and I have had my say in the *Origin*."[5]

Darwin was the only one of the protagonists to be absent. The *Origin*, then six months old, had been acknowledged by enemy and friend alike (although not by one, like Owen, who was less enemy or friend than rival) as the most important event in the recent history of science, and it was scheduled as the subject of two papers at the meeting. Public interest was great and the audience large and varied, consisting of undergraduates, women, and clergymen, as well as amateur and professional scientists. At the Zoological Section, on June 28, the Oxford Professor of Botany, Charles Daubény, read a paper on "The Final Cause of the Sexuality of Plants, with particular reference to Mr. Darwin's work on the Origin of Species," which was largely favorable to Darwin. Huxley, called upon for comment, pleaded that "a general audience, in which sentiment would unduly interfere with intellect, was not the public before which such a discussion should be carried on." Owen, however, would not be dissuaded, and "in the spirit of the philosopher," as he claimed, undertook to instruct the public in the facts that would permit it to judge Darwin's theory. The most damaging fact, he asserted, was the gorilla's brain, which "presented more differences, as compared with the brain of man, than it did when compared with the brains of the very lowest and most problematical of the Quadrumana."[6] Huxley, in rebuttal, contented himself with a flat denial of this assertion, promising to elaborate upon it at a more appropriate occasion.

For the moment the matter rested there, the subject not being alluded to at the meetings on the following day, Friday. Although it was scheduled again for Saturday, the form which the discussion promised to take seemed so inauspicious that both Huxley and Hooker decided independently not to attend. Bishop Wilberforce was to reply to the main address, and Huxley, tired and anxious to rejoin his wife, told Robert Chambers that he had no intention of being "episcopally pounded."[7] Chambers vehemently remonstrated with him against deserting the cause, until Huxley finally agreed to stay on. Hooker, for his own part, changed his mind at the last moment, when sheer boredom drove him to attend the meeting.

The most curious aspect of the affair is the paper which sparked the conflict and which has since been so completely ignored: "The Intellectual Development of Europe considered with reference to the views of Mr. Darwin" by John William Draper. It is curious, first, that a paper on this subject should be regarded as suitable for the Zoological Section of the British Association; second, that of the two papers bearing on Darwin, both should have been largely in his favor—which would suggest that the Church scientific was not so overwhelmingly prejudiced against Darwin as his friends later made out; and third, that Hooker and Huxley should have been as distressed by Draper's message as were Wilberforce and his friends.

The attitude of Darwin's friends was particularly revealing, for it epitomizes the difference in character between the two parties, the anti-Darwinians being willing to enter into any alliances that might be to their immediate advantage, and the Darwinians, strait-laced and pure-minded, determined not to compromise their principles. Draper's theme was the analogy between science and society: as the progression of species was determined by immutable law, so was the intellectual progression of mankind. Later, in his famous *History of the Conflict between Religion and Science*, this was to be spelled out as the progression of mankind from religious barbarism to scientific enlightenment. Huxley was repelled by these sweeping and uninformed generalizations, and Hooker did not try to hide his contempt for the "Yankee donkey": "For of all the flatulent stuff and all the self-sufficient stuffers, these were the greatest; it was all of a pie of Herbert Spencer and Buckle without the seasoning of either."[8]

The lecture was delivered to a packed audience. Although it took place during the long vacation and was closed to the public, so many members had turned up that it had had to be adjourned from its scheduled hall to a larger one, in which over seven hundred people assembled before the principals arrived. The discussion afterward, which was what everyone was waiting for, was opened by an economist who quarreled with Draper and Darwin on the religious question, as did his successor, a clergyman. Both men were shouted down by the undergraduates massed in one part of the room, who showed their impartiality by also shouting down the next speaker, who rose in Darwin's defense. Trying to improve upon Darwin with a mathematical demonstration of the theory, he had the misfortune to accompany his blackboard diagram with the explanation, "Let this point A be man, and let that point B be the mawnkey," whereupon the students delightedly took up the cry of "mawnkey." Professor Henslow, as chairman, ineffectually tried to confine the discussion to the scientific issue, but the audience, impatient for the main event, called for the Bishop. With a great show of modesty, Wilberforce referred to his friend, the histologist Professor Beale, who had nothing to say except to appeal for a fair hearing. His audience's appetite whetted, Wilberforce finally condescended to take the floor.

Wilberforce's speech was a rehearsal of his *Quarterly Review* article to appear a few days later. Although he ridiculed Darwin and Huxley, it was all done in "such dulcet tones, so persuasive a manner, and in such well-turned periods" that the chairman could not object, only Darwin's partisans complaining of the "ugliness and emptiness and unfairness" of it.[9] The climax came when he gallantly appealed to his audience, asking whether woman, as well as man, was supposed to be derived from a beast. Then, turning to Huxley, he put the memorable question: Was it through his grandfather or his grandmother that he claimed descent from a monkey? After this, even the eloquence of his peroration, a solemn indictment of Darwinism as contrary to the word of God, failed to make of it more than an anticlimax.

Earlier in the speech Huxley had begun to suspect that Wilberforce was not the formidable opponent he had taken him to be. When the direct question was put to him about his ancestry, he instantly saw the polemical advantage it gave him, and he startled his neighbor, the venerable surgeon Sir Benjamin Brodie, by whispering to him: "The Lord hath delivered him into mine hands."[10] Waiting patiently until Wilberforce finished and the audience called for him, he opened with a sober defense of the *Origin* as a legitimate scientific hypothesis. Still gravely, he went on to explain that it was not Darwin's intention to establish a direct relationship between man and the ape but only the descent of both, through thousands of generations, from a common ancestor. He then delivered himself of the fatal blow:

> I asserted—and I repeat—that a man has no reason to be ashamed of having an ape for his grandfather. If there were an ancestor whom I should feel shame in recalling, it would rather be a *man,* a man of restless and versatile intellect, who, not content with an equivocal success in his own sphere of activity, plunges into scientific questions with which he had no real acquaintance, only to obscure them by an aimless rhetoric, and distract the attention of his hearers from the real point at issue by eloquent digressions and skilled appeals to religious prejudice.[11]

Bishops, however great the provocation, were not often treated so disrespectfully, and the excitement was tremendous. One lady fainted and had to be carried out, while undergraduates leaped from their seats and shouted. Other speakers followed, adding to the confusion and uproar. An Oxford don disputed the theory of development by pointing out that Homer, the greatest of the poets, had lived three thousand years ago and his like had not been seen since. Sir John Lubbock defended it against some of the frauds used against it: he told of a specimen of wheat that had been sent to him as having come from an Egyptian mummy, ostensibly demonstrating that wheat had not changed since the time of the Pharaohs; upon examination the wheat proved to be made of French chocolate. Admiral FitzRoy got up to describe how he had often expostulated with his old comrade of the Beagle; lifting an immense Bible over his head, he solemnly implored the audience to believe God rather than man, "to reject with abhorence the attempt to substitute human conjecture and human institutions for the explicit revelation which the Almighty has himself made in that book of the great events which took place when it pleased Him to create the world and all that it contained."[12] Hooker, unable longer to contain himself, determined to "smite that Amalekite, Sam, hip and thigh if my heart jumped out of my mouth," as he later reported to Darwin, and proceeded to "hit him in the wind at the first shot in ten words taken from his own ugly mouth."[13] He declared that Wilberforce could never have read the *Origin* nor have been familiar with the rudiments of botany, cited his own experience and conversion, and concluded with some remarks on the relative merits of the old and new theories. His was the last word. Wilberforce did not reply and the meeting was dissolved.

Recounting these events to Darwin, Hooker implied that he had intervened only because he was afraid that Huxley's voice had not carried well and that his arguments had not been entirely effective. Certainly Huxley was drowned out at the end, but because of the very excitement he had generated; and most of the reports agree in making him the hero of the day. Yet if Hooker was being somewhat ungenerous to Huxley, history may prove to be too generous. Huxley's retort to Wilberforce on the merits of their respective ancestries has been taken to be not only the climax of the day but also the climax of the entire debate, as if with this one riposte the battle was won. To some extent, this is a defect of reporting. Since there was no full and objective contemporary account of the event, we are at the mercy of the "lives and letters" later published by eyewitnesses. And history having subsequently decided the issue in favor of Darwin, those who chose to dwell on the affair are naturally the victors rather than the vanquished. Thus it is a partisan view that has prevailed: the idea that an initially hostile audience was won over when the ignorant and coarse imputations of the Bishop were exposed by the "cool, quiet, scientific" demeanor of Huxley.[14] In fact, most of the clergy remained unmoved; the ladies who had waved their handkerchiefs in gentle acclamation of Wilberforce's wit were repelled by Huxley's impiety; while the undergraduates, having earlier hooted the bores on both sides, were now aroused more by the contest of wit than by the contest of ideas.

The professional scientists in the audience were probably less affected by the verbal sparring than the others. The majority had come to the meeting with an unfavorable opinion of Darwin's theory, and if they were abashed, as Darwin's party liked to think, by the knowledge that "the Bishop had forgotten to behave like a perfect gentleman,"[15] their discomfiture was probably not so great as to induce in them a change of heart on the subject of species. On the other hand, those scientists who had already committed themselves to Darwin were tremendously elated by Huxley's coup, and while it did not affect their scientific positions either, it did give them an encouraging premonition of triumph. Not the most sanguine of them supposed it to be an outright victory, but the fact that an open resistance to authority had resulted even in a drawn battle, as Huxley's son later described it, was gratifying.

Probably the effect of the meeting was less to shift sentiment than to harden it, to intensify party strife among those already endowed with party spirit. Huxley's followers recalled the "looks of bitter hatred" bestowed on them as they passed through the still predominantly hostile crowd after his speech.[16] If the majority, however, made them feel like despicable outcasts, among themselves they were exultant. At a triumphal party held in Daubény's rooms that evening, they congratulated Huxley upon his performance. One more naive than the others and carried away by the collective enthusiasm, was heard to wish that "it could come over again"—whereupon Huxley solemnly recalled them to the proprieties. Intimating that he had not enjoyed being forced to take so personal a tone, he assured them that once in a lifetime was enough, if not too much.[17] The congratulations of others he received with less-concealed satisfaction, particularly those of the small group of clergymen who either wished to express their disapproval of the debating tactics of their superior or who were actually attracted to Darwin's theory. (A more piquant clerical note came from the marked physical resemblance, in spite of the difference of age, between the two protagonists, so that the next day, resting at the house of his brother-in-law, Huxley was actually mistaken by one of the other guests for "the son of the Bishop of Oxford."[18])

The impact of the *Origin* was felt immediately. The *Illustrated London News* was probably the only organ of opinion to be so misguided as to conclude its appraisal of 1859 with the pronouncement, "The year 1859 will not form a remarkable year in the annals of English literature," and to deplore the losses suffered by science (the deaths of Brunel and Stephenson) without thinking to remark upon the *Origin* as either a gain or a loss.[19] In a society where science was a normal subject of discourse, where no theory seemed too difficult or recondite for the intelligent layman, the *Origin* vied with the Italian Revolution in dinner-party conversation. If it was more often cursed than praised, the very volubility of the response testified to the importance attached to it. And after a few months it became clear that it was to be no short-lived sensation, that it would remain a staple of the lecture room as of the drawing room. Those who most bitterly deplored it were quickest to concede its importance.

Even before the British Association meeting, Hooker had begun to worry about the time, in the near future, when Darwin's theories would be as generally accepted as they were then generally execrated: "There will be before long a great revulsion in favour of Darwin to match the senseless howl that is now raised, and . . . as many converts on no principle will fall in, as there are now antagonists on no principle."[20] After that meeting, Huxley, who also took pride in fighting the good battle against great odds, sensed that the ammunition and zeal of the opposing forces were running out. "The platoon firing," he reported, "must soon cease."[21]

The first, and in Darwin's opinion the most important, contingent to be won over were those "young and rising naturalists" to whom he had addressed himself with such confidence in the *Origin*[22] (and for which bit of arrogance he had been rebuked by Sedgwick). Here his predictions

were entirely borne out, for it was the young scientist at
the start of his education and career—rather than the
older, established expert—who tended to be the more receptive. Apart from the obvious social conservatism that
gave the older men a stake in the conventional theories,
there were good scientific reasons for their reluctance to
take up a new and unproved hypothesis. Even Darwin did
them the justice of realizing that if he could not hope for
much from the experienced naturalists, it was because
their minds were "stocked with a multitude of facts all
viewed, during a long course of years, from a point of
view directly opposite to mine."[23] The younger men, on
the other hand, with a less refractory stock of facts and
theories, were more open to influence. And as their elders
might be inhibited by social conservatism, so they were
emboldened by social rebelliousness. In the perennial battle of the generations, the "ancients" versus the "moderns,"
Darwin gave them the opportunity to assert themselves
over their elders and superiors.

The war of the generations was fought most strenuously
in geology. Archibald Geikie was twenty-four when the
Origin appeared; later he recalled his impatience with his
teachers for not seeing it as a "new revelation of the manner in which geological history must be studied."[24] The
American geologist Nathaniel Shaler described his student
days at Harvard, when he and a friend used to debate the
Darwinian theory in secret, "for to be caught at it was as
it is for the faithful to be detected in a careful study of a
heresy." Summoning up the courage to ask Agassiz how
species had originated, he had received the uncompromising reply: by a "thought of God."[25] While Lyell was the
only one of the older geologists to be so much as shaken in
his views of immutability, so many of the younger men
were rallying to Darwin's side that by 1860 geologists out-
weighed botanists and zoologists among his supporters.
But the other sciences also had their youthful rebels. The
mathematician and biologist Karl Pearson recalled "the joy
we young men then felt when we saw that wretched date
BC 4004, replaced by a long vista of millions of years of
development."[26] And August Weismann, who was twenty-

five when the *Origin* was published, described the antipathy of the reputable biologists of his youth to all general problems and far-reaching, deductive theories, in reaction against the romantic speculations of *Naturphilosophie:* "Darwin's book fell like a bolt from the blue; it was eagerly devoured, and while it excited in the minds of the younger students delight and enthusiasm, it aroused among the older naturalists anything from cool aversion to violent opposition."[27]

For the same reason that Darwin looked for converts among the young scientists rather than the old, so he looked for approval among laymen rather than professional scientists:

> I have made up my mind to be well abused; but I think it of importance that my notions should be read by intelligent men, accustomed to scientific argument, though *not* naturalists. It may seem absurd, but I think such men will drag after them those naturalists who have too firmly fixed in their heads that a species is an entity.[28]

Here, too, he was proved to be largely right. If the lay converts did not actually "drag after them" the professionals, they did help create the climate of opinion in which both laymen and professionals worked. What he did not properly appreciate, however, was that it was less as intelligent men "accustomed to scientific argument" that they judged and approved the *Origin* than as intelligent men susceptible to philosophical prejudice. And it was this philosophical prejudice, more than any scientific argument or knowledge, that told in its favor. The historian James Bryce remembered his undergraduate days at Oxford at the time the *Origin* appeared, when fellow students who had never so much as opened any serious book, let alone a scientific one, outside the prescribed line of their studies, read and discussed the *Origin* with an avidity that was in inverse ratio to their scientific knowledge.[29] Such young enthusiasts as Thomas Hardy, Leslie Stephen, and John Morley did not pretend to have been won over by a sober appraisal of the scientific evidence. Morley spoke for all of them when he said that the words "evolution," "adaptation," "survival," and "natural selection" appeared as "so many patent pass-keys that were to open every chamber."[30]

The *Origin* commended itself not only to the young seeking a new and freer philosophical universe but also to those who had already discovered such a universe. Henry Buckle, whose rationalist, evolutionary theory of history required an analogous rationalist and evolutionary theory of science, had indicated, in the first volume of his *History of Civilization,* his readiness to give up the "old dogma" of the fixity of species.[31] When the *Origin* appeared two years later, he naturally found it to be "full of thought and of original matter."[32] For similar reasons, and knowing even less of history than Buckle did of science, Darwin approved of Buckle's *History,* the first volume of which he judged to be, if somewhat sophistical, nevertheless "wonderfully clever and original and with astounding knowledge."[33] By the time he read the second volume, perhaps warmed by the report of Buckle's praise of the *Origin,* he had abandoned all reservations. Convinced that Buckle was animated by a "noble love of advancement and truth," he said he no longer cared whether his views were right or wrong.[34]

John Stuart Mill had as good reason as Buckle to welcome Darwin's book. It is all the more to his credit, therefore, that he did not respond to it as to a set piece of naturalistic philosophy, but rather judged it on its own merits and in its own terms. As an exercise in scientific method, he found it entirely acceptable and said as much to one of his disciples, Henry Fawcett, who wrote a notice of the *Origin* defending the method of the "probable hypothesis" as a legitimate scientific procedure.[35] Fawcett later gratified Darwin by telling him that Mill had declared the reasoning of the *Origin* to be not only "in the most exact accordance with the strict principles of logic" but also the only mode of investigation proper to the subject.[36]

As a theory, however, Mill found it exciting but not

persuasive. His first reaction to the book was genuine astonishment that so much could have been done with so fantastic an idea. Darwin had not, to be sure, proved the truth of his theory, but he had proved that it might be true, which Mill took to be "as great a triumph as knowledge and ingenuity could possibly achieve on such a question." "Nothing," he continued, "can be at first sight more entirely unplausible than his theory, and yet after beginning by thinking it impossible, one arrives at something like an actual belief in it, and one certainly does not relapse into complete disbelief."[37] Ten years later he was still at the same point midway between belief and disbelief, and could say no more on its behalf than that it was "not so absurd as it looks, and that the analogies which have been discovered in experience, favourable to its possibility, far exceed what any one could have supposed beforehand." Nevertheless, it "is still and will probably long remain problematical." Moreover, he added—reducing still further the validity of the theory as Darwin understood it—even if proved, it would not be inconsistent with Creation; he himself, on the basis of the present state of the evidence, believed in "creation by intelligence."[38]

George Eliot, who had long done battle for positivism, naturalism, and agnosticism, might also have been expected to be an enthusiastic convert. Instead, she was only a very reluctant one. Where Mill accepted the theory partially and tentatively, feeling that it had not yet been adequately proved, Eliot accepted it completely but reluctantly. She did not doubt its truth; she only disliked its implications. It was sentiment, not science, that estranged her.

A good friend of Herbert Spencer and entirely *au courant* in intellectual affairs, Eliot was sufficiently alert to the importance of the *Origin* to order a pre-publication copy. Reading it before the notices appeared, she was inclined to belittle it: "Though full of interesting matter, it is not impressive, from want of luminous and orderly presentation." The next day it found its place in her diary among the occupations of the evening: "music, 'Arabian Nights' and Darwin."[39] Subsequent reading, and perhaps also the first intimations of its public success, made it seem more impressive, and a fortnight later she was writing to a friend that the *Origin* marked an epoch in the history of the doctrine of development, as witnessing the adhesion not of an anonym like the author of the *Vestiges* but of a reputable naturalist. The book had its failings, to be sure: it was "sadly wanting in illustrative facts," and this, she predicted, would prevent it from becoming as popular as the *Vestiges*. But it would be effective in the scientific world in opening up the discussion of a question about which men had been regrettably timid. "So the world gets on step by step towards brave clearness and honesty!" Yet she herself was not entirely happy with it: "To me the Development theory, and all other explanations of processes by which things came to be, produce a feeble impression compared with the mystery that lies under the processes."[40] Having sought in ethics the same principles of necessity and law that she so envied in science, she found herself dissatisfied with those principles when they confronted her in science itself.

If some of Darwin's supporters were less enthusiastic than they might have been, others were more so. Thus he made converts not only of naturalists and rationalists but also of some clergymen. In the second edition of the *Origin*, protesting that he had said nothing inherently offensive to religion, Darwin quoted a "celebrated author and divine" who had written in praise of his book that it was "just as noble a conception of the Deity to believe that He created a few original forms capable of self-development into other and needful forms, as to believe that He required a fresh act of creation to supply the voids caused by the action of His laws."[41] One critic found it so incredible that any clergyman could have "penned lines so fatal to the truths he is called upon to teach"[42] that he challenged Darwin to produce his name. Although Darwin was incensed that anyone should have doubted his deliberate word—it was "the act of a man who has not the soul of a gentleman in him"[43]—and although he had been authorized to quote the statement, he did not wish to ask

permission to use his name and so had to suffer the asper-
sion in silence.

The author and divine who had volunteered his good
opinion of the *Origin* was Charles Kingsley. Kingsley was
also an amateur naturalist, and it was as such and as a
suspected sympathizer that Darwin had sent him a copy of
his book. In acknowledgment, Kingsley wrote that on the
basis of his own observations of domesticated plants and
animals, he had already come to disbelieve the "superstition" of the permanence of species, as he also disbelieved
the superstition that the dignity of God required a series
of special acts of creation. What the *Origin* revealed to
him was that he would probably be obliged to give up
much else that he had believed and written. But for that,
he assured Darwin, he cared little. "Let God be true, and
every man a liar!" The "villainous shifty fox of an argument," he said bravely, "had to be followed up no matter
what bogs and brakes it might lead men into."[44] It was not
long before he discovered that the bogs and brakes were
not so formidable after all, and by 1863 this image had
given way to the placid one in the *Water Babies*, where the
child approaches Mother Nature expecting to see her busy
and instead finds her sitting with folded hands. "I am not
going to trouble myself to make things," explains Mother
Nature. "I sit here and make them make themselves."[45]

Kingsley's response to the *Origin*, resembling that of Asa
Gray, was shared by other clerics—not by many, but by
enough to suggest that it was a possible interpretation and
might, in time, become a common one. Its most sophisticated theological expression came in the controversial *Essays and Reviews* published a few months after the *Origin*.
There Baden Powell, in his study of the "Evidence of
Christianity," cited Darwin's work as definitive proof "of
the grand principle of the self-evolving powers of nature."[46] For Powell, this doctrine was not a reluctant concession to science but rather confirmation of a basic and
prior religious belief. It was not religion that was following
the dictates of science but science that had finally caught
up with the principles of true religion.

The typical clerical reaction, however, was not that of Kingsley and Powell nor even of Darwin's vicar at Down who believed that natural history should be pursued without reference to the Scriptures, but rather that of the country clergyman who said: "I cannot conceive how a book can be written on the subject. We know all there is to know about it. God created plants and animals and man out of the ground."[47] A more sophisticated variant of this was the attitude of Monsignor Manning, then provost of the Catholic diocese of Westminster and later Archbishop and Cardinal. To combat "science falsely so called," Manning founded the "Academia" in 1861, where he preached against the new "brutal philosophy" of Nature: the idea that "there is no God and the ape is our Adam."[48] And while Manning was urging his Church to rescue such fragmentary remains of Christian belief as still existed in heretical England, the Anglican priesthood was attempting to cope with Darwinism in its own way. Whewell, Master of Trinity and philosopher of science—who wrote to Darwin that although he could not "yet at least" become a convert, there was "so much of thought and of fact in what you have written that it is not to be contradicted without careful selection of the ground and manner of the dissent"[49]—took the precaution of removing not only the ground and manner of dissent but also the possibility of assent from his undergraduates, by refusing to allow a copy of the *Origin* in the Trinity College Library.

Thomas Carlyle gave a special twist to the religious opposition with the admission that the theory was not demonstrably untrue, that, in fact, it might well be true. But it was irrelevant and unimportant. "If true, it was nothing to be proud of, but rather a humiliating discovery, and the less said about it the better."[50] If such things as the origin of species had to be spoken of, it should be reverently, as mysteries. "An irreverent mind," he rebuked Darwin, "is really a senseless mind."[51] Mrs. Carlyle, as was her wont, put the matter even more sharply when she ridiculed her neighbor who unaccountably preferred the latest geological treatise to the *Iliad:*

Even when Darwin, in a book that all the scientific world is in ecstasy over, proved the other day that we are all come from shell-fish, it didn't move me to the slightest curiosity whether we are or are not. I did not feel that the slightest light could be thrown on my practical life for me, by having it ever so logically made out that my first ancestor, millions of ages back, had been, or even had not been, an oyster.[52]

Darwin found Carlyle's attitude difficult enough to understand without the malice that seemed to accompany it. He assumed that, as a friend of his brother, Carlyle would be amiable to him as well, so that when Carlyle once mockingly asked him whether men might not soon turn into apes again, Darwin took it as a good-natured pleasantry. He was dismayed, therefore, to read a statement attributed to Carlyle in the *Times* which was as unflattering to him personally as to his doctrine. The statement had originated as a letter in an obscure paper, had gained wide currency in the United States where it was supported by someone who actually claimed to have heard Carlyle make it, and eventually found its way into the *Times*:

A good sort of man is this Darwin, and well meaning, but with very little intellect. Ah, it's a sad, a terrible thing to see nigh a whole generation of men and women, professing to be cultivated, looking around in a purblind fashion, and finding no God in this universe. I suppose it is a reaction from the reign of cant and hollow pretence, professing to believe what, in fact, they do not believe. And this is what we have got to. All things from frog spawn; the gospel of dirt the order of the day. The older I grow—and I now stand upon the brink of eternity—the more comes back to me the sentence in the Catechism which I learned when a child, and the fuller and deeper its meaning becomes, 'What is the chief end of man?—To glorify God, and enjoy him forever.' No gospel of dirt, teaching that men have descended from frogs through monkeys, can ever set that aside.[53]

Carlyle instantly disclaimed authorship of the letter. Hastening to Erasmus Darwin, he expressed his regret that a forged letter of such a character should appear over his name, and assured him that, in fact, he had the best opinion of Darwin, knowing him to be a "noble, generous, good man" and his intellect "of the highest scientific order."[54] Since the first was not denied in the letter and the second counted so little in Carlyle's scale of values, and since the statement had his inimitable ring, it may be assumed to be genuine enough, although probably said in confidence and never intended for publication.

The same range of responses, the same permutations of belief and disbelief, acceptance and rejection, were to be found on the continent as in England, although with peculiar national variations. Thus, while in England it was the naturalistic, "scientistic" tradition of a Spencer, Buckle, or Mill that might predispose men in Darwin's favor, the same effect was produced in France by the materialistic, deistic, anti-clerical tradition of the Enlightenment, and in Germany by the *Naturphilosophie* associated with romanticism and idealism. Each of these countries had, in fact, a different version of the *Origin*, the German and French translators having emended it as they saw fit. The German translation had been supervised by Dr. H. G. Bronn, a distinguished zoologist and paleontologist, who produced an infelicitous translation that was excessively literal when it was not excessively liberal; its liberality consisted in the omission of passages of which the translator did not approve, such as that alluding to the origin of man, and the addition of an appendix elaborating upon difficulties which the translator, although not Darwin, found in the theory. (This appendix was done with Darwin's consent.) The French version was slanted in the opposite direction. The translator, Mlle. Clémence Royer, a disciple of Comte, who saw in natural selection a metaphysics, ethics, and political philosophy capable of superseding Christianity, was impatient of Darwin's more modest claims. In footnotes she chided the author for expressing doubts she did not share or posing difficulties which she found readily explainable.

As the party favoring him had its national stamp, so did the party opposing him. In France this sometimes degenerated into unashamed chauvinism, the opposition being divided between evolutionists who resented Darwin's usurpation, as they saw it, of the honors rightfully due to Lamarck, and anti-evolutionists who based the pre-eminence of French science on the work of Cuvier. In Germany the opposition was less chauvinistic but no less distinctive. It consisted of the two extremes to which German intellectual life was prone: the exact scientists—research workers, classifiers, and analysts—to whom the kind of general theory represented in the *Origin* was mystical, metaphysical nonsense; and the metaphysicians who thought it meaninglessly empirical or dangerously materialistic. In the latter camp was Schopenhauer. Although he had earlier approved of such evolutionists as Geoffroy St. Hilaire and Goethe on the grounds that they were concerned with the philosophical relationship of natural events, and had himself interpreted nature's indifference to individuals and solicitude for species as confirmation of the doctrine that only ideas and not individuals have reality, he was repelled by what he regarded as Darwin's materialism. Shortly before his death, he read an extract from the *Origin* in the *Times* which convinced him that this was yet another one of those frothy, insubstantial "soapsud or barber" books produced by superficial scientists. It was, he protested, "downright empiricism," with no sense of the inner, hidden force, the idea behind the struggle.[55]

For the most part, the Darwinians were triumphant in Germany, the anti-Darwinians in France. While the German edition was selling briskly, three leading French publishers rejected the proposal of a French translation. Thus the first French edition did not appear until the middle of 1862, after three English, two American, and one German edition had been sold out. And while French naturalists continued to ignore the work as long as they could, becoming hostile when indifference failed, German naturalists deluged Darwin with pamphlets, articles, and even books on the subject, most of them favorable. Later it was to be said that Darwinism, born in England, found its real home in Germany—and, it might be added, in those countries sharing the German intellectual tradition. When the geologist Geikie visited Austria in 1869, he was astonished to find how generally accepted Darwinism was among natural scientists, one of whom proudly informed him: "You are still discussing in England whether or not the theory of Darwin can be true. We have got a long way beyond that stage here. His theory is now our common starting point."[56]

In the United States too, where pirated editions of the *Origin* vied with the authorized, the controversy assumed a distinctively local color. Here its most exotic feature was racial; its most typical, religious. With the appearance of Social Darwinism, the occasions for controversy increased, and Darwinism became a national political issue.

The advance of the *Origin*, at home and abroad, was more rapid than might have been thought possible of a doctrine that was, after all, abstruse, technical, and fairly difficult, and that was vigorously opposed by most of the men who were in a position to understand and judge it—and, presumably, to influence those less qualified to understand and judge. In almost a matter of months it progressed from sensationalism, through heresy, to a modicum of respectability. Some more years were required for respectability to be converted into acceptance, but even this last stage was achieved far more quickly than was expected.

As it made its way through the succession of stages that Whewell had described as the fate of all great discoveries—the first, in which people said, "It is absurd"; the second, "It is contrary to the Bible"; and the third, "We always knew it was so"[57]—Darwin's own attitude toward it changed. At first he protested that his intention was only to open up the subject for discussion and suggest a possible but by no means final solution to the problem of species. Thus, after the meeting of the British Association in 1860, he modestly conceded that if he had not "stirred up the mud" of the controversy, someone else would soon have done so, that the subject could not be settled in his own lifetime, and that his only contribution was to provide a theory on which other naturalists might try out new facts.[58] Later, with the growing acceptance of the *Origin*, the pretense of modesty disappeared, and he and his followers began to act as if he had not only opened up the subject but had also closed it. By 1863 Kingsley was commenting on the "most curious" state of the scientific world, with Darwin "conquering everywhere and rushing in like a flood."[59] The following year the citadel of English science fell to him: the Copley Medal of the Royal Society, the highest scientific honor in England, available to scientists in all fields and of all nationalities. Nominated for it the previous year, he had then failed to receive it. When it was given to him in 1864, it was against the will of the president, General Sabine, who tried to minimize its significance by suggesting, in his presidential address, that the award was a tribute to him as a working naturalist and not as a controversial theorist. When it came to the printing of the address, however, Darwin's friends prevailed, and the objectionable sentence was stricken out.

One after the other, the leading scientists of the older generation and the august heads of societies had to confess the disaffection in the ranks. In 1865 Sedgwick deplored the spread of heresy in what was once his undisputed bailiwick: "The Geological Society is partly in fetters. It is not the honest independent body it once was; and some of its leading men are led by the nose in the train of a hypothesis."[60] One by one, the professional societies succumbed. In 1866 Hooker, addressing the British Association, could complacently describe the famous meeting of 1860 as an assembly of primitive tribes in which the Sachems, presiding over the ignorant savages, tried to suppress the civilizing missionaries; against those who taught the true theory of the moon's motions, they had invoked the ancient tribal belief that each month the gods ate the old moon and created a new one to demonstrate their power and glory. Now, Hooker was happy to report, the new doctrine was the accepted gospel. Two years later it was as Sachem himself that he appeared before the Association. His presidential address, widely reported in the press, was a eulogy of Darwin. Huxley, describing the event to the ever-absent Darwin (Darwin had also been absent from the Royal Society meeting at which he had been presented with the Copley Medal), spoke of the "Darwinismus" which crept up everywhere, insinuating itself even into a lecture on Buddhist temples. "You will have the rare happiness," he wrote, "to see your ideas triumphant during your lifetime." He added a typical Huxley postscript: "I am preparing to go into opposition; I can't stand it."[61]

It was, in fact, not in his lifetime but in a single decade that Darwin saw his ideas triumph. Indeed, scientific opinion had come so far by 1868 that Darwin was beginning to fear a reaction setting in against what was by now almost the new orthodoxy. When the *Athenaeum* ventured to suggest that the belief in natural selection was already passing away, Darwin was able to retort that at least the *Origin* had brought about the "now almost universal belief" in evolution.[62] The occasional reports of the demise of his theory proved to be premature, and Darwin soon had the satisfaction of witnessing the conversion of the most indomitable journals. In 1869 the Church paper, *The Guardian*, commended to its readers the tenth edition of the *Principles*, in which Lyell had declared his adherence to Darwinism. And the *Quarterly Review* invited Wallace to write on the same subject. "Really," Wallace exclaimed, "what with the Tories passing Radical Reform Bills and the Church periodicals advocating Darwinism, the millennium must be at hand."[63] The following year the *Athenaeum* itself succumbed. With the appointment of a new editor, the magazine's "curious antipathy" to Darwinism, as it by then appeared, was abandoned; the change was made, the editor later recalled, "with great tact and apparently without doing violence to anyone's susceptibilities."[64]

This tactful and inoffensive change in the policy of the *Athenaeum* brought the decade, and with it an epoch, to a close. There still remained, to be sure, a band of resisters, but although they included some of the most eminent scientists and clergymen, they were regarded as a relic of the past. In the sixth edition of the *Origin*, published in 1872, Darwin apologized for retaining several sentences implying that most naturalists still believed in the separate creation of species. He had been much censured for these references, which were by now obviously untrue. For although, he explained, "this was the general belief when the first edition of the present work appeared . . . now things are wholly changed, and almost every naturalist admits the great principle of evolution."[65] The witticisms that had been so entertaining in the sixties—Disraeli's "Is man an ape or an angel? I, my Lord, am on the side of the angels,"[66] and *Punch's* many variations on the joke: "'I could a tail unfold.' Could you? Then lose not a moment, but go instantly to Mr. Darwin. He will be delighted to see you"[67]—were to be briefly revived in the seventies with the publication of Darwin's *Descent of Man*. But they had lost much of their wit and more of their sting.

History makes prophets of us all. Because Darwin's success was so spectacular, it is assumed to have been inevitable as well. Yet there were times, early in the controversy, when contemporaries had every right to think (and some of the shrewdest, like Owen, did think) that Darwinism would fail to be accepted. And had it failed, the historian would have been well provided with reasons and explanations as persuasive as those he now calls upon to account for its success. Such a rationale of failure is not entirely hypothetical, its elements being present in the initial hostility with which the *Origin* was received and in which it might well have foundered. At that time, most responsible scientific authorities objected to Darwin's theory, as they had objected, with good reason and with success, to similar theories in the past. Theological canon and religious sentiment told against it, and so forcefully that scientists no less than clerics were susceptible to their promptings. Vested interests, both in science and the Church, opposed it; and not only institutional interests—the schools, societies, organs of opinion, and sources of professional status—but also intellectual interests, pride in ideas that had been carefully nurtured and long cherished. The most important of these was not so much an idea as an emotion, a primitive and pervasive revulsion against the idea that man was nothing more than the last stage in a natural order embracing barely animate organisms and all-too-animate beasts. No amount of philosophical analysis of the "fallacies" of anthropomorphism, solipsism, and the like could put the mind at ease among such alien ideas. Many more than admitted to it must have shared George Bernard Shaw's sentiment, that although Darwinism could not be disproved, no decent-minded person could believe it.

The success of the *Origin* is made all the more striking by the possibility of its having been a failure. It is to the *Origin* then—the analysis of its ideas, its logic, and its meaning—that we must turn to appreciate both the potentiality of failure and the magnitude of success, and to find the true measure of Darwin's achievement in the conversion of a near failure into so considerable a success.

ANALYSIS OF
THE THEORY

BOOK V

CHAPTER 15

THE ARGUMENT OF THE *ORIGIN*

ANY analysis of the *Origin*, like the history of its background, is plagued by the confusion between the theory of evolution (what Darwin called the "theory of descent") and the theory of natural selection. It has generally been assumed that the first has logical priority and that the second—natural selection, or the "how" of evolution—is secondary and subordinate, that the fact of evolution should be properly settled before the mechanism of evolution could be discussed. This was not, however, as Darwin intended it or as the argument was presented in the *Origin*. There both theories were supported by a single structure of facts and reasons, a structure so intricate that evolution could not be separated from natural selection. Indeed, if there was any question of logical priority, it was resolved in favor of natural selection. Without natural selection, Darwin declared, the theory of descent was unintelligible and unprovable; natural selection, in showing *how* species descended from others, also showed that they *did* descend from others. Natural selection was not a corollary of the theory of descent; it was descent that was a corollary of natural selection. The full title of his book, "On the Origin of Species by Means of Natural Selection, or the Preservation of Favoured Races in the Struggle for Life," was not only an attempt on his part to stress what was original in the theory but was also a fair description of what was at the heart of it.

The idea of natural selection, indeed the very term "selection," derived from the familiar practices of horticulturists and breeders, of whom it had been said that they seemed to have "chalked out upon a wall a form perfect in itself, and then had given it existence"; while a famous pigeon fancier of the time boasted that "he would produce any given feather in three years, but it would take him six years to obtain head and beak."[1] What Darwin did was to apply this experience to nature: as man adapted plants and animals to his needs or whims, so nature adapted them to their needs and environment. And both exploited the same natural fact: the great variability of individuals in any species. Man accomplished his ends by deliberately interbreeding those varieties which would preserve the characteristics he desired, and by destroying or condemning to die without issue those he did not desire; nature by seeing to it that, in the universal struggle for life, those varieties which were better suited to the struggle survived and reproduced while the less favored died. For both nature and man, the final arbiter determining the "survival of the fittest" was death.[2]

In this selective process, Darwin found, nature was as superior to man as life is to art. Where man could only perceive the grossest of variations and act upon them in the crudest of fashions, nature responded to the most subtle and invisible variations, consulting not some passing whim or private need but the good of the species itself, and having not a single lifetime or even the lifetime of a civilization but entire geological periods in which to exercise her discrimination. "Can we wonder, then, that Nature's productions should be far 'truer' in character than man's productions; that they should be infinitely better adapted to the most complex conditions of life, and should plainly bear the stamp of far higher workmanship?"[3]

Natural selection was the chief means by which individuals were elevated into varieties and varieties into species, but not, Darwin maintained, the only one. Another means at nature's disposal was "sexual selection." Sexual selection accounted for those secondary sexual characteristics—the plumage or singing of birds, for example—which had no apparent utilitarian function other than to differentiate the sexes and make them attractive to each other. Here variations had the effect of favoring individuals not in the struggle for existence but in the struggle for females. Yet death was still, ultimately, the penalty for failure. The unsuccessful competitor, instead of dying, left few or no offspring so that the extinction of his kind was accomplished gradually, in the course of generations.

For Darwin, sexual selection was not merely a subcategory under natural selection, as so many later commentators would have had it; it was an alternative or complementary theory to natural selection.[4] It was intended not only to cover the obvious case of the stag's horns or cock's spurs which permitted them to do battle for the possession of the females, but also all distinctions between the sexes, the theory being that all such distinctions, all the organs of display, must have originated in the attempt to attract, if not actually to catch, the opposite sex.

Darwin's critics, however, have objected that many of these organs of display serve no apparent purpose either in attracting or catching the opposite sex. Experimenters have painted the wings of butterflies and attached male wings on females without affecting their breeding habits. They have shown that the colors of moths breeding at night are as striking as those of butterflies, and that although male fish are vividly colored in the breeding season, the female does not even see the particular male fertilizing her eggs. Nor has it been proved that in all species where the male is more brilliantly colored than the female, the male is also more plentiful in number; but without this disparity there would be no competitive motive, no reason why the male should vie for the favors of the female. And even where the males do compete for the females, the victors do not always produce more offspring than the vanquished, who may have to content themselves with less colorful but no less prolific mates. These are not trivial matters, for it is precisely such sexual characteristics—so difficult to account for in terms of either natural or sexual selection—that are often the primary marks of distinction among the species.

Although sexual selection cannot be subsumed under natural selection, it does share some of the assumptions and difficulties of natural selection. It focuses attention upon one of the most significant and least appreciated aspects of Darwin's theory: the location of the struggle for existence primarily *within* species rather than *between* species. Superficially, the opposite might have been expected, the most common image conjured up by the struggle for existence being of one species preying on another or of a number of species competing among themselves for a limited supply of food. The point of the theory, however, which is dramatically illustrated by sexual selection, is that it is precisely among individuals of the same species or among varieties of the same species, who "frequent the same districts, require the same food, and are exposed to the same dangers,"[5] that the struggle is most intense. (Similarly, the struggle among the species of one genus is more intense than among species of different genera.) It is this conflict within the species that Darwin took to be the mechanism for the origin of species. By favoring the strong and eliminating the weak, individuals of superior endowment perpetuated themselves and became established first as varieties and then as species. Thus are old species transformed and new ones originated.

The difficulty here is that such a struggle for existence need not necessarily terminate in the survival of the fittest; it might well terminate in the survival of the unfittest, or at least of the unfit. The process of sexual selection, by virtue of those very qualities of display which permitted a species or variety to establish itself, might actually leave it weakened for the ordinary struggle for existence. Thus the peacock's train, so useful in courtship, might be a serious hindrance in competing for food or warding off enemies. In the same way, elaborate courting patterns have been said to make some species more vulnerable to their environment and more prone to genetic disturbance. It might be argued in rebuttal that if the organs of display or habits of courtship had, in fact, proved a hindrance in the struggle for existence, they would not have been perpetuated. But this is to reassert the hypothesis without advancing the argument. The fact remains that by the ordinary canons of evidence and common sense, such display, in its more extravagant forms, would seem to be as much conducive to extinction as to survival.

This problem is basic to Darwin's theory and is by no means confined to sexual selection. There are other areas of life in which the principle of the survival of the fittest is questionable, or at most meaningful only in the tautological sense that the survivors, having survived, are thence judged to be the fittest. One Darwinian, confronted with situations in which the survivors were not visibly or conspicuously the fittest, has suggested a modification of the principle to read merely that the "fit" survive[6]—a revision that may be more in keeping with the evidence but which destroys the very foundation of the theory. For if it is only the fit rather than the fittest who necessarily survive, there is no occasion for the operation of natural selection, which depends precisely upon just those small differences and advantages making one individual or variety more fit than the other, until some one form emerges as the most fit. Moreover, even if the original formula is retained, the fittest being presumed in some sense to survive, this itself is not sufficient to insure natural selection and evolution. The struggle for existence may as readily have a retarding as an advancing effect upon the species, the fittest at the end of the struggle being less fit than those at the beginning, thus giving no opportunity for the emergence of more fit forms that will constitute new varieties or species. The most familiar example is a war which leaves both victor and vanquished enfeebled.

Not only the concept of the survival of the fittest but even that of the struggle for existence has been found to be less obvious and more dubious than might seem to be the case at first sight. For if competition is a primary fact of nature, so is cooperation. Darwin was obliged to recognize this in the two most dramatic cases of cooperation, ants and bees, to account for the sterility of the workers. He tried to get around this difficulty by saying that what was here to the obvious disadvantage of the individual was to the clear advantage of the community. Even the sting of the bee, resulting in the insect's death, was amenable to the same explanation: "For if on the whole the power of stinging be useful to the community, it will fulfil all the requirements of natural selection, though it may cause the death of some few members."[7] Yet, neither the arbitrary shifting of the biological unit from the individual to the group, nor the ingenious attempt to subsume cooperation under the larger category of competition, by making the cooperation of individuals a means to the better competition of groups, succeeds in resolving the difficulty. The problem remains of reconciling the motive and meaning of cooperation, in which the primary measure of utility is the species, with the struggle for existence, the survival of the fittest, and natural selection, in all of which the primary measure is the individual.

It is in its most individualistic sense, as the competition of individuals rather than groups, that the struggle for existence was crucial to Darwin's theory, for only thus was he able to assert his independence from Lamarck. Unlike Lamarck's theory of adaptation, which depended primarily upon the struggle of organisms with their environment, Darwin's depended upon their struggle with each other for the best exploitation of their environment. Ultimately, to be sure, what Darwin's theory, like Lamarck's, had to account for were the "exquisite adaptations" of the different parts of the organism to each other and the whole to its "conditions of life."[8] But where Lamarck arrived at his end directly, by having the organism adapt itself, as by an act of will, to its environment, Darwin arrived at it circuitously, by engaging the organism in competition with its neighbors so that only the best adapted survived.

It is interesting to observe how far into the background the "conditions of life," or environment, receded in the first edition of the *Origin*—and this not only because of Darwin's wish to distinguish himself from Lamarck, but also because there often seemed to be little evidence of any adaptation, however circuitous, to the environment.

The geographical distribution of living species, as of fossils, showed both greater uniformities and greater diversities than could have been expected on any familiar theory of adaptation. Under a variety of physical conditions there has persisted a conspicuous uniformity of type, while elsewhere a diversity of types have flourished in the face of identical physical conditions. Similarly, while one species was undergoing the same change over a wide range of conditions, another might be transported into a new and different set of conditions without undergoing any change. From these facts Darwin deduced that the organism responded far less to its habitat than to its neighboring inhabitants. It was not so much its suitability to its physical environment that determined the success of a newly imported species, as it was a host of more remote and subtle conditions: the fact, for example, that the species, dominant elsewhere, would already have established itself in great numbers and with many variations; this variability alone, quite apart from the particular character of the variations, would constitute an advantage relative to a neighboring species, which, however indigenous and suitable to its habitat, was less numerous and variable. Thus, for Darwin, adaptation was a necessary and limiting, but by no means sufficient or directing, condition for the evolution of species.

Not being bound by adaptation in any direct or active sense, Darwin was liberated from the mechanical rigidity of the Lamarckian theory. He could indulge a freedom of inquiry and hypothesis that was not possible to Lamarck. But the greater scope afforded his imagination was itself dangerous. For if it liberated him from the definite, prescribed limits within which Lamarck was obliged to function, by the same token it set him free to be irresponsible as Lamarck was not. Lamarck was bound by a causal conception which, however it might violate the facts, was at least definite and direct; in Darwin's theory cause and effect were related in such a devious way as to permit almost any conjecture and to resist all control or verification.

The undisciplined nature of Darwin's concept of adaptation may be seen in his reply to those critics who objected that the same process that might be thought to account for the long neck of the giraffe might also have been expected to produce long necks in other species, the ability to browse upon the high branches of trees being of as much apparent advantage to one quadruped as to another. In a later edition of the *Origin* Darwin attempted to meet this objection, first by explaining that an adequate answer to this as to so many other questions was impossible because of our ignorance of all the conditions determining the number, range, size, and structure of species; and then by suggesting possible reasons why the giraffe alone developed a long neck, such as that only in that one species were all of the necessary correlated variations present in precisely the right degree and at the right time. He frankly admitted that these reasons were "general," "vague," and "conjectural."[9] In fact they were as hypothetical as the hypothesis they were intended to support.

There would be no objection to such hypothetical reasons, if they merely served to establish the consistency of the hypothesis. But what they establish is less its consistency than its plasticity, the ease with which it can be bent into any desired shape. If some animals had long necks, Darwin could summon up enough general, vague, and conjectural reasons to account for this peculiar fact; if others did not, he had at hand a different but equally general, vague, and conjectural set of reasons to account for that. Why, indeed, were not all species evolving in all different directions, ostriches acquiring the useful faculty of flying, other terrestrial animals of swimming, and so on? Because, Darwin replied, the appropriate variation must appear at the optimum moment in the creature's development, correlated with other variations also appearing at the optimum moment, each being of particular advantage in the particular economy and habitat occupied by the animal—an answer so general as to beg the question. The critic who protested that Darwin's theory was too *teres atque rotundus* and discredited itself by trying to explain too much[10] was not being as unreasonable as Darwin thought. Nor was a later critic who remarked upon the "labour-saving" quality of Darwin's concept of adaptation:

By suggesting that the steps through which an adaptive mechanism arose were indefinite and insensible, all further trouble is spared. While it could be said that species arise by an insensible and imperceptible process of variation, there was clearly no use in tiring ourselves by trying to perceive that process.[11]

Darwin himself, after a while, came to regret the precipitancy with which he had abandoned Lamarck—not because of any suspicion of methodological fault on his own part, but because, having securely established his own theory, he could afford to be more tolerant of Lamarck's and exploit its greater dramatic appeal. Natural selection, being a more subtle conception, was also more tenuous, and it was sometimes desirable to reinforce it by the hardier Lamarckian notion of adaptation.

This was all the more tempting in that Darwin did not share the common modern objection to the idea of the inheritance of acquired characteristics. In his *Variation of Animals and Plants under Domestication,* published in 1868, he gave examples of variations acquired in one generation and inherited in succeeding ones: a cow, having lost its horn by suppuration, gave birth to three calves, each with a small, bony lump in place of a horn; guinea pigs transferred to their progeny epilepsy which had been produced in the parent by injury to the spinal cord, malformed ears and eyelids originally produced by nerve injuries, or missing toes when the parent had bitten off theirs as a result of gangrene; horses commonly inherited the bony growths on the legs developed by their parents accustomed to traveling on hard roads; a man who had his little finger partially cut off had sons all of whom had a similarly crooked finger on the same hand. Darwin considered, only to reject, the possibility that these were coincidences. All, he decided, were authentic cases of the inheritance of acquired characteristics.

It was this discussion in the *Variation* that he cited, in a later edition of the *Origin,* as evidence that he had never believed natural selection to be the only agency of modification. And he added other such examples: the giraffe's neck which was the combined result of natural selection and increased use, and organs of other animals which were developed by the "inherited effects" of use and disuse and only "strengthened" by natural selection.[12] When one of his critics demonstrated statistically how unlikely it was that even the most favorable variation occurring only once in a large population could be perpetuated by natural selection, he conceded the point, granting that variations were not the completely chance and individual phenomena he had once thought them, but a recurrent phenomenon "owing to a similar organisation being similarly acted on." Indeed, he admitted, "the tendency to vary in the same manner has often been so strong that all the individuals of the same species have been similarly modified without the aid of any form of selection."[13]

As time went on, he became ever more receptive to the Lamarckian theory, until by 1875 he acknowledged that each year he came to attribute more and more to the agency of use and disuse.[14] Thus the older theory of adaptation was thrown into the breach when natural selection was found wanting.

If in one respect natural selection may be criticized for trying to explain too much; in another it may be thought to explain too little. Even at the time of its publication, a common charge brought against the *Origin* was its failure to establish a *vera causa* for evolution. As Samuel Butler later put it: "The 'Origin of Variation,' whatever it is, is the only true 'Origin of Species.' "[15] Natural selection, critics complained, might account for the persistence of some variations and the disappearance of others, but it did not account for the origin of the variations themselves. And only an explanation of the origin of the variations would constitute a *vera causa*. One critic compared Darwin unfavorably, in this respect, with his predecessors, Lamarck and the author of the *Vestiges,* who, however benighted, at least had the forthrightness to propose specific explanations for the origin of the variations. And even in his own camp Asa Gray and others confessed themselves troubled by this inadequacy in the theory.

Provoked by the criticism of friends and enemies, Darwin fell back upon historical precedent: if Newton was not obliged to show what gravity is, apart from how it manifests itself, so he was not called upon to go behind the mechanism of natural selection to the variations through which it operated. Yet he appreciated the difficulty of those who took natural selection to be a *vera causa* in the same sense in which Lamarck's theory was. In the sixth edition of the *Origin* he tried to clarify the difficulty. He admitted that "in the literal sense of the word, no doubt, natural selection is a false term." He had not intended, he said, to imply that natural selection was an "active power" inducing variability, but only that it was the process by which such variations as may arise were preserved. Still less had he meant to imply a conscious choice on the part of the organisms that were modified—which would have exempted from the workings of his theory at least the whole range of plants. Nor had he meant to personify nature as the selecting agent; he had intended it only to represent "the aggregate action and product of many natural laws." Natural selection was, in short, a metaphor or a shorthand expression, and he assured his readers that such "superficial objections" as they might now have would disappear with familiarity.[16] Privately, however, he was less confident, more ready to admit that natural selection was an unfortunate expression. "Natural preservation" or "survival of the fittest," he reflected, would have been less ambiguous.[17]

"Natural preservation" or "survival of the fittest" might indeed have been preferable, in making it clear that what he had in mind was a process that did not bring into being the crucial variations but merely acted upon them once they were in existence. The problem was thus clarified, but the objection, as he admitted, remained:

> The laws governing inheritance are quite unknown; no one can say why the same peculiarity in different individuals of the same species, and in individuals of

different species, is sometimes inherited and sometimes not so; why the child often reverts in certain characters to its grandfather or grandmother or other much more remote ancestor; why a peculiarity is often transmitted from one sex to both sexes, or to one sex alone . . .[18]

He also confessed that if he sometimes spoke as if variations were due to chance, this was only a loose way of saying that they were due to causes of which we are ignorant.

Although Darwin did not pretend to have an original theory of inheritance, he did share with his contemporaries the assumption that inheritance involved a blending of parental traits—a tall and a short parent, for example, producing neither tall nor short but medium-size offspring. In the essay of 1842 he had written:

> Each parent transmits its peculiarities. Therefore if varieties allowed freely to cross, except by the *chance* of two characterized by same peculiarity happening to marry, such varieties will be constantly demolished. . . . Free crossing great agent in producing uniformity in any breed.[19]

Initially, the blending theory of inheritance seemed to give to natural selection its strength and, almost, its validity. For if blending had the effect of diminishing variations, only the intervention of some such process as natural selection could account for the persistence of any variations at all. By seizing upon a favorable variation as soon as it appeared, natural selection might stabilize and preserve it, thus counteracting the destructive effect of blending. In this way, natural selection might be held responsible not only for the particular variations that became dominant— for the transformation of varieties into species—but for the very persistence of variation itself. In the 1844 essay this was suggested by the remark that without selection the free crossing of individuals would have the "unimportant result"—meaning the negative result—of eliminating variations.[20]

Yet, in spite of the greater significance given to natural selection by virtue of a blended inheritance, this argument does not appear in the *Origin*. Sometime between 1844 and 1859 Darwin must have realized that the idea of a blended inheritance raised as many difficulties for his theory as it seemed to solve. The main difficulty was the enormous quantity of variations that would be required in order to counteract the effect of blending and provide natural selection with the materials upon which it might operate. And not only must the stock of variations be far greater than anything observed in nature; it must also be constantly renewed, the variations available for natural selection at any one time having to be of fairly recent origin. The geneticist Sir Ronald A. Fisher has calculated that under the blending theory, in which half of the variations are lost in each generation, if the total variation is to remain constant, one-half must be new in each generation, one-quarter must be one generation old, one-eighth two generations old, and so on, with less than one-thousandth as much as ten generations old.[21] Even without these precise calculations, Darwin must have been aware that the blending theory threatened to deprive him of two of his most important props: the great expanse of time, measured not in mere generations but in eons, during which variations gradually and almost imperceptibly accumulated (this is response to those who protested that they could not find much evidence of new variations in their own time); and the ability of species to revert to an earlier type and so resurrect a variety that had been lost or dormant (this in order to enrich still further the stock of variations upon which selection might draw). Time was of little significance, and reversion inexplicable (if not impossible), with the blending theory.

By 1857 Darwin was beginning to despair of the conventional blending theory. He wrote to Huxley:

> I have lately been inclined to speculate, very crudely and indistinctly, that propagation by true fertilization will turn out to be a sort of mixture, and not true fusion, of two distinct individuals, or rather

of innumerable individuals, as each parent has its parents and ancestors. I can understand on no other view the way in which crossed forms go back to so large an extent to ancestral forms.[22]

Yet, the *Origin* contains no hint of these speculations. Not only did he abstain from offering any alternative to the blending theory, but he ignored the issue entirely—a remarkable omission, considering the decisive role of inheritance in his scheme. The principle of reversion, however, was essential to his thesis, and without arguing the point, without indicating how it was at variance with the conventional theory, he incorporated it into his argument. There must be present, he surmised, in each successive generation a "tendency" to reproduce the remotely inherited character, a tendency which finally gains "ascendancy" under some "unknown favorable conditions";[23] in the sixth edition of the *Origin* the character is described as "lying latent" until the unknown favorable conditions permit it to emerge.[24] If this description of reversion sounds a modern note, it is because it is in keeping with the currently accepted "particulate" theory of inheritance, according to which characters are inherited as "particles" or units which remain true to type and are not blended. But Darwin was certainly not aware of it in this sense.

Darwin's only attempt to formulate a systematic theory of inheritance was in the form of "pangenesis." He had conceived it many years earlier, but had had so little faith in its ability to convince others that he had excluded it from the *Origin*. He finally summoned up the courage to publish it in the *Variation*, and then only toward the end of the second volume as a "provisional hypothesis or speculation."[25] To Huxley he apologized for this "very rash and crude hypothesis," whose only justification was that it permitted him to "hang on it a good many groups of facts"; to another friend he spoke of the relief it gave him not to have a host of facts "floating loose in my mind."[26] The theory advanced so tentatively he reluctantly christened "pangenesis"—reluctantly, because he was more than half persuaded by his wife's objection that it sounded wicked, like pantheism; and also because he felt that other names, such as "cell-genesis" or "atom-genesis," would have been more precise. Euphony, however, and years of thinking of it under that name spoke in favor of pangenesis, and so it remained.

The theory held that cells threw off minute atoms or granules, termed "gemmules," which circulated through the system, multiplied by self-division, joined with each other by "elective affinity," and eventually developed into cells like those from which they originated. It was these gemmules, uniting in the sexual elements or reproductive organs, that he took to be the basic agents in the generation of new organisms. The idea was not particularly original. Speculations much like it have circulated since antiquity; Aristotle is known to have argued against a similar theory. Nor was Darwin's notably superior to those of his predecessors, not only because of such archaisms as "elective affinity" but also because of its assumptions about reversion and the inheritance of acquired characters. Like most breeders of the time, he believed that an inferior mating episode could damage the offspring of all later crossings. He gave the example of a "nearly" pure-bred Arabian chestnut mare whose first mating with a quagga (kin to the zebra) resulted in a hybrid, and whose later mating with a black Arabian horse produced a striped colt with distinctively quagga-like hair, from which Darwin deduced that "there can be no doubt that the quagga affected the character of the offspring subsequently begot by the black Arabian horse." This and "many similar and well-authenticated" cases testified to the "influence of the first male on the progeny subsequently borne by the mother to other males." Darwin considered, and rejected, other explanations for the remarkable influence thus exerted by the step-father, so to speak, on the physical character of his step-children—that, for example, the mother's "imagination" had been affected by the earlier union. He himself preferred the explanation suggested by pangenesis that "the male element acts directly on the reproductive organs of the female, and not through the crossed embryo."[27] Thus the quagga affected the character of all the mare's subsequent offspring by affecting permanently the character of the mare herself. It was the insatiable appetite for variations, the desperate need to create and preserve variations, that motivated his theory of pangenesis, prompting him to elevate all relations to the rank of blood relations and acquired characteristics to the status of hereditary ones, and to resurrect, by alleged reversion, whatever varieties were once, however remotely, present.

Pangenesis was never accepted by most of Darwin's supporters, let alone his opponents. And modern genetics has discredited it still further. The study of genetics had started even before the appearance of the *Origin,* when the Austrian monk and botanist Gregor Mendel began his famous experiments on peas. The results of these experiments were published in 1865, three years before the *Variation* containing Darwin's theory of pangenesis. It is not, however, to Darwin's reproach that he should have been ignorant of Mendel's paper. It appeared in the obscure publication of a local scientific society, and although the title and author were duly recorded in the Royal Society Catalogue of Scientific Papers, no one recognized in the "Versuche über Pflanzenhybriden" the work that was to revolutionize biological studies. Even so estimable a botanist as Karl von Nägeli, when the monograph was presented to him by Mendel, failed to appreciate its importance. It was not until the Dutch botanist Hugo de Vries rediscovered it in 1900 (at the same time that two other scientists independently came upon it) that Mendel's work received the recognition it deserved and that the science of genetics was established on secure grounds.

Mendel, and de Vries after him, confirmed one part of Darwin's theory, only to refute another. They confirmed Darwin's suspicion that inheritance is a discrete mixture of characteristics rather than a diffusion or blending. Each gene, as the unit of heredity is now known, is inherited independently of every other, and recessive genes (equivalent to Darwin's "latent" ones) become effective under specified conditions. Thus genetics came to the assistance of natural selection (not of pangenesis, which it completely exploded) by insuring that variations would not be lost, and by accounting for their origin and transmission. However, it solved the problem of the origin of variations only at the expense, it has been argued, of Darwin's theory of the origin of species. As genetics made pangenesis obsolete, so, in the opinion of some, it made natural selection superfluous.

Genetics may be thought to supersede natural selection by making the explanation of the origin of variations do service for the origin of species. In the experiments of Mendel and de Vries new species appeared suddenly in the form of mutations. No intermediate steps or gradual transitions, no struggle for existence or survival of the fittest was required for the creation of new and stable forms. Nor were the mutations responsible for new species related to the variations present in the old: between the shortest specimen of a new giant mutant and the tallest specimen of the old form, there was a distinct gap. De Vries conducted experiments to see whether he could create a permanent change of type by the selection of already-existing variations, as distinct from mutations, and discovered that he could not. Although he did succeed in creating an ear of corn with extra rows of kernels, the form persisted only as long as he continued to cross his best specimens—that is, as long as he continued to practice selection. Once he ceased to exercise control over it, the corn reverted to its normal condition. This reversion to type would not have occurred had the new form been the result of mutation instead of selection. August Weismann discovered the reason for this in the distinction between germ cells and somatic cells. Only the germ cells concerned in reproduction are inherited; somatic cells, making up the rest of the body, can be modified by environment and use, but unless these modifications affect the germ cells, which they do not necessarily do, they are not passed on. Thus not all variations, however favorable, are transmitted.

More recently, the neo-Darwinians have offered a rebuttal in the fact that the only effective mutations are not large but small, that most mutations are fatal to the organism, and the only ones at all likely to be favorable are those which depart least from the parental type. Thus, it is claimed, mutations are little more than Darwin's variations, and the Darwinian theory may be reinstated by the simple verbal amendment of "variation" to "mutation." Natural selection therefore remains what it always was: a "court of last appeal" determining which mutations, or variations, survive and which disappear.[28]

And not only a court of last appeal, but also an agency for defying, so to speak, the laws of probability, or, as it has been said, for "generating an exceedingly high degree of improbability."[29] For it is now discovered that favorable mutations are not only small but exceedingly rare, and the fortuitous combination of favorable mutations such as would be required for the production of even a fruit fly, let alone a man, is so much rarer still that the odds against it would be expressed by a number containing as many noughts as there are letters in the average novel, "a number greater than that of all the electrons and protons in the visible universe"[30]—an improbability as great as that a monkey provided with a typewriter would by chance peck out the works of Shakespeare. The strength of natural selection, Darwinians now argue, may be measured by the difficulties it overcomes, the odds against which it is pitted. For it is natural selection alone that transforms a manifest improbability into a fact. Only natural selection, by insuring the survival of those rare mutations that are favorable, and the still rarer concurrence of necessary and favorable mutations, can bring into being "the most apparently improbable adaptations."[31]

The neo-Darwinians, it is apparent, are as adroit as Darwin in making a virtue of necessity and in converting difficulties into assets. When Darwin's critics had objected that there were not enough variations in nature to satisfy the inordinate appetite of natural selection, Darwin had replied that nature was replete with variation, that variability was as basic a principle as heredity itself. When a second generation of critics, the early geneticists, pointed out that the only "variations" stable enough to provide material for natural selection were mutations—not the "infinitesimally small inherited modifications" that Darwin had insisted upon, but rather the "great and sudden modifications" he so distrusted[32]—and that these mutations, by being the sufficient cause of the origin of species, made natural selection redundant, the neo-Darwinians came along with the rebuttal that only the smallest mutations could be favorable and that such favorable mutations were, in fact, so rare a phenomenon that without natural selection not even a fruit fly, let alone a man, could have developed. Thus it became the very paucity of variations, the very improbability of their concurrence that was now made to tell in favor of natural selection.

Natural selection, in fact, has become the *deus ex machina* rescuing nature from the impossible situation in which the Darwinians had put her. Long before Darwin, men had recognized the improbability that nature, working blindly and by chance, could have evolved the universe as we know it. The triumphant discovery of the neo-Darwinians is, after all, only a feeble echo of an ancient cry. The laborious calculations of probability—the number represented by an infinity of noughts, the monkey pecking out the works of Shakespeare—are at least as much an argument in favor of the creationist theory as of natural selection, insofar as they can be said to be an argument in favor of anything.

The same mode of reasoning, the same technique of turning the tables may be seen in Darwin's treatment of another major difficulty: the case of the geological record. Why, he anticipated his critics, were there not fossil remains of those intermediate varieties or transitional forms which, according to his theory, must have linked up the species? The very process of extermination and transformation which he presumed to have taken place throughout history on such a vast scale should have produced a correspondingly vast number of intermediate varieties. Every geological stratum, it might be thought, would turn up many such fossils. Geology, however, has been notably unforthcoming, and, instead of being the chief support of Darwin's theory, it is one of its most serious weaknesses. The explanation of this anomaly Darwin found in the "imperfection of the geological record."[33] The record is imperfect in two respects: first, in the sense that our paleontological collection is faulty, representing only a small proportion of the fossils that may be presumed to exist; and second, in the sense that the process of fossilization itself is faulty. The inevitable decay of some parts of organisms and conditions unfavorable to the preservation of the rest have conspired to prevent the fossilization of some animals and to destroy such fossils as did once exist.

The particular circumstances responsible for these gaps in the record are elaborated in great detail, so that it comes as a shock to find Darwin admitting to further difficulties not accounted for by this formidable array of explanations. It might have been expected, for example, that in those cases where the geological record is more or less complete —and there are such cases—we would find, in any single geological formation, closely graduated varieties of species existing at the beginning and at the close of the period. Yet, even here we do not find such a graduated series. Nor is it easy to understand the abrupt way in which whole groups of species suddenly appear in certain formations, as if they had "started into life all at once."[34] Nor why whole groups of species should suddenly and abruptly appear in the lowest fossiliferous strata, with none of their expected progenitors in the beds beneath them. These difficulties inspired Darwin to further exercises in explanation, each more hypothetical and ingenious than the last. The entire discussion of the imperfections of the geological record is thus an exercise in hypothetical, "imaginary" (as he put it) reasoning:

> Geological research, though it has added numerous species to existing and extinct genera . . . yet has done scarcely anything in breaking down the distinction between species, by connecting them together by numerous, fine, intermediate varieties; and this not having been effected, is probably the gravest and most obvious of all the many objections which may be urged against my views. Hence it will be worth while

to sum up the foregoing remarks under an imaginary illustration. . . .[35]

The case at present must remain inexplicable; and may be truly urged as a valid argument against the views here entertained. To show that it may hereafter receive some explanation, I will give the following hypothesis. . . .[36]

The inconclusiveness of Darwin's argument escaped neither his friends nor his critics. Huxley summed up the matter precisely. "What," he asked, "does an impartial survey of the positively ascertained truths of paleontology testify in relation to the common doctrines of progressive modification?" To which he frankly replied:

It negatives these doctrines; for it either shows us no evidence of such modification, or demonstrates such modification as has occurred to have been very slight; and, as to the nature of that modification, it yields no evidence whatsoever that the earlier members of any long-continued group were more generalized in structure than the later ones.[37]

Although several years later Huxley thought he had found positive evidence of Darwin's theory in the fossil history of the horse, his reconstruction of that history, as he admitted, contained serious gaps, at least one crucial link being still missing and the length of time which must have intervened between the origin of vertebrates and their first known appearance as fossils being, even for him, "appalling to speculate upon."[38]

In raising the question of time in connection with geology, Huxley had put his finger on perhaps the most vulnerable part of Darwin's case. The length of time that must have intervened between the origin of vertebrates and their first appearance as fossils was appalling to contemplate, not only because it was difficult to understand why there were no fossil remains in the earlier stages of development, but also because there was no convincing evidence that such a length of time was in fact available for the slow development envisaged by Darwin. In the first edition of the *Origin* Darwin had calculated that "a far longer period than three hundred million years has elapsed since the later part of the Secondary period"[39]— which would have given natural selection enough time to perform its laborious task. But this estimate proved to be so immensely above that of the reputable geologists and physicists of the time that later editions of the *Origin* omitted it. Instead, Darwin substituted the findings of one geologist who calculated that a thousand feet of solid rock might be worn down in six million years; applied to the thirteen-thousand-feet thickness of the secondary strata, mentioned by him on a previous page, this would mean that only seventy or eighty million years had elapsed since the beginning of the secondary period, and much less since the later part of that period. Darwin did not explicitly combine these two figures to draw this conclusion, nor confess to having abandoned his earlier considerably higher estimate. Nor did he indicate how the new figures presented new difficulties for natural selection. Instead, he pretended that the retreat was of no consequence, since in any case "we have no means of determining, according to the standard of years, how long a period it takes to modify a species."[40]

At one point in his autobiography Darwin objected to the criticism that he was a good observer but a poor reasoner. The *Origin*, he protested with justice, was "one long argument from the beginning to the end" and could only have been written by one with "some power of reasoning."[41] He also remarked that he had a "fair share of inventiveness"—which erred only in being too modest. For his essential method was neither observing nor the more prosaic mode of scientific reasoning, but a peculiarly imaginative, inventive mode of argument.

It was this that Whewell objected to in the *Origin*:

> For it is assumed that the mere possibility of imagining a series of steps of transition from one condition of organs to another, is to be accepted as a reason for believing that such transition has taken

place. And next, that such a possibility being thus imagined, we may assume an unlimited number of generations for the transition to take place in, and that this indefinite time may extinguish all doubt that the transitions really have taken place.[42]

What Darwin was doing, in effect, was creating a "logic of possibility." Unlike conventional logic, where the compound of possibilities results not in a greater possibility, or probability, but in a lesser one, the logic of the *Origin* was one in which possibilities were assumed to add up to probability.

Like many revolutionaries, Darwin embarked upon this revolutionary enterprise in the most innocent and reasonable spirit. He started out by granting the hypothetical nature of his theory and went on to defend the use of hypotheses in science, such hypotheses being justified if they explained a sufficiently large number of facts. His own theory, he continued, was "rendered in some degree probable" by one set of facts and could be tested and confirmed by another—among which he included the geological succession of organic beings. It was because it "explained" both these bodies of facts that it was removed from the status of a mere hypothesis and elevated to the rank of a "well-grounded theory."[43] This procedure, by which one of the major difficulties of the theory was made to bear witness in its favor, can only be accounted for by a confusion in the meaning of "explain"—between the sense in which facts are "explained" by a theory and the sense in which difficulties may be "explained away." It is the difference between compliant facts which lend themselves to the theory and refractory ones which do not and can only be brought into submission by a more or less plausible excuse. By confounding the two, both orders of explanation, both orders of fact, were entered on the same side of the ledger, the credit side. Thus the "difficulties" he had so candidly confessed to were converted into assets.

This technique for the conversion of possibilities into probabilities and liabilities into assets was the more effec-

tive the longer the process went on. In the chapter entitled "Difficulties on Theory"[44] the solution of each difficulty in turn came more easily to Darwin as he triumphed
over—not simply disposed of—the preceding one. The
reader was put under a constantly mounting obligation;
if he accepted one explanation, he was committed to accept the next. Having first agreed to the theory in cases
where only some of the transitional stages were missing,
the reader was expected to acquiesce in those cases where
most of the stages were missing, and finally in those where
there was no evidence of stages at all. Thus, by the time
the problem of the eye was under consideration, Darwin
was insisting that anyone who had come with him so
far could not rightly hesitate to go further. In the same
spirit, he rebuked those naturalists who held that while
some reputed species were varieties rather than real species,
other species were real. Only the "blindness of preconceived opinion,"[45] he held, could make them balk at going
the whole way—as if it was not precisely the propriety
of going the whole way that was at issue.

As possibilities were promoted into probabilities, and
probabilities into certainties, so ignorance itself was raised
to a position only once removed from certain knowledge.
When imagination exhausted itself and Darwin could de-
vise no hypothesis to explain away a difficulty, he resorted to the blanket assurance that we were too ignorant
of the ways of nature to know why one event occurred
rather than another, and hence ignorant of the explanation
that would reconcile the facts to his theory. When one
botanist argued that his theory was contradicted by the
fact that some forms remained unaltered through long
periods of time and wide expanse of space, Darwin admitted the objection to be "formidable in appearance, and
to a certain extent in reality." But this did not deter him:

> Does not the difficulty rest much on our silently
> assuming that we know more than we do? I have
> literally found nothing so difficult as to try and always
> remember our ignorance. I am never weary, when
> walking in any new adjoining district or country, of

reflecting how absolutely ignorant we are why certain old plants are not there present, and other new ones are, and others in different proportions. . . . Certainly *a priori* we might have anticipated that all the plants anciently introduced into Australia would have undergone some modification; but the fact that they have not been modified does not seem to me a difficulty of weight enough to shake a belief grounded on other arguments.[46]

Somehow the fact that no adequate explanation suggested itself today seemed a warrant for the belief that such an explanation would suggest itself in the future, and that the explanation, moreover, would be bound to vindicate his theory. Thus the argument from ignorance was made the prelude to a confident affirmation:

We are far too ignorant, in almost every case, to be enabled to assert that any part or organ is so unimportant for the welfare of a species that modifications in its structure could not have been slowly accumulated by means of natural selection. But we may confidently believe . . .[47]

It may be objected, however, that in the logic of science, as in the logic of grammar, three negatives do not normally constitute a positive.

To be sure, a scientific theory that explains equally well a variety of contradictory phenomena may still be true; there are reputable theories that cannot, in this sense, be falsified,[48] and hypothetical reasoning is a legitimate, even necessary, scientific technique. The difficulty with natural selection, however, is that if it explains too much, it also explains too little, and that the more questionable of its hypotheses lie at the heart of its thesis. Posing as a massive deduction from the evidence, it ends up as an ingenious argument from ignorance.[49]

CHAPTER 16

MECHANISM AND TELEOLOGY

THE same supple and yet aggressive tactics are displayed in Darwin's efforts to capture that traditional stronghold of teleology: the argument from perfection. What all his critics assumed to be a major difficulty in his theory, he blandly took as confirmation of that theory. While they objected that perfection implied design, that complex and intricate organs could not have evolved by the slow process of selection acting upon chance variations, he insisted that such organs could never have been created in a perfect state: "Almost every part of every organic being is so beautifully related to its complex conditions of life that it seems as improbable that any part should have been suddenly produced perfect, as that a complex machine should have been invented by man in a perfect state."[1]

At one time he had been considerably less confident of this than he later professed to be. "I remember well the time," he recalled, "when the thought of the eye made me cold all over."[2] The eye, as one of the most complex organs, has been the symbol and archetype of his dilemma. Since the eye is obviously of no use at all except in its final, complete form, how could natural selection have functioned in those initial stages of its evolution when the variations had no possible survival value? No single variation, indeed no single part, being of any use without every other, and natural selection presuming no knowledge of the ultimate end or purpose of the organ, the criterion of utility, or survival, would seem to be irrelevant. And there are other equally provoking examples of organs and processes which seem to defy natural selection. Biochemistry provides the case of chemical synthesis built up in several stages, of which the intermediate substance formed at any one stage is of no value at all, and only the end product, the final elaborate and delicate machinery, is useful—and not only useful but vital to life. How can selection, knowing nothing of the end or final purpose of this process, function when the only test is precisely that end or final purpose?

The question arises in the case not only of organs or processes that are nothing until they are complete, but also of organs or processes in which the variations in the initial stages must have been so small and insignificant relative to the final complex product as to have no discernible survival value. The characteristics distinguishing species are often trivial enough, but they must have been still more trivial and insignificant in the first stages of their evolution. The appearance of a white spot in a coat of hair, or even of a faintly lighter color all over, could hardly have been of such value to the bear as to encourage the development of an entirely different colored species. Nor could an extra inch in the forerunner of the giraffe have been of much use to him in reaching those top boughs of the trees which presumably gave him his superiority over his neighbors. The case of the giraffe, the classical exemplar of evolution, is made even more problematical by the fact that the extra inch of neck would have been particularly useless to the young offspring of the new variety, who, being too small to attain even to the height of the adult of the old form, would have died in a drought as readily as the others.

Darwin was quick to see the problem, but not so successful in resolving it. His technique here, as elsewhere, was first to assume that by acknowledging the difficulty, he had somehow exorcised it; and second, if this act of confession did not succeed in propitiating his critics, to bring to bear upon the difficulty the weight of authority of just that theory which was being called into question. His discussion of the eye opened disarmingly with the admission that even to him the idea that natural selection could have fashioned so intricate and perfect an organ seemed at first "absurd in the highest degree." Yet reason soon assured him that the difficulty, "though insuperable to our imagination, can hardly be considered real." So long as there were gradations of eyes among different organisms, even though there was no evidence of gradation among the lineal members of any one species, he saw "no very great difficulty (not more than in the case of many other structures)" in supposing natural selection to have converted the simplest optic nerve into the most complex and perfect instrument.

> He who will go thus far, if he find on finishing this treatise that large bodies of facts, otherwise inexplicable, can be explained by the theory of descent, ought not to hesitate to go further, and to admit that a structure even as perfect as the eye of an eagle might be formed by natural selection, although in this case he does not know any of the transitional grades. His reason ought to conquer his imagination; though I have felt the difficulty far too keenly to be surprised at any degree of hesitation in extending the principle of natural selection to such startling lengths.[3]

Thus, when the actual evidence in the case proved to be insufficient for his purpose, Darwin referred back to the theory which he had found adequate in other cases—although it was precisely the adequacy of the theory that was being challenged.

The argument was too evasive and the reasoning too circular to satisfy all of Darwin's champions, let alone his critics. Weismann addressed himself to the question of whether natural selection could operate in the initial steps of the development of an organ: "To this question even one who, like myself, has been for many years a convinced adherent of the theory of selection, can only reply: 'We must assume so, but we cannot prove it in any case.'" He then candidly remarked: "It is not upon demonstrative evidence that we rely when we champion the doctrine of selection as a scientific truth; we base our argument on quite other grounds."[4]

Not only on this count—that natural selection could not explain the evolution of complex and "perfect" forms—did the argument from perfection fail Darwin. It also failed him because, having held out the promise of perfection, having subjected every part of every organism to a ruthless struggle for existence in which imperfect forms necessarily yield to less imperfect and ultimately perfect ones, natural selection had not, after all, succeeded in eliminating imperfection. Indeed, the argument from perfection, as Darwin stated it, was irreconcilable with the fact of imperfection. Yet he not only admitted the fact of imperfection but even embraced it as additional proof of his theory. While at one point of his argument he invoked the perfection of the eye as evidence of natural selection, at another point, in proof of the same theory, he quoted Helmholtz on the "inexactness and imperfection" of the eye.[5] Helmholtz was more forthright than most in condemning the ineptitude with which nature fashioned an organ having "every possible defect that can be found in an optical instrument, and even some which are peculiar to itself."[6] A biologist put the point more mildly when he said that many organisms were not, after all, particularly well fitted to their environment. "Natural Selection is stern, but she has her tolerant moods."[7]

The idea of imperfection is common enough, an entire genre of fiction being occupied with little else than the conceptualization of new and improved species of human beings with superior physiological, mechanical, and intellectual powers. And whole branches of biology and medicine are devoted to the analysis of man's infirmities and imperfections. One evolutionist, trying to defend natural selection against the charge that no material process could account for the wonderful perfection of nature, had no difficulty showing the remarkable imperfection of nature. He cited only three of the many failings with which man was uniquely endowed: his susceptibility to certain blood diseases, the mechanical shortcomings of an upright carriage, and the ineptitude of wound healing in injuries of his skin.[8] Any surgeon devoting his life to the correction of nature's mistakes could extend the list indefinitely. Nor are these the disadvantages arising out of man's advantages, the unfortunate by-products of otherwise sound arrangements. Some are gratuitous, uncompensated evils.

Credited with such marvels as the origin of new species, natural selection might have been expected to accomplish such lesser miracles as the elimination of their grosser imperfections. It is true that natural selection, as Darwin said, aspired not to absolute perfection but rather to a "standard of perfection" in which each organic being is "as perfect as, or slightly more perfect than, the other inhabitants of the same country with which it comes into competition."[9] Yet even this standard of perfection is not always achieved. The persistence, through the whole span of a species, of grave imperfections not shared by others in its neighborhood—and even, in some cases, the aggravation of these imperfections in the course of time—is as difficult to reconcile with natural selection as is the real or presumed evidence of nature's perfection.

Not only the persistence of imperfections but the persistence without change of any forms over a long period of time is difficult to explain by natural selection. If natural selection is intended to account for the development of species from the simple to the complex and from a low to a higher order of organization, how can it also account for the persistence of the simple and low? How can it be that variants, often numerically small compared with the total population, are frequently found existing side by side, and over long periods, with the typical forms? Why have not the superior or higher forms supplanted the inferior or lower?

The problem may be illustrated by the case of the hive bee, which Darwin discussed at some length as demonstrating the ability of natural selection to create a perfect instinct. He described how, slowly and irrevocably, selection developed the "inimitable architectural powers" by which the hive bee instinctively constructed cells containing the maximum amount of honey with the minimum expenditure of wax.[10] What he failed to explain was why all species of bees did not arrive at this optimum solution of their problem. That all bees do not create cells of such perfection—bumblebees, for example, being content with extremely irregular and uneconomical structures—he cited only as evidence of "the great principle of gradation" by which nature shows her hand.[11] Natural selection, that is, had left behind its visible traces in the several stages it had passed on its road to perfection. But why should there be these living, not dead, remains? Why had not natural selection itself eliminated these imperfect and superseded forms? Elsewhere, when Darwin was obliged to account for the gaps in the geological record—the absence of those transitional forms which his theory led him to expect—he had recourse to just this argument: that natural selection had eliminated the imperfect transitional forms before they became plentiful enough to leave traces in the shape of fossils. But here, in the case of the bumblebee and the numerous other inferior species which are both plentiful and persistent, he had quite another argument at hand. Natural selection, he now reasoned, does not imply absolute perfection or the "progressive development" of all organisms but only that level of perfection and development induced by the particular "relations of life" of any given organism.[12] Thus, by a judicious shifting of argument, natural selection could be made to account for any stage or degree of perfection or imperfection arrived at by any organism.

Some of the difficulties of Darwin's theory came from a metaphysical confusion. What kind of nature was it that was celebrated in the theory of natural selection? For it was not enough to protest that he had really meant nothing by "nature" apart from the workings of natural laws. One can have a philosophy of nature without being guilty of personifying nature.

Many of Darwin's defenders and more of his critics, both among his contemporaries and afterward, took his theory to be materialistic or mechanistic, an affirmation of the self-sufficiency and omnipotence of nature. Sedgwick spoke of the "cold atheistical materialism" that inspired the *Origin;*[13] another critic discussed the book under the title of "The Materialists' Stronghold";[14] and still another found the "Fallacies of Darwinism" to consist of the ideas that life evolved out of non-life, that spirit and matter were not fixed and distinct, and that spirit did not direct and give purpose to the movements of nature.[15] What were fallacies to some were virtues to others. John Dewey eulogized Darwin for liberating science from the shackles of teleology by destroying the old idealistic notion of a species as a fixed form or final cause. The new logic of Darwin, he rejoiced, "forswears inquiry after absolute origins and absolute finalities in order to explore specific values and the specific conditions that generate them."[16] This has been the most common reading of the *Origin,* and it is this that has been taken to be its great philosophical import.

Yet there are anomalies and ambiguities that do not bear out this interpretation, and there are many who have found the *Origin,* whether for good or bad, to be preeminently a teleological doctrine. Several scientists have accused Darwin of refurbishing a discredited and dangerous myth in his assumption of the "greatest possible adaptability to prevailing conditions"—an assumption that revives the search for "finality" in nature, thus encouraging the return to the romantic tradition of *Naturphilosophie,* and retarding the development of biology into an exact science.[17] Others have objected that the idea that every organ and characteristic has some utility or survival value upon which selection operates ignores the fact that the most important features of plants, and those which are crucial to their identification, have no discernible purpose, and that their morphological study can only proceed upon entirely different and unteleological lines. Huxley himself, surprisingly, was not disturbed by the charge that Darwinism was essentially teleological in character:

> There is a wider teleology which is not touched by the doctrine of evolution, but is actually based on the fundamental proposition of evolution. This proposition is that the whole world, living and not living, is the

result of the mutual interaction, according to definite laws, of the forces possessed by the molecules of which the primitive nebulosity of the universe was composed. That acute champion of teleology, Paley, saw no difficulty in admitting that the "production of things" may be the result of mechanical dispositions fixed beforehand by intelligent appointment and kept in action by a power at the centre.[18]

If Huxley could find evidences of Paley in the *Origin*, the idealists were even quicker to seize upon such evidence, praising Darwin for explaining the structures and habits of organisms in terms of the ends or purposes they served. The *Origin*, they reported, by establishing the idea of the good, the idea of a "final cause," had restored teleology to its ancient noble role—not, perhaps, in the cruder theological sense that all organisms tend to the good of man, but that they tend to the good, the realization of themselves. In the same vein, the French philosopher Henri Bergson distinguished between the older teleology with its doctrine of "internal finality," in which organs were related to the end or good of the organism, and Darwinian teleology with its doctrine of "external finality," in which organs were related to the end or good of the species.[19] It was impossible, he found, to read Darwin without this teleological understanding:

> If the accidental variations that bring about evolution are insensible variations, some good genius must be appealed—the good genius of the future species— in order to preserve and accumulate these variations, for selection will not look after this. If, on the other hand, the accidental variations are sudden, then . . . all the changes that have happened together must be complementary. So we fall back on the good genius again, this time to obtain the convergence of simultaneous changes.[20]

There was much in the *Origin* to justify this teleological interpretation, starting with the frontispiece featuring quotations from Bishop Butler and William Whewell, oddly juxtaposed with one from Francis Bacon.[21] Whewell was quoted to the effect that material events are brought about "not by insulated interpositions of Divine power, exerted in each particular case, but by the establishment of general laws"; Butler to the effect that the only meaning of "natural" is fixed or settled, so that "what is natural as much requires and presupposes an intelligent agent to render it so, i.e., to effect it continually or at stated times, as what is supernatural or miraculous does to effect it for once"; and Bacon to justify the study of God's works equally with his words. Later in the book Paley was added to this august company as authority for the view that natural selection can only act for the good of the creature, no organ being formed for the purpose of paining or injuring its possessor.[22] Darwin did not, to be sure, quote any of Paley's more extreme statements of the teleological method, although he did sometimes seem to echo the passage in which Paley exhorted the inquirer not to be dissuaded from his opinion by the charge that he knew "nothing at all" about the matter in question: "He knows enough for his argument. He knows the utility of the end; he knows the subserviency and adaptation of the means to the end. These points being known, his ignorance of other points, his doubts concerning other points, affect not the certainty of his reasoning."[23]

The teleological bias revealed itself in the bypaths of argument as well as in the central thesis of the *Origin*. Darwin often spoke as if nature, in its supreme wisdom, not only contrived by selection to insure the best possible adaptation of organisms to their conditions, but also contrived to arrange things so that selection would have the best possible conditions in which to function. If there are, he reasoned, so few genuine hermaphrodites in nature, it is not only because hermaphroditism is unfavorable to the individual and to the species in reducing their vigor, but also because it is unfavorable to natural selection itself in limiting the development of varieties and diminishing the opportunities for selection. Thus nature was credited first with exploiting whatever circumstances happened to be favorable to the development of a particular individual or species, and second, with promoting the development of "circumstances favorable to natural selection."[24] Natural selection, from being a means to other ends—the ends of individuals and species—had become an end in itself, an object of primary solicitude on the part of nature.

Sometimes it was not even the end of the individual or species that Darwin seemed to have in mind but an end conceived in human terms, teleology betraying him into a crude anthropomorphism. In the description of the "slave-making instinct" exhibited by some species of ants, he prefaced his observations with the remark that he had approached the subject in a skeptical frame of mind, "as any one may well be excused for doubting the truth of so extraordinary and odious an instinct as that of making slaves."[25] Elsewhere he confessed that he could not see how certain sexual peculiarities, such as the tuft of hair on the breast of the turkey cock, could be regarded as either useful or ornamental; if this feature had appeared under domestication, he added, it would have been termed a "monstrosity."[26] Indeed, his entire discussion of sexual selection is anthropomorphic in its basic conception, for whether the coloring of a bird is judged to be either beautiful or monstrous, it is by human standards that the judgment is made.

Conventionally, teleology has been optimistic. If nature has been credited with an end or purpose, it has commonly been a happy end and a purpose congenial to man. This is not, to be sure, a necessary consequence of the teleological mode of thought. The end can conceivably be—and has been so interpreted by some strong-minded thinkers—as not happiness but misery, the design or purpose of the universe tending not to the elevation or glorification of its inhabitants but to their denigration and humiliation. Darwin himself occasionally toyed with such a pessimistic view of the ways of nature. "What a book a devil's chaplain might write," he suggested to Hooker, "on the clumsy, wasteful, blundering, low and horribly cruel works of nature!"[27] Whether it was because this view would hardly have endeared him to his readers or because he himself could not contemplate such unredeemed misery, he chose in the *Origin* to dwell on a happier vision. As theologians have looked upon evil as the prelude and even the necessary cause of good, so Darwin now saw the cruelty of nature as part of a larger beneficent design. His chapter on the struggle for existence concludes with the soothing thought: "When we reflect on this struggle, we may console ourselves with the full belief, that the war of nature is not incessant, that no fear is felt, that death is generally prompt, and that the vigorous, the healthy, and the happy survive and multiply."[28] The final words of the book were an apostrophe to teleology in its most optimistic sense:

> As all the living forms of life are the lineal descendants of those which lived long before the Cambrian epoch, we may feel certain that the ordinary succession by generation has never once been broken, and that no cataclysm has desolated the whole world. Hence we may look with some confidence to a secure future of great length. And as natural selection works solely by and for the good of each being, all corporeal and mental endowments will tend to progress towards perfection.
>
> . . . From the war of nature, from famine and death, the most exalted object which we are capable of conceiving, namely, the production of the higher animals, directly follows. There is grandeur in this view of life, with its several powers, having been originally breathed by the Creator into a few forms or into one; and that, whilst this planet has gone cycling on according to the fixed law of gravity, from so simple a beginning endless forms most beautiful and most wonderful have been, and are being, evolved.[29]

Although this teleological interpretation comes with the authority of the *Origin* itself, privately Darwin was in great doubt about it. When Asa Gray interpreted him as saying that the system of nature had "received at its first formation the impress of the will of its Author, foreseeing the varied yet necessary laws of its action throughout the whole of its existence, ordaining when and how each particular part of the stupendous plan should be realized in effect,"[30] Darwin protested that this was not at all what he meant. To find such evidences of design not only in the end product of natural selection but also in each stage of it was to deny his theory altogether. For if each variation was predetermined so as to conduce to the proper end, there was no need for natural selection at all, the whole point of his theory being that, out of undesigned and random variations, selection created an evolutionary pattern. "If the right variations occurred, and no others, natural selection would be superfluous."[31]

Nor was he even certain, in his private exchanges with Gray, that there was evidence of a larger, final design or beneficent providence in nature, although the *Origin* had said so unambiguously enough. "There seems to me too much misery in the world," he protested. Would a beneficent, omnipotent God have so arranged matters that parasitic insects should feed upon the living bodies of caterpillars, or that cats should play with mice? On the other hand, he could not quite bring himself to look on this wonderful universe, and especially man, as the result of mere brute force. He was, he confessed, in a hopeless muddle.

His other friends conspired to keep him in a muddle. In an address to the British Association, Hooker found evidence of design in the general fact of variation: "By a wise ordinance it is ruled, that amongst living beings like shall never produce its exact like. . . . A wise ordinance it is, that ensures the succession of being, not by multiplying absolutely identical form, but by varying these. . . ."[32] But although Darwin had said much the same thing in the *Origin*, Hooker hastened to assure him that he had not meant a word of it; it was all "bosh and unscientific" and was intended only to show those who had to have a providence that Darwin's was the true one.[33] Lyell, too, when finally he declared himself on Darwin's side, did so on the understanding that it was according to a "preconceived plan" that evolution had taken place, and that the "power, wisdom, design and forethought" required for this evolution was at least as great as that required for a multitude of separate creations.[34] All this made the muddle worse, and by 1874, when Gray paid him the tribute of having restored teleology to natural science, Darwin thanked him earnestly: "What you say about Teleology pleases me especially, and I do not think any one else has ever noticed the point."[35] Yet only a few years earlier, in the *Descent*, he had explained that if the *Origin* erred in putting too great an emphasis on natural selection, it was because he had not yet entirely thrown off the prevailing teleological habit of mind that was a vestige of the old theory of creation.[36]

There is a danger, in an analysis such as this, of inflating details and exaggerating their import. It is the critic's conceit to think that what has been criticized has been destroyed. In fact, however, the *Origin*, so far from being destroyed, still dominates the thinking of most men today, and for the same reasons that it captured the minds of a considerable number of Darwin's contemporaries.

Much of the attractiveness and influence of the *Origin* may be ascribed to the very qualities that have here been featured as its flaws. The first of these was the simplicity of the theory, at once strikingly bold and yet apparently so self-evident as almost to rank as a truism. Thus, while some readers were shocked to attention by its audacity, others were coerced into assent by its obviousness. "How extremely stupid not to have thought of that," was Huxley's first reaction, reflecting that Columbus' companions had probably felt the same way when he made the egg stand on end.[37] The same thought suggested itself to the ornithologist Alfred Newton, who did not know whether to be "more vexed at the solution not having occurred to me, than pleased that it had been found at all," particularly since it was "a perfectly simple solution" of the problem that had been plaguing him for months.[38]

But simplicity alone would not have been enough. It was also the impression of a massive structure supporting this simple skeleton that gave men confidence in its stability and durability. The paper read before the Linnean Society probably failed to attract notice because it was too slight either to shock or to persuade. The *Origin* did not make this mistake. Even those familiar and sympathetic with the thesis confessed that they did not completely appreciate it or experience its full effect until they had read it in the final version. "How different the *book* reads from the manuscript," Hooker was startled to find,[39] while Huxley was so moved by the reading that he declared himself ready to go to the stake for it. And Wallace, who had, after all, thought it out independently, doubted whether anyone, reading it only once, could gain "any clear idea of the accumulated argument"; he himself, he said, had to read it five times before he fully appreciated its strength.[40] And where the facts or argument seemed insufficient, there was the reminder that this was only the "abstract" of an even weightier document yet to come. Darwin had, in fact, hit upon precisely the right format: he gratified his readers by the assurance that there was more behind the book, without burdening them with the need to read any more. He was thus able to win their assent without unduly trying their patience.

It was probably less the weight of the facts than the weight of the argument that was impressive. The reasoning was so subtle and complex as to flatter and disarm all but the most wary intelligence. Only upon close inspection do the faults of the theory emerge. And this close inspection, by the nature of the case, was rarely vouchsafed. The points were so intricately argued that to follow them at all required considerable patience and concentration—an expenditure of effort which was itself conducive to acquiescence. Only those determined in advance to be hostile were likely to maintain a vigilant and hence critical attitude. In his rapid volley of explanations, where one might fail, another would hit the mark, and where one line of defense had to be abandoned, another was hastily erected. And there were few to point out that in the strategy of reason, as in the strategy of warfare, the cause was not better served by a succession of feeble defenses than by a single strong one.

More important, however, than any assets which Darwin's theory might be thought to possess was the bankruptcy of his opponents. The only serious rival, as a general theory, was creation. (As a general theory, because the idea that some species had originated by variation and others by creation committed its adherents, ultimately, to the theory of creation.) And the theory of creation was no more satisfactory than the theory of evolution. Able to defy natural laws whenever it chose, creation was obviously not bound by the usual canons of scientific proof. Any particular hypothesis might be disproved, such as Archbishop Usher's calculation that the world was created at 8 P.M. on Saturday, October 22, 4004 B.C.; but the general theory of creation was at least as difficult either to prove or disprove—which come to the same thing—as Darwin's theory.

Moreover, the theory of creation had its own problems. The geological record that was a difficulty for the one was also a difficulty for the other. If the evidence of a gradual succession of species was far from complete, there were nevertheless enough indications of such a succession to embarrass the conventional theory, for why should God have willfully chosen to create species in such an order as sometimes to approximate a "natural" order of descent? If evolution found it difficult to account for the persistence of separate and well-defined species, the theory of creation found it difficult to account for the multitude of varieties that merged imperceptibly into each other. If perfection was more readily understood as the product of an omniscient God than of a blind, natural process, imperfection might more easily be ascribed to nature than to God. If God sometimes elected to work through natural laws, why and when did he personally intervene in this natural order? And evolution, rather than creation, accounts for the presence of organs that can best be understood as vestigial.

But these were difficulties for the theory of creation only when confronted with the theory of evolution, not that of natural selection. Many who were ready to grant the case for evolution boggled at natural selection. And it was natural selection that was at issue. Darwin himself, in spite of his protestations to the contrary, helped contribute to the confusion between the two. Having opened the *Origin* with the assertion that evolution without natural selection was meaningless, that he meant to confine his argument to natural selection since evolution would naturally follow in its wake, he had then proceeded to reverse the process, citing evidence in favor of evolution to bolster the case for natural selection. The natural order of beings, the facts of morphology, embryology, and rudimentary organs—all the traditional arguments of evolutionists—were presumed to bear witness to natural selection. Thus the theory of natural selection surreptitiously drew upon the strength of the theory of evolution. And the doctrine of creation, which might have been able to hold its own against natural selection, was faced with the combined forces of a more redoubtable enemy.

Sedgwick tried to brazen the matter out. He did not claim that the doctrine of creation had been scientifically demonstrated, or even that it was scientifically demonstrable. Creation, he said, was "a power I cannot imitate or comprehend; but in which I can believe, by a legitimate conclusion of sound reason drawn from the laws and harmonies of Nature."[41] What he and others resented was the pretense that Darwin's theory was anything more than this.

CHAPTER 17

THE ORIGIN OF MAN

THE great debate of 1859 sometimes seemed to be less about the origin of species than about the origin of man. While Darwin was discoursing upon the varieties of pigeons, his readers were wondering, with Lyell, whether they were being inveigled into going "the whole orang."[1]

Darwin himself had never doubted that what was true of pigeons was also true of orangs and men. More than twenty years earlier, when the theory of natural selection had not yet occurred to him, he had toyed with the idea that animals and men "partake our origin in one common ancestor," that "monkeys make men, men make angels."[2] And in the sketch of 1842 he had said that his theory necessarily extended to all mammals, however weakened his reasons became in the process. He would have liked to avoid the subject in the *Origin* because he thought it would prevent his book from getting a fair hearing, but he could not help implicating man in his conclusion that all animals were probably descended from "one primordial form"; his readers did not need after this to be warned that "light will be thrown on the origin of man and his history."[3] To Lyell, who frankly confessed that the origin of man stood in the way of his accepting the theory of the origin of species, he was more forthright: "I believe man is in the same predicament with other animals. It is in fact impossible to doubt it." And in a postscript he prepared him for the worst: "*Our* ancestor was an animal which breathed water, had a swim bladder, a great swimming tail, an imperfect skull, and undoubtedly was a hermaphrodite. Here is a pleasant genealogy for mankind."[4]

In the years after the publication of the *Origin* the figure of man continued to hover behind the abstraction of species. It was the subject of man's genealogy that aroused the strongest fears and passions, and it was a measure of Darwin's triumph that more and more men responded to the challenge, "Is man an ape or an angel?" by declaring themselves on the side of the ape. The growing popularity of his cause relieved him of any sense of urgency in stating his case, so that it was not until the two volumes of the *Variation* were finished that he finally turned his attention to man. In February 1867, in his spare time while awaiting the proofs of *Variation*, he started work on what was first intended to be a "chapter," then a "short essay," then a "very small volume" on man,[5] and which ended up as the two-volume work entitled *The Descent of Man, and Selection in Relation to Sex*. Because of the usual number of interruptions on account of illness, the work was completed only in August 1870 and published on February 24, 1871.

Even before he started work on the *Descent,* he confessed that he would have little new to contribute to the discussion. Wallace, Huxley, and Lyell had already written on the subject, and to much the same purpose. If he now entered the lists, he said, it was to vindicate himself against the taunt that he was concealing his opinions. By the time he finished it, so many others had joined the ranks that he felt more than ever deflated. The first edition of Haeckel's *History of Creation* had appeared in Germany in 1868 and a second edition in 1870, covering the same ground as his own book.[6] Little more than a decade after the *Origin,* Darwin was obliged to apologize that there was little left for him to say on the subject of man.

The circumstances attending the publication of the *Descent* recall the publication of the *Origin.* Darwin voiced the familiar self-doubts: "The work half killed me, and I have not the most remote idea whether the book is worth publishing."[7] Nor, apparently, had his publisher, Murray, any more confidence in it. For all their cordial, and profitable, relations, Murray continued to be as little an admirer of Darwin in 1871 as he had been in 1859. As before, he consulted Whitwell Elwin, who had given Darwin the memorable counsel to scrap the *Origin* and write instead about pigeons. Still unabashed and unregenerate, Elwin advised Murray that the *Descent* was "little better than drivel," unreadable and not worth refuting. He predicted that with the first appearance of a "really eminent naturalist," this latest bubble would be pricked and exploded.[8]

On this dubious recommendation, Murray ordered a first edition of 2500 copies. This was immediately sold out, and reprints were issued bringing the total to 7500 by the end of the year. Two thousand five hundred—or even three times that number—would hardly seem to be an impressive figure for a book by the author of the most sensational work of the age, on a subject that promised to be even more sensational than the first. As with the *Origin*, however, the publication figures give little indication of its notoriety. Hooker, who found evolution discussed at every dinner table, was naive in supposing that those ladies who thought it improper to talk about it were ordering it "on the sly" and reading it with clandestine delight.[9] Fewer people bought the book, and possibly still fewer read it, than talked about it.

What was most surprising was the way people talked about it. Hooker's experience—"I dined out three days last week, and at every table heard evolution talked of as an accepted fact, and the descent of man with calmness"[10] —was shared by Darwin himself, who was surprised at the general assent and absence of abuse with which the book was received: "Everybody is talking about it without being shocked."[11] This amiability, however, was often more a matter of tone than of substance. Professional scientists and critics were not outraged, but neither were they placated. Although Darwin confessed himself astonished to find "in how extraordinary a manner the judgment of naturalists has changed since I published the *Origin*,"[12] the change was not yet as thoroughgoing as he would have liked. Perhaps in ten years, he ventured to predict, the progress of thought on the subject of evolution would extend to man. In the meantime, he admitted elsewhere, the *Descent* had "met the approval of hardly any naturalists as far as I know."[13]

This was the general tone of the reviews: unshocked, but also, for the most part, unconvinced. The *Edinburgh Review* was being overdramatic when it said that the *Descent* had raised "a storm of mingled wrath, wonder and admiration."[14] The storm was more in the nature of a drizzle, and it was not wrath so much as sorrow that descended upon Darwin. The *Saturday Review* was nearer the truth when it said that no one need pretend to be startled by ideas that had for so long been familiar.[15] Although the general thesis was familiar, however, the details still had the power, if not to shock, at least to offend. It was one thing to say that in some remote past and by some collateral ancestor, man and the ape were related. But to spell out in detail the physical affinities between man and various animals, and to derive man's moral and mental faculties from those of the animals, was to exacerbate wounds that twelve years of familiarity had not entirely healed.

The *Times* reacted most sharply. Perhaps because it had not the chance to expend its indignation upon the *Origin* (Huxley having appropriated that review), it was all the more outraged now. Darwin's ideas, it solemnly advised, were as mischievous as they were unscientific. Should they ever gain wide acceptance, "morality would lose all elements of stable authority," and uncontrollable passions would rule. Indeed, at that very moment, such a "loose philosophy" was contributing to the disorganization of French society. While Paris was being consumed in the flames of the Commune, Darwin was exploiting the "authority of a well-earned reputation" to advance the "disintegrating speculations of this book." That he should have done so on the basis of cursory evidence and hypothetical arguments was not only unscientific, the *Times* complained, but positively reckless.[16] It was this review that John Morley, turning it on its head, later used as an object lesson of the degraded moral sense prevailing in England: to invoke the Commune in its judgment of the *Descent*, he said, was typical of the lack of principles that reduced everything, including even scientific truth, to the level of political expediency.[17]

Most critics took a less haughty tone. Ridicule rather than outrage, the tedium of familiarity in place of shock, were the accents in which they registered their disapproval. The *Athenaeum* was debonair: "No man will ever develop religion out of a dog or Christianity out of a cat." And Darwin's tales of peculiarly gifted and sensitive animals were greeted with arch incredulity: as there was no such thing as a legal conscience, neither was there a canine conscience, and the rare case of a conscientious dog no more disproved the rule than did the case of a conscientious lawyer.[18]

Everywhere criticism was tempered with praise. As the *Times* had conceded Darwin's well-earned reputation while charging him with immoral and unscientific conduct, so the *Edinburgh Review* carefully balanced displeasure and tribute: "Mr. Darwin appears to us to be not more remarkable for the acuteness and ingenuity of his powers of observation of natural phenomena, than he is for the want of logical power and sound reasoning on philosophical questions."[19] And the *Spectator*, where in 1859 Sedgwick had done his worst, was now markedly deferential, extolling Darwin for his sincerity, his theory for its lucidity, and his style—most improbably—for its dispassion and intellectual vitality.[20] Even the *Dublin Review* found matter for praise. The subject of sexual selection, it was pleased to note, was handled "in a way that entirely strips it of all offensiveness."[21] And while it did not conceal its conviction that some of the implications of the *Descent* no orthodox Christian could accept—the idea that the human soul developed out of animal life flatly contradicted the Scriptural statement that Adam had received his soul from the breath of God—it also found much with which it could agree. It quoted Saint Augustine in favor of "reason and experiment" in natural questions, said that the idea of evolution was not in itself heretical, cited authority to prove that the six days of creation need not be interpreted literally, declared "*Natura non facit saltum*" to be a scholastic principle, and was confident that Darwin had exalted, not debased, God in seeking the evidence of design and natural law. And even on the vexed question of the origin of man's soul, it made its obeisance to science, claiming the Christian doctrine to be more naturalistic and scientific than Darwin's. Was it not more natural to conceive of the soul as simply being breathed into the body of man, rather than to invoke a double miracle by which the animal soul was first breathed out of him and the human soul then breathed in?

Even more than the reviews of the *Origin* were those of the *Descent* occupied with religion, and for obvious reasons. The *Descent* dealt with the most provocative question of all, the question of man, in the most provocative possible way: by deducing his moral and mental, as well as physical, qualities from the lower animals. Yet, even now, critics refused to divide neatly on this issue. In spite of Huxley's warning that no man could be both "a true son of the Church and a loyal soldier of science,"[22] many persisted in thinking of themselves as both. The *Contemporary Review,* where Huxley delivered this challenge, had earlier been the host to just such a conciliator of religion and science. Sir Alexander Grant opened his article on "Philosophy and Mr. Darwin" with the assurance that Darwin was "worthy of all respect" and that his latest book need occasion no nervousness: "There is nothing atheistical in Mr. Darwin's work; on the contrary, it might be described as a system of Natural Theology founded on a new basis."[23] The *Spectator* was no less enthusiastic, declaring the *Descent* to be a "far more wonderful vindication of Theism than Paley's *Natural Theology*," adding, as an afterthought, "though we do not know, so reticent is his style, whether or not he so conceives it himself."[24]

It was in the name of a higher and purer morality that Darwin was exonerated from the charge of atheism. *Macmillan's Magazine* argued that the basic religious beliefs—the nobility of conscience, our power of communion with

God, and our hopes of immortality—were in no way impugned by Darwinism, since no matter how far back Darwin succeeded in tracing the evolution of man, the laws governing that evolution must still ultimately be ascribed to a Creator. In one respect, indeed, Darwin could be credited with refining and exalting the Christian sense of morality. By deriving morality from a social instinct and making the good of the community the end and aim of our moral nature, he gave a firm, biological basis to utilitarianism while redeeming it from the reproach of selfishness.[25] The *Saturday Review* was also delighted with the way Darwin managed to root morality in the most elementary instincts and at the same time elevate it above selfish calculations. It assured its readers that they would be more than compensated for any shock his conclusions might give them by the "intellectual pleasure of following so exquisite a chain of philosophical deductions."[26]

Not all readers were as easily conciliated as the *Saturday Review* had assumed. The strongest criticism, on the score of both religion and science, came from the Roman Catholic biologist St. George Mivart in the *Quarterly Review*. The article was anonymous, but the author's identity was exposed by a comparison with Mivart's *Genesis of Species,* published shortly before the *Descent.* Although the controversy with Mivart came to rival in bitterness the earlier ones with Owen and Wilberforce, it started out mildly enough. When Darwin first read the *Genesis* (which was a critique of the *Origin*), he went out of his way to affirm his faith in Mivart's good intentions. He also declared himself pleased to find that Mivart agreed with him in locating man, at least in his physical structure, in a single series with the animals; indeed, Mivart had gone even further than he in this respect.[27] Only later, when the *Quarterly Review* article had appeared and his friends had persuaded him that Mivart's criticisms were not only unjust but also influential, did Darwin have second thoughts about him. He had himself observed that Mivart had twice neglected to complete quotations from the *Descent,* but now he was told that the omitted words were essential to the argument, upon which he sorrowfully concluded that "though he means to be honourable, he is so bigoted that he cannot act fairly."[28] Later still, even this measure of honorableness was withdrawn, and when Mivart wrote expressing his friendship and respect, Darwin dismissed this as a bit of double-dealing inspired by an "accursed religious bigotry."[29]

This hostility to Mivart is difficult to understand, except perhaps as resentment of a critique that was both more extensive and more effective than Darwin would have cared to confess. Certainly the charges of gross unfairness are difficult to sustain. One of Darwin's American champions, Chauncey Wright, hinted darkly, in the *North American Review*, of partial and distorted quotations.[30] Yet only once did he find Mivart out in such a distortion, and even then the questionable passage was not a direct quotation but a paraphrase—and an accurate one, as Wallace pointed out at the time to Darwin.[31] Nor was the "unjust and unbecoming" treatment of which Huxley complained in the *Contemporary Review*[32] any more unjust or unbecoming than is customary in polemics, least of all in those Huxley engaged in. On one point Huxley and Darwin were right to take offense: this was Mivart's innuendo that only Wallace's reticence permitted Darwin to take sole credit for the theory of natural selection. For the rest, their outrage seems excessive, as when Huxley complained that Mivart catalogued Darwin's changes of opinion in the spirit of an Old Bailey barrister trying to prejudice a jury. Himself a master of Old Bailey practices, Huxley must have known that the contradictions and irresolutions in Darwin's theory were legitimate points of evidence.

Huxley himself was not the most fastidious opponent. He opened his attack on Mivart with some reflections on the progress of public opinion as revealed by a comparison of the *Quarterly Review's* present treatment of the *Descent* with that of the *Origin* a decade before. He might also have taken the occasion to reflect upon how much more sensitive to criticism, how much quicker to take offense the Darwinians had become during that time, for compared with Wilberforce, Mivart was almost a convert; yet Huxley treated him with the special contempt reserved not for the enemy but for the traitor. In his *Genesis of Species* Mivart had advanced an evolutionary theory of his own, "specific genesis," according to which species, while fixed at any one time, contained an "internal innate force" for the generation of new species, these new species being produced suddenly as "harmonious self-consistent wholes."[33] Even in the *Quarterly Review,* where this theory was not discussed (the pretext of anonymity prohibiting it), Mivart managed to make his evolutionary sympathies clear: he accepted natural selection as a partial explanation of evolution and admitted that man, as a physical creature, had evolved from the animals. He insisted, however, that God must have intervened for the creation of man's soul and that it was this that distinguished man radically from all the other creatures. He concluded by saying that there was nothing prejudicial to religion in the theory of evolution (not even, he had said in the *Genesis,* in the Darwinian theory), and that some of the greatest authorities of the church—Augustine, Aquinas, and Suarez—had spoken in its favor.[34]

It was upon this last point that Huxley pounced. Obtaining a copy of Suarez's works, he proceeded to compile a dossier of anti-evolutionary statements. For five thousand words he went on about Suarez, quoting at length in Latin, analyzing, comparing, and interpreting texts, and in general, as he mischievously wrote to Darwin, defending the true Catholic orthodoxy against Mivart. It was the kind of disputation he gloried in, but it was not strictly relevant to the issue or, indeed, of the slightest interest to Huxley, except as a stick with which to beat Mivart. (He may have felt vindicated twenty years later when Mivart was denounced by the church for liberalism and excommunicated, although not on this issue.)

The Mivart affair prompted Darwin to one of his few interventions in controversy. As usual, he did not make a personal rebuttal, but he did take the initiative, and incur the cost, of having Chauncey Wright's article published as a pamphlet in England, although he privately thought it badly written and ill informed. (Huxley's article had not then been written.) He also wrote a new chapter for the sixth and last edition of the *Origin*, taking account of Mivart's criticisms. With the publication of Huxley's piece in the *Contemporary Review*, he declared himself satisfied with the state of the controversy, confident that the pendulum would soon swing again in his favor and that Mivart could write his worst without mortifying him. It later appeared, however, that Mivart still had the power to upset him, although in an entirely unexpected way. In 1874 Darwin's son George wrote an article in the *Contemporary Review* advocating eugenics, whereupon Mivart replied (again in the *Quarterly Review*) that this was to encourage tyranny and sexual license. Darwin was so incensed as to consider issuing a public statement, but on Huxley's advice he contented himself with informing Mivart that he would never speak to him again. Two years later, in writing his autobiography, he forgave all his critics except Mivart. He had often, he wrote, been misrepresented and ridiculed, but generally "in good faith" —with the single exception of that "pettifogger" and "Old Bailey lawyer," Mivart.[35]

The progress of opinion can be seen not only in the conversion of such magazines as the *Spectator* and *Saturday Review* (more than compensating for the backsliding of the *Times*) and in the friendlier tone of even the critical notices, but in the fact that this was despite the greater provocation given by the *Descent*. The most obvious cause of provocation was its conclusion: that man was descended from a "hairy, tailed quadruped, probably arboreal in its habits," and ultimately from "an aquatic animal, provided with branchiae, with the two sexes united in the same individual, and with the most important organs of the body (such as the brain and heart) imperfectly or not at all developed."[36] And lest anyone have too inflated an idea of this remote ancestor, Darwin specified that it was like the larvae of the existing marine Ascidians.

This was provocative enough. What enhanced the provocation was the fact that the *Descent* was, both as literature and as polemic, distinctly inferior to the *Origin*. That criticism should have receded just when it had reason to advance testifies to the general acceptance of the doctrine. Where the *Origin* is one long, intricate, and often subtle argument, the *Descent* is a loose amalgamation of ideas which might better have appeared as two separate books. The original edition was in two volumes of about four hundred pages apiece. Of these, only two hundred and fifty pages of the first volume were properly on the subject of the descent of man. The rest—on sexual selection—constituted, in effect, a postscript or addendum to the *Origin;* four hundred pages of this did not relate to man at all, but gave a detailed account of sexual selection among insects, fish, birds, and mammals.

Sexual selection had a far more important role in the *Descent* than in the *Origin.* It assumed much of the burden for the origin of species that had earlier been carried by natural selection—so much so that it would no longer be accurate to describe the theory as that of natural selection. Darwin himself had taken to referring to his theory as "the principle of evolution."[37] This principle included several explanations for the origin of man, of which natural selection was only one, the others being sexual selection, the inherited effect of use and disuse, the direct action of the environment, the correlation of growth, and one unspecified cause. Of these, sexual selection, not natural selection, was the most important, as Darwin himself now admitted: "For my own part I conclude that of all the causes which have led to the differences in external appearance between the races of man, and to a certain extent between man and the lower animals, sexual selection has been by far the most efficient."[38]

The extent of Darwin's retreat from natural selection may be measured by his differences with Wallace on the subject of race. In periodical essays first printed in 1864 and republished in 1870,[39] Wallace gave what might have been taken to be the orthodox Darwinian doctrine of the origin of human races. He argued that races must have developed early in history, when men, ranging far from their original homes, confronted new physical conditions, and when the combined processes of natural selection and correlation of growth (one change bringing in its wake other correlated ones) caused them to evolve peculiarities of skin color, hair texture, eye formation, and the like. At this stage of man's development, his mental power must have reached the point where further physical adaptation was superfluous. Once he learned to use and invent tools, there was no longer any incentive, so to speak, for nature to evolve special kinds of hands; when he started to wear clothes, hairiness was no longer of any particular advantage or disadvantage, so that the accentuation of hairy and non-hairy varieties was brought to a halt. Thus the ascendancy of mind, by putting an end to the evolution of man's physical structure, fixed the different races in their different and permanent features.

Some such reconstruction of human evolution based on natural selection could plausibly be held to account for the difference of races. Yet Darwin did not avail himself of it. Instead, he chose to explain these differences as the result of sexual selection. It was not, according to the *Descent,* by a process of the survival of the fittest that men of a particular skin color, hair texture, and so on came to dominate in any particular area; but rather because each isolated tribe formed for itself "a slightly different standard of beauty"[40] which then, by a process of sexual selection, fixed the characteristics of the tribe and eventually of the race. That primitive tribes must have had such standards of beauty he deduced from the fact that savage tribes had them today, to judge by their elaborate practices of self-decoration and self-mutilation. Moreover, these ideals must have been an intensification of their natural characteristics, just as savage tribes—he quoted Humboldt to this effect—admire and exaggerate the features peculiar to themselves. Thus, by selecting those variations which exaggerated their peculiarities, they succeeded in differentiating themselves from their neighbors. He sympathized with those readers who might think it "a monstrous supposition that the jet blackness of the negro has been gained through sexual selection,"[41] but he was convinced that it must have been so.

The same process of sexual selection that served to

differentiate the races Darwin also saw as a major influence in the differentiation of man from the animals and, in their secondary characteristics, the male from the female. In the case of male and female, the greater size and strength of the male might be attributed to natural selection; but such other characteristics as the beard he assumed to have been acquired by sexual selection alone, "as an ornament to charm or excite the opposite sex."[42] What charmed and excited the female, however, did not seem to have charmed or excited the male. For while our male ape-like progenitors were developing beards, our female ape-like progenitors, at the same time and for the same reasons of sexual attractiveness, were being denuded of hair. Simultaneously, while the males were transmitting their beards to their male descendants only, the females were transmitting their hairlessness to descendants of both sexes, thus serving to distinguish not only male and female but also humans and animals. By a similar process of reasoning, Darwin was able to explain how speech, music, and other traits conventionally regarded as being uniquely human derived from the courtship patterns of these ape-like ancestors, the "sweeter voices" of women, for example, having been acquired to attract the male.[43]

Such explanations were not only exceedingly contrived and hypothetical; they were also contradictory and illogical. Did the voice of the female become "sweeter" than that of the male because the female population was larger and the women were therefore obliged to compete for the favors of the male? Darwin did not prove any such disparity of numbers; and Wallace confounded the thesis further by pointing out that savages do not value fine voices and that savage women rarely sing at all. Even if it was claimed that each sex, regardless of relative numbers, competed for the possession of the most attractive members of the other sex, there was no assurance that the most desirable couples were more prolific than the least desirable, and thus no promise that the tribal standard of beauty would predominate and its distinctive character be fixed. Nor is it certain that the tribe's standard of beauty was merely an expression or exaggeration of its natural traits,

as Darwin's theory required. Are not the practices of mutilation and decoration described by Darwin as much a
denial of nature? The black savage who paints his body
red, stains or removes his teeth, tattoos his face, gashes
his temples, tortures his legs, dangles rings from his nose,
ears, and lips—is he not violating rather than celebrating his
natural character? And this standard of beauty that is so
capricious among savages must have been even more so
among prehistoric men, to favor a patch of hair around
the chin of man and to discourage it on woman. To com-
plicate matters, this capriciousness must have remained
constant for an untold number of generations, if the species was to evolve at the slow pace Darwin set for it.

It was a bold experiment to make so tenuous and
hypothetical an idea as the esthetic standards of our ape-
like progenitors bear the burden of such weighty matters
as the evolution of man from the animals and the distinctions of sex and race. Sexual selection has all the faults
of natural selection and more: the suspicious facility with
which it can be made to explain anything and everything,
the manipulation of evidence for whatever purposes are
convenient, and the invocation of ignorance when all else
fails. Ignorance is resorted to even in so crucial a matter
as the intellectual disparity between man and the apes.
We cannot, Darwin protested, be expected to know exactly why the ape, having so many of the other advantages
of man, did not develop man's intellect, because of "our
ignorance with respect to the successive stages of development through which each creature has passed."[44]

In view of the many difficulties confronting sexual selection, Darwin's choice of it in preference to natural selection, in which he had both an emotional and professional
stake, may at first seem strange. It was not, however,
without cause that he abandoned natural selection. What
forced his hand was the realization that natural selection
was untenable as the main explanation either for the development of man from the animals or for the distinctions
of race and sex. Natural selection assumed that beneficial
variations alone would be preserved. The difficulty was
that "the races of man differ from each other and from

their nearest allies amongst the lower animals, in certain characters which are of no service to them in their ordinary habits of life."[45] The advantage of sexual selection was that it did not have to prove utility; what was patently not useful, and therefore not subject to natural selection, could be regarded, by a sufficient exercise of imagination, as sexually attractive. Indeed, sexual selection could not only account for those characteristics that were useless; it could also account for those that were actually injurious. Darwin now admitted that sexual selection could work at cross purposes with natural selection, the objects of sexual admiration interfering with the more elementary struggle for existence.

Having dispensed with natural selection when there was no evidence of utility, he soon came to dispense with it even where he might have made out a case for utility. More and more, the Lamarckian principle of the inherited effects of use and disuse came to replace natural selection. A variety of phenomena were now attributed to this cause: the smallness of the tail in some monkeys and its absence in man, the development of the vocal organs and power of speech, the thin legs and thick arms of Indians who spent most of their lives in canoes, the larger hands of English laborers compared with those of the gentry, the hardened skin on the soles of the feet, the inferiority of Europeans compared with savages in sight and other senses, customs such as the deliberate eradication of hair and other mutilations,[46] even the virtuous habits inculcated in youth. Where once he would have re-interpreted these findings to make them conform to natural selection—and they are amenable to such re-interpretation—he was now easily persuaded of the simpler Lamarckian idea.

In a remarkable confession Darwin explained how and why he formerly erred in giving too much prominence to natural selection:

> A very large yet undefined extension may safely be given to the direct and indirect results of natural selection; but I now admit . . . that in the earlier editions of my "Origin of Species" I probably attrib-

uted too much to the action of natural selection or the survival of the fittest. . . . I had not formerly sufficiently considered the existence of many structures which appear to be, as far as we can judge, neither beneficial nor injurious; and this I believe to be one of the greatest oversights as yet detected in my work. I may be permitted to say as some excuse, that I had two distinct objects in view, firstly, to show that species had not been separately created, and secondly, that natural selection had been the chief agent of change, though largely aided by the inherited effects of habit, and slightly by the direct action of the surrounding conditions. Nevertheless I was not able to annul the influence of my former belief, then widely prevalent, that each species had been purposely created; and this led to my tacitly assuming that every detail of structure, excepting rudiments, was of some special, though unrecognized, service. Any one with this assumption in his mind would naturally extend the action of natural selection, either during past or present times, too far. . . . If I have erred in giving to natural selection great power, which I am far from admitting, or in having exaggerated its power, which is in itself probable, I have at least, as I hope, done good service in aiding to overthrow the dogma of separate creations.[47]

The confession is fascinating, not only for what Darwin said but how he said it—the alternating rhythm of self-recrimination and self-extenuation. In the second edition, perhaps smarting as a result of Mivart's triumphant citation of this passage, he tried to undo the damage by recanting, in effect, some of his earlier recantation. He now declared himself convinced "that very many structures which now appear to us useless, will hereafter be proved to be useful, and will therefore come within the range of natural selection."[48]

Even more interesting, however, than the confession itself was what followed it: the admission of a new factor in the variation of species, more momentous in its implications for his theory than even sexual selection, and which he did not afterward elaborate upon or so much as refer to again (except in the conclusion, where the point was repeated in almost exactly the same words). Falling under none of the other categories that he recognized as responsible for evolution—natural selection, sexual selection, the direct action of the environment, the effect of use and disuse, and correlation of structure—the variation induced by this new factor was of no service to the organism, either in its inception or in its later development. And the nature of its cause was unknown. Darwin could only assume that, whatever its cause, so long as it continued to act "uniformly and energetically" over a long period, the result would be the production not of "mere slight individual differences, but well-marked constant modifications."[49] Not only did these variations arise "spontaneously," in the sense he had used the term in the *Origin,* but—and here he went far beyond the *Origin*—having so arisen, they were not subject to any selective process, natural or sexual, since they were in no way beneficial to the organism (although injurious variations would be eliminated by selection). And these variations would persist so long as either the original conditions producing them persisted or as the free crossing of individuals insured the normal operation of heredity. The latter, he suspected, was more important than the former: "They relate much more closely to the constitution of the varying organism, than to the nature of the conditions to which it has been subjected."[50] Darwin had come far indeed from the doctrine of natural selection.

As Mivart and others were quick to point out, this admission of an unknown cause as one of the means by which man attained his present state—"aided perhaps by others as yet undiscovered"[51]—dangerously undermined Darwin's enterprise. What before was said to be entirely explained was now, at a critical point, left unexplained. The admission that "strange and strongly-marked peculiarities of structure"[52] could arise from unknown causes, and could be perpetuated without reference to any principle of selection or adaptation, opened the way for those

sudden leaps of nature—and of God—which had tradition-
ally been invoked to explain the origin of species, and
especially the origin of man. That Darwin soon became
aware of this danger is suggested by his attempts to modify
his concession and minimize its effects. Thus, in the second
edition of the *Descent*, the "well-marked and constant
modification" brought about by this unspecified cause was
qualified as being "of no physiological importance"[53]—a
qualification that was not explained and that made nonsense of the whole discussion. Nor did he try to justify
another important emendation: in the first edition the
change resulting from the operation of this unknown cause
was described as being "perhaps a large one";[54] in the
second edition this phrase was deleted. When he complained so bitterly of Mivart's omissions, he might have
recalled his own, which like Mivart's were judiciously left
unexplained and unremarked.

The *Descent* was far less rigorous than the *Origin* in
argument and more relaxed in tone—a consequence, to
some extent, of its subject matter. The origin of man raised
problems which were not amenable to the systematic treatment accorded to the origin of pigeons, and the systematic
treatment which they might have received Darwin was
not capable of providing.

To establish the evolution of man from the lower animals, he had to show that the mental and moral faculties of
man differed only in degree and not in kind from those
of animals. This contention was facilitated by his including
among mental faculties such things as the experience of
pleasure and pain, terror, suspicion, curiosity, and affection,
as well as the powers of imitation, attention, memory,
imagination, and reason. The discussion was less a reasoned
argument than a catalogue of examples. From personal
experiences, the observations of friends, and reports in
learned journals, he drew his illustrations. His own dog,
after an absence of five years, had responded to him instantly, demonstrating both memory and faithfulness. The
birds on the Galapagos Islands who avoided men were behaving intelligently; only after their reactions had become

instinctive and inherited did they cease to qualify as intelligent. Monkeys, who notoriously dread snakes, seemed almost to have "some notion of zoological affinities"[55] because some of them were also suspicious of lizards and frogs. Birds, when they dream, give evidence of a "power of imagination."[56] Sleigh-bearing dogs diverge when they approach thin ice; Darwin was not certain whether this was an instinct inherited from their ancestors, the Arctic wolves, or whether it testified to the experience and reason of each individual dog. A retriever, confronted with the problem of one wounded and one dead duck, killed the wounded so as to be able to retrieve both; this was one of Darwin's prize exhibits, but he could not help reflecting that a more perfect exercise of reason would have led the dog first to return the wounded duck and then the dead one. Chimpanzees crack nuts with the help of stones, demonstrating their ability to use (although not fashion) tools. And a baboon had been known to protect himself from the sun by throwing a straw mat over his head; here "we probably see the first steps towards some of the simpler arts, namely rude architecture and dress, as they arose amongst the early progenitors of man."[57] Even language was not unique to man: one species of monkey utters as many as six sounds, "which excite in other monkeys similar emotions,"[58] while dogs bark in four or five distinct tones.

Darwin conceded that between the most precocious animal and the most primitive savage there was far greater difference than between the lowest barbarian and the most civilized man. A Fuegian was remarkably like an Englishman in disposition and mental faculties after only a few years' residence in England, whereas the most advanced animal could not approach the lowest savage in articulate speech. Nevertheless, he insisted that the wide range in mental power among both humans and animals proved that the difference between them was of degree only. The many gradations by which the lowest fishes were separated from the highest apes, and the savage from a Shakespeare, made him confident that the two orders must also have been linked by a series of fine gradations.

But although he could give examples of the links within each order, and even some explanation of how one link evolved from the next, at the crucial point of the transition between man and animal both illustration and explanation petered out.

The lapse is most noticeable where he had taken most pains, as in the discussion of language. Here may be found scattered hints of explanation: the derivation of language from the inarticulate cries and gestures of animals; the theory of sexual selection suggesting that speech, like music, developed out of the emotional stress of courtship; the "not . . . altogether incredible" theory that "some unusually wise ape-like animal should have thought of imitating the growl of a beast of prey, so as to indicate to his fellow monkeys the nature of the expected danger";[59] and the "principle of the inherited effects of use," which would account for the strengthening and perfecting of the vocal organs and, in turn, the improvement of speech. Having come this far, however, he was obliged to retreat, with the admission that even the most imperfect form of speech could not have come into being without the prior development of a brain superior to that of the ape. Thus there must have existed at the very outset "some early progenitor of man" with mental equipment distinguishing him from the ape.[60] But it was precisely to account for the origin of that early progenitor that he had introduced such considerations as the development of speech. If, after all, a species of man must be presumed to have antedated the origin of speech, his enterprise had come to naught, the evolution of man from the animals remaining unexplained. It did not compensate for the failure of that enterprise— although it may have distracted attention from it—to insist, as he did vigorously and at length, upon a proposition no one was inclined to question: that the continued use of the power of speech would, in time, have helped develop and perfect the mind; nor upon another more dubious proposition: that the different languages all grew out of each other and that within each language, by a process of natural selection, "the better, the shorter, the easier" words and forms came to prevail.[61]

The same lack of rigor in argument, the same collapse at the critical moment, may be seen in his controversy with Mill on the subject of moral sense. It was in a footnote that he rebuked Mill for supposing that the moral sense of man was acquired and not innate: "It can hardly be disputed that the social feelings are instinctive or innate in the lower animals: and why should they not be so in man?"[62] Since it was not only the difference between man and animals that was at issue but also the difference between morality and sociability, the point was hardly decisive. Later in the discussion, indeed, he seemed to be arguing that morality and sociability were not identical, that morality arose out of sociability. Among animals, savages, and presumably primeval men, he said, actions were judged as right or wrong depending upon how they affected the welfare of the herd or tribe. Where infanticide was encouraged, it was because it was taken to be for the good of the tribe; when intemperance was sanctioned, it was because the "insufficient powers of reasoning" of primitive men blinded them to its evil consequences.[63] But what kind of instinct was this, it might be asked, that depended upon the power of reasoning? Moreover, considering the number of primitive practices which could be deemed either irrelevant or opposed to the best interests of these tribes—"their utter licentiousness," Darwin was shocked to report, "not to mention unnatural crimes, is something astounding"[64]—he would have had to resort suspiciously often to this catch-all explanation of their "insufficient powers of reasoning." When there are more exceptions to the rule than exemplifications of it, it would seem time to abandon the rule.

Perhaps even more disturbing than any particular faults of argument was Darwin's tendency to resolve all issues at their lowest level. This was apparent in his discussion of religion. As he had earlier denied that language was a unique attribute of man, so he was also constrained to deny that the religious impulse was unique to man. He conceded that if religion be taken to mean a "belief in unseen or spiritual agencies," then it would appear to be almost universal among men.[65] He also conceded that the elements that went into the making of a religious sense—love, submission, fear, reverence, gratitude—required at least a moderate development of the intellectual and moral faculties. Yet he professed to find "some distant approach to this state of mind" in the love of a dog for his master or of a monkey for his keeper; and he cited a German professor who held that "a dog looks on his master as on a god."[66] Thus, as he earlier reduced language to the grunts and growls of a dog, he now contrived to reduce religion to the lick of the dog's tongue and the wagging of his tail.

Indeed, he seemed to find the "religious" sense of the dog more acceptable than that often displayed by men. In the very sentence after he had approvingly quoted the German professor on the religious instinct of the dog, he summoned all the indignation of a proper Victorian gentleman to denounce the outlandish superstitions of human religion. Some of these superstitions, such as human sacrifice and trial by ordeal, he found "terrible to think of";[67] others were less abhorrent but equally unworthy of respect; these included the "senseless practice of celibacy," rules of caste, and injunctions against the eating of certain foods. He did not quite understand, he confessed, how such "absurd rules of conduct" and "absurd religious beliefs" could have come to prevail in large parts of the world, "in complete opposition to the true welfare and happiness of mankind."[68]

It is a common habit of the scientist not only to reduce the complicated and sophisticated to the simple and primitive but also to prefer the simple and primitive. He will cheerfully contemplate the reduction of the most exalted belief to a mechanical or instinctive act, but will not tolerate a belief that, in all its complex reality, happens to conflict with his own rationalist prejudices. Darwin, who could find religious meaning in a dog's devotion to his master, had not the faintest appreciation of the religious significance to be found in such rites as sacrifice, celibacy, caste, or diet. His sensibility was of that inverted order that is unable to extend to human beings the same sympathy and respect it has for animals.

This practice of seeking explanations in the lowest common denominator—morality in terms of instinct, human motives in terms of animal impulses, and civilized conduct in terms of primitive customs—was perhaps a professional failing. As a zoologist, Darwin was naturally more at home in the realm of animal behavior than of philosophy. This may be why so much of his discussion of religion, morality, and esthetics seems painfully naive. Although he apologized for presuming to dispute with Mill, he was not aware of his presumption in venturing upon some of the most ancient, profound, and thoroughly controverted of philosophical questions. In typical auto-didact fashion, he seemed to feel that with the aid of a few simple and obvious principles of his own devising, he could bring light where there had too long been darkness. What little reading he did in philosophy was parochial in the extreme. Apart from a few stock references to Kant and several quotations, surprisingly, from Marcus Aurelius, all his other authorities were British, and none dated back before the eighteenth century. Paying homage to the "many writers of consummate ability" who had applied themselves to the problems of morality, he appended a list (not compiled by him) of twenty-six British philosophers whose names, he felt assured, would be familiar to all his readers—and that are today, quite as assuredly, entirely unknown.[69] It is difficult to take seriously a discussion that had, as its most frequently cited moralist and philosopher, the historian William Lecky.

Darwin's failures of logic and crudities of imagination emphasized the inherent faults of his theory; a finer, more subtle mind would only have obscured or minimized them. The theory itself was defective, and no amount of tampering with it could have helped. Strangely enough, it was Wallace who went to the heart of the matter. Neither natural selection nor any of Darwin's other explanations, Wallace reluctantly concluded, could account for the emergence of man as an intellectual and spiritual creature. Even if there were no other evidence of a leap between the highest animals and the most primitive men, there would be the inexplicable anomaly that the brain of savage man was not only larger than the brain of his nearest animal kin but also vastly disproportionate to his own talents and needs. The unique and at the same time disproportionate size of the savage brain only ceased to be an anomaly if it was assumed that some special agency—analogous, Wallace suggested, to that which had first produced organic life and then consciousness—had created a species not once removed from the ape but, at least potentially, many times removed from him. This was teleology on a scale never intended by Darwin: it presumed a knowledge of the end not only of the individual or even of the species as it then existed but of the species as it might exist many epochs later. Moreover, Wallace argued, so long as research failed to uncover evidence of man's presence far back in the history of mammals, it must be presumed that he came into existence much later and hence much more rapidly than Darwin's theory allowed.

The problem of the development of the human brain—of man, in short—has been and still is the critical point in the evolutionist case. It was this that inspired the harshest attacks of Darwin's contemporaries and the most earnest labors of scientists until this day. For a time it seemed as if the difficulty had been resolved in favor of the evolutionists—and not by theory but by incontrovertible fact. The discovery in 1912 of the Piltdown skull, which combined a man-like brain with an ape-like jaw, seemed to provide the missing link required for Darwin's theory. It was also the occasion for experiments that appeared to vindicate Darwinism in other respects. These experiments were less dramatic than the discovery of the missing link, but they were perhaps quite as important in bolstering up the confidence of the Darwinians and confirming them in their methods and theories.

One of the confirmations apparently provided by the Piltdown man was of the paleontological method itself, the technique of reconstructing a skeleton from a few fragments of bones. In the case of Piltdown, the difficulty of reconstruction was enhanced by the dissimilarity of jaw and brain case. Moreover, there were conflicting opinions on the crucial question of the size of the brain case. Working with only a few fragments, anthropologists varied in their estimates of the size of the brain from one thousand cubic centimeters, which would have put it in a subhuman category, to one thousand five hundred centimeters, which would have made the Piltdown man a more likely candidate for humanity and a far more significant missing link. At one point in the controversy, Sir Arthur Keith, who held out for the larger estimate, was challenged to put his methods to the test by performing a similar feat of reconstruction on the basis of a few equally incomplete fragments of a modern skull. When he succeeded in arriving within twenty cubic centimeters of the correct size (although he mistook its sex and provided it with a masculine brow ridge), his reconstruction of the Piltdown skull seemed triumphantly vindicated. After some adjustment and compromise, most scientists, with the exception of some unreconciled skeptics, agreed that Piltdown man must have had a brain capacity of about one thousand four hundred cubic centimeters, which made him a respectable human progenitor.

Another opportunity for experiment was provided by the heavily ridged brows which gave Piltdown man his ape-like appearance. The ridges themselves were not in question; the fragments clearly suggested these. What was in question was how the species could have evolved, in the relatively short time now allotted him (Piltdown man was estimated to be only fifty thousand years old), from the ridged brow of Piltdown man to the smooth brow of modern man. It was calculated that if each such difference between our progenitors and ourselves was based upon a separate genetic variation, it was mathematically impossible for ape man to have been transformed into modern man in the time at his disposal. An American anthropologist then undertook to show that this could have happened given Darwin's theory of correlated variations, according to which one genetic variation would necessarily bring other variations in its wake without the intervention of new mutations. Experimenting with newborn rats, he discovered that if he removed the jaw muscle on one side, the skull of the fully grown rat would present the appearance on one side of the deep ridges customarily found in rats (as in Piltdown man), and on the other, the operated side, a smooth brow such as is found in modern man. Thus the one change in the jaw muscle could have been enough to account for so momentous and rapid a development.

It was all the more unfortunate, then, for the Darwinians that after all these experiments apparently confirming their method and theory, Piltdown man should have been exposed as a fraud. The ape-like jaw and brow and the man-like brain, which seemed almost tailor-made to fit the theory, proved to be all too literally tailor-made. While it is true that evolution neither stands nor falls with Piltdown man—some Darwinians were skeptical of it all along—nevertheless, its exposure as a fraud leaves the theory, after a century of search, without the much-desired missing link, and also without even such antiquity as Piltdown offered. The anthropologist Loren Eiseley has vindicated Wallace on this important point: "Man and his rise now appear short in time—explosively short."[70]

Nor can it be maintained, as some Darwinians have done, that the exposure of Piltdown man leaves them no better and no worse off than they were before. It does, in fact, weaken their position in regard to both their theory and their methods. The zeal with which eminent scientists defended it, the facility with which even those who did not welcome it managed to accommodate to it, and the way in which the most respected scientific techniques were soberly and painstakingly applied to it, with the apparent result of confirming both the genuineness of the fossils and the truth of evolution, are at the very least suspicious.[71] However earnestly scientists may now dissociate themselves and their theory from Piltdown man, they cannot entirely wipe out the memory of forty years of labor expended on a deliberate and not particularly subtle fraud. And not forty years in the remote past, but forty years which came to an end only as recently as 1953.

DARWINISM

BOOK VI

CHAPTER 18

DARWINISM, RELIGION, AND MORALITY

IDEAS have a radiation and development, an ancestry and posterity of their own, in which men play the part of godfathers and godmothers more than that of legitimate parents. In this dictum of Lord Acton may be seen the relationship between Darwin and Darwinism, between the theory of natural selection and the theories of religion, morality, society, and politics presumably derived from that theory.[1] Of the ancestry of his ideas or the climate of opinion in which they were nurtured, Darwin had little real knowledge or appreciation; of their posterity or the climate which they created, he had, necessarily, even less. It has already been observed how much narrower in extension and more specific in intention were his theories than either the background from which they emerged or the response which they evoked; the image is of an hourglass, with Darwin comprising the narrow waist.

Even Darwin's own opinions on religious and social subjects belong to the public realm of Darwinism. They are as much a deduction from his doctrine as are the views of any outsider, and he is as little an authority on the legitimacy of this deduction as anyone else. Indeed, his views often carry less weight than those of others who were able to find more subtle, complex, and profound significances in Darwinism than he himself ever suspected. If, nevertheless, attention persists in focusing upon him, it is because of a natural curiosity to know what even a godfather makes of his godchild.

Darwin's early religious orthodoxy has already been re-marked upon. It showed itself in his willingness to join the ministry and in his irreproachably orthodox behavior on the Beagle. Disbelief, by his own account, did not begin to trouble him until after his return from the voyage. He then began to doubt not only the historical authenticity of the Bible—the Gospels, he realized, differed among them-selves and must have been written long after the events they purported to describe—but also its teaching. The mir-acles, he felt, were not credible to any "sane man," and fixed laws could explain everything in nature. He tried to resist these doubts, even going so far as to invent day-dreams about the discovery of old manuscripts which would confirm all that was said in the Gospels. But the most liberal exercise of imagination could not restore his faith in Christianity as a divine revelation. "Thus," he con-fided, "disbelief crept over me at a very slow rate, but was at last complete. The rate was so slow that I felt no dis-tress, and have never since doubted even for a single sec-ond that my conclusion was correct."[2]

The progress of his disbelief must have been sufficiently advanced by 1838, when he became engaged to be mar-ried, to provoke his father's warning about the advisability of concealing one's doubts from one's wife. Darwin was too candid to act upon this advice, however, and Emma was from the first aware of his waning belief. Things never came to the disagreeable pass predicted by his father, in which each illness of the husband would provoke the wife to loud laments about his eternal salvation; in Darwin's in-valid state this would have been serious indeed. But his father was not entirely wrong, and in spite of Emma's resolution not to "bother" him, as she once put it,[3] she could not resist occasionally raising the subject. During one particularly ailing period she confided her feelings in a letter:

> I am sure you know I love you well enough to be-lieve that I mind your sufferings, nearly as much as

I should my own, and I find the only relief to my own mind is to take it as from God's hand, and to try to believe that all suffering and illness is meant to help us to exalt our minds and to look forward with hope to a future state. When I see your patience, deep compassion for others, self-command, and above all gratitude for the smallest thing done to help you, I cannot help longing that these precious feelings should be offered to Heaven for the sake of your daily happiness.[4]

Darwin respected, even if he did not share, his wife's sentiments, and he permitted her to set the tone of their domestic life. Pious and given to prayer in moments of stress, she attended church regularly, took the Sacrament, read the Bible to her children, and was assailed by doubts as to the propriety of knitting or playing patience on Sunday. Although herself Unitarian, she had the children baptized and confirmed in the Church of England. Darwin apparently did not object to the baptism or confirmation, but by a curious religious scruple drew the line at the ceremony which would have bestowed upon them godparents—"not," he explained, "from any objection to their having such, but as we should in that case have been obliged to have stood proxies, and we both disliked the statement of believing anything for another."[5] Even in public he was governed by his solicitude for his wife's feelings, to the extent of refraining from open pronouncements on religious matters. For what disturbed her even more than his personal salvation was the thought of public scandal. She was so upset by the attacks on the *Origin* that she refused to let the children see the worst of them. And when the *Descent* was published, she confessed to her daughter: "I think it will be very interesting, but that I shall dislike it very much as again putting God further off."[6]

What was said in the *Descent* was unavoidable, but elsewhere Darwin tried hard not to give offense. When the editor of an American rationalist journal solicited his views for publication, he declined, pleading ill health and the lack of time for sustained thought on the subject of "religion in relation to science, or on morals in relation to society."[7] That this was not the whole story—he had, after all, just completed a major work in which the question of morals and society figured prominently—appears in the unpublished portion of this letter:

> Many years ago I was strongly advised by a friend never to introduce anything about religion in my works, if I wished to advance science in England; and this led me not to consider the mutual bearings of the two subjects. Had I foreseen how much more liberal the world would become, I should perhaps have acted differently.[8]

And when Marx proposed to dedicate to him *Das Kapital*, he firmly refused the honor, explaining that it would pain certain members of his family if he were associated with so atheistic a book.[9]

It was only in his autobiography that Darwin gave free expression to his religious opinions. And it was when his son prepared to publish the autobiography in the *Life and Letters* that Emma Darwin revealed the true measure of her conventionality. Having succeeded in maintaining a modicum of discretion in his lifetime, she objected to having the floodgates of scandal opened after his death, and solemnly warned her son that unless he deleted some of the franker passages, her life would be made unendurably miserable. She was supported by Henrietta on the grounds that Darwin had expressed himself badly in the autobiography and that its publication would "lower him in the eyes of the world."[10] Although the sons were all in favor of printing it *in toto*, Francis reluctantly agreed to the deletions out of respect for their mother's wishes. William, the eldest, protested strongly against this concession. He did not believe that his mother could be so grievously affected by the publication of a few sentences, and he was frankly disgusted by his sister's parade of piety. "I have little doubt," he wrote, "if the *Descent* had been left unpublished, Henrietta would have been for cutting out passages as unworthy of his mind."[11]

The full extent of Darwin's disbelief, therefore, can be seen neither in his published work nor even in his published autobiography, but only in the original version of that autobiography. Where the edited version[12] stated simply that he had come to see "that the Old Testament was no more to be trusted than the sacred books of the Hindoos," the original added: "or the beliefs of any barbarian." It also specified what it was in the Old Testament that he found so objectionable: "its manifestly false history of the world, with the Tower of Babel, the rainbow as a sign, etc., etc."

If he found the Bible an untrustworthy source, neither could he be persuaded of the existence of God by "the deep inward conviction and feelings which are experienced by most persons." He himself, he confessed, had once had such feelings; in the grandeur of the Brazilian forest he had been possessed by the conviction that there must be more in man than "the mere breath of his body." But later even the grandest scenes could not evoke such thoughts in his mind. It might be argued, he realized, that he was like a man who had become color-blind and who alone among his fellow men could not see red when confronted with it. Such arguments, however, failed to move him. Too many men had equally deep but different convictions and feelings for these to inspire confidence:

> It cannot be doubted that Hindoos, Mohammedans and others might argue in the same manner and with equal force of the existence of one God, or of many gods, or with the Buddhists of no God. There are also many barbarian tribes who cannot be said with any truth to believe in what we call God; they believe in spirits or ghosts, and it can be explained, as Tylor or Herbert Spencer have shown, how such a belief would be likely to arise.

The fact was, as he told his cousin: "I look upon all human feeling as traceable to some germ in the animals."[13] This being so, all religious doctrines were equally pretentious and spurious. The most exalted convictions and feelings could be explained as the simple effect of custom:

May not those be the result of the connection between cause and effect which strikes us as a necessary
one, but probably depends merely on inherited experience? Nor must we overlook the probability of
the constant inculcation of a belief in God on the
minds of children producing so strong and perhaps an
inherited effect on their brains, not as yet fully developed, that it would be as difficult for them to
throw off their belief in God, as for a monkey to
throw off its instinctive fear and hatred of a snake.

Darwin gave as little credence to the moral argument
as to the psychological one: "Beautiful as is the morality
of the New Testament, it can hardly be denied that its
perfection depends in part on the interpretation which we
now put on metaphors and allegories." And sometimes that
morality, no matter how liberally the metaphors and allegories might be interpreted, was considerably less than
beautiful or perfect:

I can indeed hardly see how anyone ought to wish
Christianity to be true; for if so, the plain language
of the text seems to show that the men who do not
believe, and this would include my Father, Brother,
and almost all my best friends, will be everlastingly
punished. And this is a damnable doctrine.[14]

Nor was God as benevolent as he was reputed to be:

A being so powerful and so full of knowledge as a
God who could create the universe, is to our finite
minds omnipotent and omniscient, and it revolts our
understanding to suppose that his benevolence is not
unbounded, for what advantage can there be in the
sufferings of millions of lower animals throughout almost endless time.

For himself, Darwin preferred a morality independent of
religion and untainted by the moral defects of Christianity.
In an addendum to his autobiography, he spelled out the
derivation and implication of a naturalistic ethics:

A man who has no assured and no present belief in the existence of a personal God or a future existence with retribution and rewards, can have for his rule of life, as far as I can see, only to follow those impulses and instincts which are the strongest or which seem to him the best ones. A dog acts in this manner, but he does so blindly. A man, on the other hand, looks forwards and backwards, and compares his various feelings, desires, and recollections. He then finds, in accordance with the verdict of the wisest men, that the highest satisfaction is derived from following certain impulses, namely the social instincts. If he acts for the good of others he will receive the approbation of his fellow-men and gain the love of those with whom he lives; and this latter gain undoubtedly is the highest pleasure on this earth. By degrees it will be more intolerable to him to obey his sensuous passions rather than his highest impulses, which when rendered habitual may be almost called instincts. His reason may occasionally tell him to act in opposition to the opinion of others, whose approbation he will then not receive; but he will still have the solid satisfaction of knowing that he has followed his innermost judge or conscience.[15]

Darwin had come a long way from his earlier protestations that religion was beyond the reach of the human intellect, that "a dog might as well speculate on the mind of Newton," and that man could only "hope and believe what he can."[16] As often happens, victory, rather than abating hostility, had only served to intensify it. His cousin, Julia Wedgwood, remarked upon the curious fact that his antagonism to religion increased, "while all the apparent reasons for it were vanishing quantities." In the same proportion, she observed, as the churches approached him in docile and even eager acceptance of his teachings, so he receded from them. "He was far more sympathetic with religion when his books were considered wicked, by the religious world, than when (as was the case for some years before he died), the dignitaries of the Church were eager to pay him the highest honour."[17]

This is not, however, to say that Darwin's disbelief was ever as hard and unremitting as some might have liked. Edward Aveling, Karl Marx's son-in-law, came away from a conversation with Darwin shortly before his death complaining of his parochialism in regarding himself as an agnostic rather than an atheist. Aveling consoled himself, however, with the thought that "'Atheist' is only 'Agnostic' writ aggressive, and 'Agnostic' is only 'Atheist' writ respectable"[18]—on which Francis Darwin rightly commented that it was precisely the difference between aggressiveness and unaggressiveness that distinguished his father "from the class of thinkers to which Dr. Aveling belongs."[19]

Darwin himself, later in life, no longer doubted the wisdom of his particular brand of disbelief but only the wisdom of silence. To his son George, then at Cambridge, he confessed his doubts: "It is a fearfully difficult moral problem about the speaking out on religion, and I have never been able to make up my mind." As if in self-justification, he cited the cases of honorable men who had chosen to keep silent:

> Last night Dicey and Litchfield were talking about J. Stuart Mill's never expressing his religious convictions, as he was urged to do so by his father. Both agreed strongly that if he had done so, he would never have influenced the present age in the manner in which he has done. His books would not have been text books at Oxford, to take a weaker instance. Lyell is most firmly convinced that he has shaken the faith in the Deluge far more efficiently by never having said a word against the Bible, than if he had acted otherwise. . . .

> I have lately read Morley's *Life of Voltaire* and he insists strongly that direct attacks on Christianity (even when written with the wonderful force and vigor of Voltaire) produce little permanent effect: real good seems only to follow the slow and silent side attacks.[20]

Although Darwin is only one, and not necessarily the most authoritative, witness on the religious significance of

Darwinism, he does represent a common opinion. It is the opinion of all those who, for good or bad, find Darwinism and religion to be irreconcilable. There were many who agreed with Huxley that one of the great merits of the theory of evolution was its "complete and irreconcilable antagonism to that vigorous and consistent enemy of the highest intellectual, moral, and social life of mankind—the Catholic Church."[21] And not Catholicism alone but all religion. The German Biblical critic, David Strauss, admitted that Darwin's theory was irresistible to those who thirsted for "truth and freedom."

> Vainly did we philosophers and critical theologians over and over again decree the extermination of miracles; our ineffectual sentence died away, because we could neither dispense with miraculous agency, nor point to any natural force able to supply it, where it had hitherto seemed most indispensable. Darwin has demonstrated this force, this process of Nature; he has opened the door by which a happier coming race will cast out miracles, never to return. Every one who knows what miracles imply will praise him, in consequence, as one of the greatest benefactors of the human race.[22]

There was apparently justice in Mivart's complaint that if the *odium theologicum* inspired some of Darwin's critics, the *odium antitheologicum* possessed not a few of his supporters.

The party of the *odium theologicum* had more in common with their adversaries than they might have liked to think. Not only did they agree upon the character and import of Darwinism; they might even be found to agree as to its truth—and it was then that they were the more profoundly and bitterly antagonistic. One young man, having read the *Origin* at the age of sixteen, cursed "the rigorous logic that wrecked the universe for me and for millions of others," that gave them "a feeling of utter insignificance in face of the unapprehended processes of nature . . . a sense of being aimlessly adrift in the vast universe of consciousness, among an infinity of other atoms, all struggling desperately to assert their own existence at the expense of all the others."[23] William Hale White, whose anguished soul searching was conducted under the pseudonym of "Mark Rutherford," protested that he could no longer believe in an immortality that would perpetuate through all eternity the countless millions of barbaric, half-bestial forms that must have inhabited the earth before the final evolution of man. White gave up immortality, but others were more tenacious, clinging by the mere force of will to traditional beliefs in which they no longer truly believed, desperately resisting what they now felt to be the truth. It was an attitude that was given philosophical respectability by Bradley in the statement: "I could not rest in a truth were I compelled to regard it as hateful."[24]

In extremis, some men went even further than this, beyond the admission of a truth that could be resisted only by a deliberate exercise of will. The final act of desperation was not to resist but to submit to the truth, however hateful. This, for some, was the truly diabolical result of Darwinism: not the displacement of God entailed in the conventional loss of faith, but the substitution of Satan in the place of God, or even the Satanization of God himself. The perversion of faith rather than the loss of faith was the ultimate in heresy. Natural selection, it was said, by making creation a consequence of destruction, made of God the supreme destroyer. The poet Thomson, in his famous "City of Dreadful Night," exploring the horror of a purposeless, heartless, mindless universe, where "Death-in-Life is the eternal king," did not know whether to be more horrified at the thought of a God who was the "Creator of all woe and sin! abhorred, malignant and implacable!" or at the thought that there was indeed no God, "no Fiend with names divine" who made and tortured us, but only

> . . . Universal laws
> Which never had for man a special clause
> Of cruelty or kindness, love or hate.

It was with such sentiments in mind that George Romanes,

a devoted friend and disciple of Darwin, later came to deplore the spiritual consequences of modern science:

> Never in the history of man has so terrific a calamity befallen the race as that which all who look may now behold advancing as a deluge, black with destruction, resistless in might, uprooting our most cherished hopes, engulfing our most precious creed, and burying our highest life in mindless desolation. . . . The flood-gates of infidelity are open, and Atheism overwhelming is upon us.[25]

What drove some men to despair inspired in others new hope and joy. Looking through a microscope, Tennyson once remarked: "Strange that these wonders should draw some men to God and repel others."[26] And so it was when confronted with Darwin. The *Literary Churchman,* reading his descriptions of the "beautiful contrivances" of orchids, was moved to comment: "O Lord, how manifold are Thy works!"[27] Many found that if they had to give up their old conception of God, they were provided with a new and loftier one. Like Gray, they found in natural selection a modern, refined variant of natural theology. Gray explained his motive in associating the natural teleology of evolution with the natural theology of religion:

> The important thing to do is to develop aright evolutionary teleology, and to present the argument for design from the exquisite adaptations in such a way as to make it tell on both sides; with Christian men, that they may be satisfied with, and perchance may learn to admire, Divine works effected step by step, if need be, in a system of nature; and the antitheistic people, to show that without the implication of a superintending wisdom nothing is made out, and nothing credible.[28]

What John Dewey contemptuously dismissed as a theory of "design on the installment plan,"[29] others welcomed as an affirmation of religious faith. And, indeed, theology has always been more accommodating to the demands of science than is generally thought. Much of the time, both before and after Darwin, the relations between science and religion were entirely amicable—and genuinely so, not as an armed truce, or a grudging concession to an unpleasant reality. "Natural theology"—the search for the "evidences" of Christianity and God in the facts of nature —was not a device invented by shrewd theologians to make science subservient to religion. It was, rather, an attempt to explore nature in the only way that seemed to make nature, as well as God, intelligible—in terms of design. Paley set down the first principle of the creed: "There cannot be design without a designer, contrivance without a contriver."[30] And although he himself preferred to look for illustrations of design where he could find no evidence of natural or mechanical laws, others found design precisely in the operation of such laws. Thus scientists were able to devote themselves to their calling, confident that they could not but enhance the glory of God.

Earlier, in the seventeenth century, it had been mathematics that had been invoked to demonstrate the rationality and thus the divine providence of the universe: Newton's scientific work was inspired by the same religious mission that led him to devote so many years to the allegorical examination of the prophecies of Daniel. In the eighteenth and early nineteenth centuries it was natural science that was ransacked for "Christian evidences." Explicitly and exhaustively, the authors of the Bridgewater treatises undertook to demonstrate, as they were commissioned, "the Power, Wisdom, and Goodness of God, as manifested in the Creation; illustrating such work by all reasonable arguments, as for instance the variety and formation of God's creatures in the animal, vegetable, and mineral kingdoms; the effect of digestion, and thereby of conversion; the construction of the hand of man, and an infinite variety of other arguments; as also by discoveries ancient and modern, in arts, sciences, and the whole extent of literature."[31] The author of one of these treatises, thinking it irreverent to harp on the name of God, explained that he would use the word "nature" instead to express the same power. Prince Albert put the official stamp of piety on the scientific enterprise when he said that, in the Great

Exhibition, man was fulfilling his sacred mission: "His Reason being created after the image of God, he has to use it to discover the laws by which the Almighty governs His creation and, by making these laws his standard of action, to conquer nature to his use; himself a Divine instrument."[32] If the Great Exhibition, with its display of steam mills and shower baths, papier-mâché chairs and rococo vases, could be said to epitomize the quest for the divine laws of nature, the *Origin* could certainly be taken in the same way.

To be sure, as science advanced, so did theology, leaving Paley behind—so far behind that the more adventurous spirits thought they had transcended natural theology itself. Baden Powell, the Oxford professor whose contribution to *Essays and Reviews* succeeded the *Origin* as the great scandal of the time, declared that the modern discoveries of science made it hazardous to base Christianity on any "connexion with physical things."[33] Astronomy, geology, and the recent findings about the antiquity of man and the development of species were as undeniable in themselves, he said, as they were undeniably in contradiction with revelation. If man was to continue to live in harmony with both religion and modern science, he had to rid himself of the habit nurtured by Paley of thinking of physical facts as evidences of Christianity. A closer reading of this essay, however, suggests that even Powell was not as ruthless in divorcing religion from science as has been supposed, that ultimately he too accepted a variety of natural theology. For the "physical things" whose connection with theology he so deplored were not physical facts as such but rather the supposed physical miracles. It was these miracles, acceptable to a people "entertaining an indiscriminate belief in the supernatural," that he feared would offend and alienate an age accustomed to the reign of law.[34] He had no objection, however, to the connection between religion and natural laws. In an earlier essay he had implied that such a connection was not only desirable but necessary. The assurance of the uniformity, immutability and sufficiency of natural laws, upon which science depended, testified to a moral purpose "prior and superior" to these laws; thus science itself was "but a perpetual worship of God in the firmament of His power."[35]

The theologians assembled in *Essays and Reviews* would not have welcomed the thought that it was the reform of natural theology and not its dissolution that they were presiding over; revolutionists do not welcome being exposed as mere reformers. That they were essentially reformers, however, may be seen in the sermon delivered to the British Association in July 1860 by Frederick Temple, Headmaster of Rugby and a contributor to *Essays and Reviews*. Temple criticized the theologian's habit of drawing his arguments from science's confessions rather than from its revelations, as if theology started where science ended. Instead, he urged that since the book of science and the book of revelation were by the same author, the power of God could more properly be found in the laws of nature than in the miraculous interference with those laws:

> The fixed laws of science can supply natural religion with numberless illustrations of the wisdom, the beneficence, the order, the beauty that characterize the workmanship of God; while they illustrate His infinity by the marvellous complexity of natural combinations, by the variety and order of His creatures, by the exquisite finish alike bestowed on the very greatest and on the very least of His works, as if size were absolutely nothing in His sight.[36]

Others joined in this reformation of natural theology, agreeing that God was to be sought not in the interstices of nature but in the entirety of nature. The American clergyman Henry Ward Beecher pronounced "design by wholesale" to be more noble than "design by retail";[37] while a Scottish clergyman, Henry Drummond, mocked those who "ceaselessly scan the fields of Nature and the books of Science in search of gaps—gaps which they will fill up with God."[38] Where once men found evidence of God in the fact of "missing links," in the failures of evolution, now they found it in the links themselves, in the very process of evolution. New credos were written bearing such proud titles as *Theology of an Evolutionist:*

> The theistic evolutionist believes that God is the one Resident Force, that He is in His world; that His method of work in His world is the method of growth; and that the history of the world, whether it be the history of creation, or providence, or of redemption, whether the history of redemption in the race or of redemption in the individual soul, is the history of a growth in accordance with the great law interpreted and uttered in that one word evolution.[39]

The idea of development, it was discovered, had always been present in the Christian view of the divine economy; Darwin had only borrowed it and returned it with interest. One reviewer of the *Origin*, otherwise unsympathetic, wondered how it could be thought that the primitive "three or seven-storied structure" of the old myth was more to the "glory of God" than the grand, new, modern structure of Darwin.[40] Another observed that piety must be fastidious indeed to object to a theory which assumed all organic life to be in a "perpetual progress of amelioration."[41] Even Gladstone belatedly discovered that "the doctrine of evolution, if it be true, enhances in my judgment the proper idea of the greatness of God, for it makes every stage of creation a legible prophecy of all those which are to follow it."[42] From being a detestable scientific truth, Darwinism had emerged as an agreeable religious myth. No greater tribute can be paid to a theory.

The religious managed to find in Darwinism a variety of consolations and virtues not dreamed of even in natural theology. One distinguished botanist bewildered Darwin by declaring himself a convert on the grounds that the theory finally made intelligible the birth of Christ and redemption by grace. A clergyman was converted on the grounds that it opened up new and more glorious prospects for immortality. And theologians declared themselves ready to give up the old doctrine of "the fall" in favor of the happier idea of a gradual and unceasing progress to a higher physical and spiritual state.

It was not even necessary to interpret Darwinism teleologically in order to find religious meaning in it. Some Calvinists gloried in it precisely because it exalted chance, not design. It was this that confirmed their faith in special providence, in the arbitrary election of the chosen, and in the spontaneous, unpredictable, and often tragic nature of the universe. A prominent American theologian, holder of the chair of "The Harmony of Science and Revelation," suggested that Darwinism was to natural science as Calvinism was to theology: a foe to sentimentalism and optimism, a check on the reign of law and the trust in reason.[43]

In the Catholic Church, Darwinism had a history and literature of its own. As has already been observed, Catholics were, in some respects, better situated than Protestants to accept Darwinism. Depending rather on the authority of the Church than on Scripture, they were not bound by the literal Biblical texts; ecclesiastical authority could and did interpret the Biblical account of creation as it saw fit. It was, indeed, the purpose of Newman's theory of development, which anticipated the *Origin* by over a decade, to provide a rationale for the evolution and progressive re-interpretation of dogma. Liberal Catholics, wanting to exploit this Scriptural freedom, and at the same time wanting to endear themselves to other liberals, besought the Church not to intervene in scientific quarrels. On the eve of the publication of the *Origin*, Newman himself urged this course upon the hierarchy, not only on the plea that Catholics deserved to be spared any addition to the heavy load of controversy which they already bore, but also because, by copying "the divine wisdom in not making formally binding the old accepted cosmology," they would demonstrate to the world the "divinity of their religion."[44]

The Papacy, however, assured both of its divinity and of a measure of wisdom larger than that credited it by Newman, thought that aggression rather than discretion was called for. In 1864 Pius IX issued the "Syllabus of Errors" condemning "progress, liberalism and modern civilization." Although this was commonly understood to include Darwinism, many Catholics, taking advantage of the absence of any formal prescription (neither the *Origin* nor the *Descent* ever joined Erasmus Darwin's *Zoonomia* on the Index), continued to speculate freely with evolutionary ideas. Some went further than St. George Mivart, and perhaps even further than Darwin himself. At the centenary celebrations of Darwin's birth, Canon Dorlodot, representing the University of Louvain, praised him for revealing a creation grander and worthier of God than had before been suspected. Later he wrote a book on Darwinism, in which he declared that both the principles of Catholic exegesis and the teachings of the Church fathers made "eminently probable the theory which derives all living things from one or a few very simple types of organisms." Although he specifically left out of consideration the problem of the origin of man, he more than made up for this by his attitude toward the origin of life. Of Darwin's suggestion that a special intervention of God still seemed necessary to account for the creation of life, Dorlodot cavalierly remarked that this was probably a legitimate hypothesis —"at least for the time being."[45]

A century after the publication of the *Origin*, Catholics were able to assert with impunity the entire teachings of Darwin, even on the development of man, provided only that they did not tamper with the divine origin of the soul. (Since Darwin did not speak of a "soul," this was no great hardship.) The encyclical "Humani Generis," published in 1950, condemned the idea that evolution could account for the origin of all things, on the grounds that such an extension of the theory was conducive to a variety of heresies: monism, pantheism, dialectical materialism, idealism, immanentism, pragmatism, and existentialism—in short, all those "false opinions," as the subtitle put it, "which threaten to sap the foundation of Christian teaching." But it conceded that a more modest theory of evolution, concerning itself only with the physical development of the human body, would avoid this litany of errors and be a legitimate subject of research. Most Catholic evolutionists professed to be well satisfied with this pronouncement. An article in the Jesuit journal, the *Month*, said that while "macro-evolution does not show up too well" on the evidence, it was nevertheless preferable to "special creation," the omnipotence of God being best revealed in the laws of nature. The only qualification the Catholic need make was the belief that "God intended, from all eternity, to create Man to put him in this world as the crown of this particular order of creation, and that He took those means which He saw to be most suitable for the purpose."[46]

When Henry Adams wrote that "unbroken Evolution under uniform conditions pleased everyone—except curates and bishops; it was the very best substitute for religion; a safe, conservative, practical, thoroughly Common-Law deity,"[47] he had underestimated its appeal. The most respectable curates and bishops were even then succumbing to its charms.

The basic religious quarrel provoked by the *Origin* was not between the theists who rejected it and the atheists who favored it, as has been thought, but rather between the reconcilers and the irreconcilables, those who believed the *Origin* to be compatible with Christianity and those who thought that it was not. The irreconcilables of both parties—the one *rejecting* Darwinism because it demeaned religion, the other *embracing* Darwinism because it demeaned religion—were as contemptuous of the efforts of the reconcilers as they were hostile to each other. Huxley, engaged in debate with the elusive Gladstone, finally protested: "There must be some position from which the reconcilers of science and Genesis will not retreat."[48] The god of the reconcilers too often resembled that of Coleridge: "a something-nothing-everything which does all of which we know."[49] Agnostics and believers alike objected that such a god "who is the final reason of everything is the scientific explanation of nothing."[50] They dismissed as pat and meaningless the clever paradoxes: "providence without a god,"[51] "religion without revelation,"[52] a god who was identical with "space-filling matter."[53]

If the agnostics were aggrieved at the thought of a fruitless victory in which Darwinism was assimilated into the old religious categories, many religious men were dismayed by a defeat that might turn out to be even more disastrous than had at first appeared. To the religious irreconcilables, Darwinism threatened to suck in a multitude of well-intentioned people on the pretense of offering them a respectable compromise. But the compromise, they felt, was a sham, a façade behind which lurked skepticism and atheism. What the reconcilers saw as an aloof and noble god who did not stoop to intrude in the practical administration of his realm, the irreconcilables saw as an impoverished and impotent monarch, a ceremonial figurehead. Thus Agassiz complained of the "repulsive poverty" of the god of the evolutionists: "The resources of the Deity," he protested, "cannot be so meagre, that in order to create a human being endowed with reason, he must change a monkey into a man."[54] R. H. Hutton said that what scandalized many was the thought that God should be presumed to work in the tentative and negative fashion implied by Darwinism: "They ask how an omniscient mind which knows precisely what is wanted, can set Nature *groping* her way forward as if she were blind, to find the path of least resistance."[55]

What the religious feared was that this new god, who bore so attenuated a resemblance to the old, would become more and more attenuated until little remained to identify him as God at all. The irreconcilables, it would seem, took more seriously than their opponents the inevitability and pervasiveness of the evolutionary process. They saw that the new evolutionary theology was itself subject to the laws of evolution and that it would necessarily evolve toward an expanding Darwinism and a diminishing religion. Hutton was only one of many, in Darwin's own time, to point to the danger of reducing God to the status of a logical postulate, an impersonal First Clause, far removed from the concrete facts and processes of nature as men experience them:

> The people who believe today that God has made so fast the laws of His physical universe, that it is in many directions utterly impenetrable to moral and spiritual influences, will believe tomorrow that the physical universe subsists by its own inherent laws,

and that God, even if He dwells within it, cannot do with it what He would, and will find out the next day, that God does not even dwell within it, but must, as M. Renan says, be "organised" by man, if we are to have a God at all.[56]

The dispute about the religious significance of Darwinism was part of a long history of religious controversy. God was being "organised" by man long before the *Origin* appeared, and the hopes and fears that greeted Darwin's work were echoes of earlier exultations and laments. It is neither to deny nor denigrate the sincerity of the despair felt by so many upon reading the *Origin* that one recalls similar outbursts a century and more before. It was in 1736 that Bishop Butler complained: "It is come, I know not how, to be taken for granted, by many persons, that Christianity is not so much a subject of inquiry; but that it is now at length, discovered to be fictitious."[57]

The complaint, already familiar in Butler's time, was to become more so in the following century. In the 1830's Carlyle was to be found contrasting the easy assurance of the past, when men had accepted unquestioningly the divinity of human beings and affairs, with the "iron, ignoble circle of Necessity" now holding them in thrall or tempting them to rebellion: "The invincible energy of young years wastes itself in sceptical, suicidal cavillings; in passionate 'questionings of Destiny', whereto no answer will be returned."[58] In the same vein, James Froude recalled the spiritual anguish of the 1840's:

> All round us, the intellectual lightships had broken from their moorings, and it was then a new and trying experience. The present generation which has grown up in an open spiritual ocean, which has got used to it and has learned to swim for itself, will never know what it was to find the lights all drifting, the compasses all awry, and nothing left to steer by except the stars.[59]

In 1851 Harriet Martineau described how her loss of faith in a divine providence was succeeded by a new accession of faith in the eternal and irreversible laws of nature—whereupon Charlotte Brontë recoiled in disgust: "If this be Truth, man or woman who beholds her can but curse the day he or she was born."[60]

Through the decade the crescendo of despair mounted in the writings of such estimable witnesses as Arnold and Clough, Tennyson, Ruskin, Mill, and Newman. A catalogue of doubt, of every variety and intensity, can be compiled in the pre-Darwinian epoch. When Romanes and others later complained that they were sucked into Darwinism as into a mire of hopeless, meaningless nihilism, it may be supposed that they would have had the same experience even without Darwinism.

What the *Origin* did was to focus and stimulate the religious and nihilist passions of men. Dramatically and urgently, it confronted them with a situation that could no longer be evaded, a situation brought about not by any one scientific discovery, nor even by science as a whole, but by an antecedent condition of religious and philosophical turmoil. The *Origin* was not so much the cause as the occasion of the upsurge of these passions.

As men were provoked to take up new or exaggerated religious positions, so they were inspired to rethink the philosophical basis of morality. Darwin did this in a primitive fashion in the *Descent*, but long before then, before even the *Origin*, Herbert Spencer had set down the basic principles of evolutionary ethics. It was Spencer's belief that by basing morality on evolution, he had created not merely a philosophy of ethics but a science of ethics. Moral conduct was defined as that which contributed to man's better adaptation and to his higher evolution. Since personal happiness was also the result of a satisfactory adaptation, morality and happiness were essentially one. Thus utilitarianism was amalgamated with evolutionism, the individual was reconciled with society, and hedonism became altruism under a different guise. The perfect moral man, "the completely adapted man in the completely evolved society," was one in whom there was a "correspondence between all the promptings of his nature and all the requirements of his life as carried on in society."[61] Essentially the same argument appeared later in Leslie Stephen's *Science of Ethics*. Again what was sought was the rooting of ethics in science; like Spencer, Stephen was concerned to provide a sanction for morality that was not dependent upon the discredited dogmas of religion, particularly of immortality. And again the test of individual conduct was found in the welfare of society, and the warrant of individual happiness in social utility. Morality, happiness, and the evolutionary process were assumed to be different aspects of the same thing.

There were many, and not only among the religious orthodox, who found this multiple identity both superficial and arbitrary. Henry Sidgwick warned Stephen that evolutionism and utilitarianism were not so easily reconciled, that the evolutionary movement of society did not necessarily promote the greatest happiness of the greatest number, still less the greatest happiness of any one individual. And many others, like Morley, were distressed by the deterministic implication of Darwinism. If morality is determined by the impersonal processes of nature, "what becomes of man's voluntary agency"?[62] What is moral behavior if not a voluntary act of will? What have science and nature to do with ethics? Even if Darwinism could explain the development of the moral sense, which was doubtful enough, this did not authorize it to prescribe any particular moral code. The ethical Darwinists, their critics said, had been guilty of the ancient fallacy of confusing description with prescription, of making facts—and dubious facts at that—do service for values.

What made matters worse was the apparently immoral character of nature revealed by evolution itself, hardly qualifying it to set itself up as a preceptor. No one any longer shared Paley's faith in the benevolence of nature: "It is a happy world after all. The air, the earth, the water teem with delighted existence. In a spring noon, or a summer evening, on whichever side I turn my eyes, myriads of happy beings crowd upon my view."[63] When Paley rhapsodized over the thoughtfulness of a God who devised that clever structure in the throat to prevent choking—"Consider a city-feast, what manducation, what deglutition, and yet not one Alderman choked in a century" —the reader was now apt to reflect that this was to look at the matter too exclusively from the point of view of the alderman, and not enough from that of the sacrificial sheep or lobster.[64]

Men did not need Darwin to convince them that nature was not a reliable guide to the conscience. Matthew Arnold had given them fair warning:

> Nature is cruel, man is sick of blood;
> . . .
> Nature is fickle, man hath need of rest.
> . . .
> Man must begin, know this, where Nature ends;
> Nature and man can never be fast friends.[65]

And Swinburne learned the same lesson, not from Darwin but from that "modern pagan philosopher," the Marquis de Sade:

> Nature averse to crime? I tell you, nature lives and breathes by it; hungers at all her pores for bloodshed, aches in all her nerves for the help of sin, yearns with all her heart for the furtherance of cruelty. . . . Unnatural is it? Good friend, it is by criminal things and deeds unnatural that nature works and moves and has her being. . . . If we would be at one with nature, let us continually do evil with all our might.[66]

But although the theme of nature's cruelty is an ancient one, and texts to that effect readily accessible, it was Darwin who was generally taken as the authority. What others had vaguely sensed, he was thought to have proved: cruelty was not only a fact of nature; it was the governing force of nature, the motivating power of life.

It was left to Huxley to explore the anomalies of an evolutionary ethics: an ethics that sought guidance from nature, and a nature that repelled the conscience. But this was the mature Huxley. The young Huxley, Darwin's bulldog, seemed content to follow the lines laid down by

Spencer. In 1860, at the most tragic time of his life, when his firstborn, four-year-old son suddenly died, he wrote a letter to Kingsley which was, in effect, a testimonial of the agnostic's faith. He explained that, comforting as the thought of immortality might be, he could not bring himself to believe what could not be proved. Fortunately, he went on, it was not a belief that was essential to his peace of mind. Most men accepted it because they assumed that without it the "moral government of the world" would be imperfect and practical morality ineffective. He himself saw no reason for these pessimistic assumptions:

> I am no optimist, but I have the firmest belief that the Divine Government (if we may use such a phrase to express the sum of the "customs of matter") is wholly just. The more I know intimately of the lives of other men (to say nothing of my own), the more obvious it is to me that the wicked does *not* flourish nor is the righteous punished. . . .
>
> The ledger of the Almighty is strictly kept, and every one of us has the balance of his operations paid over to him at the end of every minute of his existence. . . .
>
> The absolute justice of the system of things is as clear to me as any scientific fact. The gravitation of sin to sorrow is as certain as that of the earth to the sun, and more so—for experimental proof of the fact is within reach of us all—nay, is before us all in our own lives, if we had but the eyes to see it.[67]

This was the classical doctrine of evolutionary—and, indeed, of any naturalistic—ethics. The moral code originated not in a supernatural dispensation but in nature, and was enforced not by a supernatural judge with all eternity at his disposal but by man himself in the here and now. Like so many other Victorian moralists—Eliot, Hardy, or Meredith—who believed in retribution on earth all the more firmly for disbelieving in it in heaven, Huxley at this time insisted upon the "very unmistakeable *present* state of rewards and punishments."[68] As late as 1886, while discoursing on "science and morals," he reaffirmed that the entire security of morals rested upon "a real and living belief in that fixed order of nature which sends social disorganisation upon the track of immorality, as surely as it sends physical disease after physical trespasses."[69]

What may have moved Huxley to reconsider the matter was his growing distrust of Spencerism and Positivism, his discovery that the religion of humanity and ethics of evolution were more immediate threats to the moral welfare of society than even the religion and ethics of the Church. Something more was required of ethics than the trinity of science, nature, and evolution. To be sure, he himself never admitted that it was a major intellectual revolution that he had embarked upon: by quoting his own remarks of thirty years earlier in support of some minor points, he encouraged the impression that he was only repeating what he had long argued.

The first intimation of his new attitude came in an article on "The Struggle for Existence in Human Society" in 1888, directed especially against Spencer. (Spencer was not mentioned by name, but Huxley's private letters frankly identify him as the enemy, and Spencer read it as a declaration of war.) It was a protest against the idea that nature, and particularly the struggle for existence, could or should be projected into society as a moral ideal. Nature, Huxley insisted, was neither conspicuously benevolent nor necessarily progressive. Even if it could be shown that in the long run the struggle made for good, this would still not absolve nature of the evils of suffering and death, the sufferings of one generation not being expunged by the satisfactions of a later generation. Moreover, the Darwinian doctrine did not entail that necessary progression from lower to higher forms which might be thought to justify suffering: retrogression was as likely a phase of evolution as progression. Before men ventured to take the struggle for existence as a model for society, he advised them to reflect upon the enormities perpetrated by nature: "The optimistic dogma, that this is the best of all possible worlds, will seem little better than a libel upon possibility. . . . From the point of view of the moralist, the animal world is on about the same level as a gladiator's show."[70]

Having once broached the subject, Huxley could not let it alone. With horrified fascination, he watched the case against nature grow out of his own testimony. In 1889 he took the occasion, in an article on "Agnosticism," to condemn both the Christian and the Comtian brands of religion for worshiping what was plainly unworthy of worship: a god who created evil, and a humanity that was itself evil. "I know no study," he announced, "which is so unutterably saddening as that of the evolution of humanity." And he proceeded to indict mankind for bearing the mark of the beast, his slow rise having been attended by "infinite wickedness, bloodshed, and misery," the best men in the best times being those who "make the fewest blunders and commit the fewest sins."[71]

By the time of his Romanes lecture in 1893, the case against nature was occupying his entire imagination. The title of the lecture was "Evolution and Ethics"; the theme, evolution versus ethics. Evolution was what it had always been for him: the cyclical movement of nature proceeding by struggle and selection, the "cosmic process" by which man and the world had arrived at their present state. Ethics, however, was something else again, for this cosmic process, so admirable a mechanism in the eyes of scientists, was profoundly deplorable in the eyes of the moralist. Not only did man and civilization advance by way of evil, but in their very advance this evil was perpetuated and accentuated. Man had acquired his leading position by virtue of his success in the struggle for existence; he excelled over the ape and the tiger in just those qualities that are commonly associated with these animals—the cunning, ruthlessness, and ferocity with which he proceeded against his enemies. As civilization developed, to be sure, these animal virtues became defects, and man would often have liked to get rid of the ape and tiger in him. But he could not so easily kick over the ladder by which he had ascended. Pain and grief, crime and sin remain to remind him of his origins. His highest qualities —memory, imagination, sympathy—serve to intensify his misery by enlarging his capacity for suffering and by providing him with new occasions for sin and pain. No thoughtful man, Huxley declared, could find in this cosmic process even the elementary concepts of justice and goodness. Certainly there was no justice in the animal world, the pleasures and pains of life having nothing to do with just deserts. And even in mankind, justice was flagrantly wanting:

> If there is a generalization from the facts of human life which has the assent of thoughtful men in every age and country, it is that the violator of ethical rules constantly escapes the punishment which he deserves; that the wicked flourishes like a green bay tree, while the righteous begs his bread; that the sins of the fathers are visited upon the children; that, in the realm of nature, ignorance is punished just as severely as wilful wrong; and that thousands upon thousands of innocent beings suffer for the crime, or the unintentional trespass of one.[72]

Brought before the tribunal of ethics, Huxley declared, the cosmos must stand condemned. Those who preach an "ethics of evolution" tend to confuse two things: a knowledge of how the good and the evil have come about, and a guide to what is good and evil, or why the good is preferable to the evil. In fact, there is no ethics of evolution, the cosmic process furnishing no guide for morality save the negative guide:

> Social progress means a checking of the cosmic process at every step and the substitution for it of another, which may be called the ethical process; the end of which is not the survival of those who may happen to be the fittest . . . but of those who are ethically the best. . . . The ethical progress of society depends, not on imitating the cosmic process, still less in running away from it, but in combating it.[73]

With this audacious proposal, that the microcosm pit itself against the macrocosm, that man set himself to subduing nature, Huxley concluded his lecture and rested his case against the ethical Darwinists.

When Huxley confided to one correspondent that his lecture was an "egg dance,"[74] he did not mean, as might be thought, that he was trying not to offend the evolutionists. He was referring to the conditions of the lecture, which forbade comments on religion or politics. It was, indeed, a most skillful egg dance, because Huxley's main intent was political. Having exerted himself not to mention politics in the lecture itself—except for one lapse about the "fanatical individualism" that would pattern society upon nature and make the individual heedless of his obligations to others[75]—he could not resist adding a prolegomenon to the published version of the lecture, making his political purpose explicit. Here it emerged that what he feared was not only the individualism of Spencer, which would leave men to the mercies of an unrestricted struggle for existence, but also the regimentation of the Comtists and eugenicists, who would try to enforce upon society the notions of fit and unfit derived from nature. If it was dangerous to forbid to intelligence any part in the organization of society, it was equally dangerous to assume that any one man or group of men could have so preternatural an intelligence as to enable them to determine the "points" of a good or bad citizen, in the way breeders judge the points of a calf; a presumed "scientific" administration of society would be as intolerable a tyranny as any yet known. The lecture was thus an attack upon those who, from opposite directions, sought to intrude science into the alien territory of society and morality.

Huxley was, for the moment, so preoccupied with the enemy within, the misguided scientists, that he almost gave the game away to the enemy without, the theologians. Only a few years earlier it was the theologians he had been busy denouncing, in a long and often hilarious polemic with Gladstone on the subject of the Gadarene swine. Gladstone had defended the Biblical miracles as moral allegories, to which Huxley had rejoined that these allegories were as questionable in morality as in history; that Jesus should have caused the demons to depart from the soul of a man and enter into a herd of swine was a "wanton destruction of other people's property," a "misdemeanour of evil example." And more serious than the injustice it inflicted on either the swine or their keepers was the "monstrous and mischievous" picture it gave of a world possessed by demons.[76] Yet, once Huxley turned his attention from the misdeeds of the theologians to those of the scientists, he found himself entertaining a conception of nature that was, if not demonic, at least very nearly Satanic. He himself described the Romanes lecture as an orthodox discourse on the text, "Satan, the Prince of this world," and he declared the superiority of the best theologians over most of their opponents to be their recognition of the realities of evil:

> The doctrines of predestination, of original sin, of the innate depravity of man and the evil fate of the greater part of the race, of the primacy of Satan in this world, of the essential vileness of matter, of malevolent Demiurgus subordinate to a benevolent Almighty, who has only lately revealed himself, faulty as they are, appear to me to be vastly nearer the truth than the "liberal" popular illusions that babies are all born good, and that the example of a corrupt society is responsible for their failure to remain so; that it is given to everybody to reach the ethical ideal if he will only try; that all partial evil is universal good, and other optimistic figments . . .[77]

The religious impulse in Huxley has often been remarked upon. What has not often been observed is that this impulse revealed itself less in his making of science a religion, or his habit of conducting scientific disputes with religious fervor, than in his basically religious, metaphysical cast of mind. As he belatedly came to see, his dualistic conception of ethics, of morality at war with the cosmos, is more akin to the conventional religious view than to that of the naturalists and materialists with whom he is commonly associated.

If the Darwinians could not agree upon a theory of ethical Darwinism, they could agree upon matters of practical morality. This, to be sure, was true of most eminent

Victorians. Whatever their scientific beliefs or religious convictions, whatever their characters and personal behavior, their moral sentiments were identical. The Darwinians are singled out here only because they, and others, often spoke as if their Darwinism should make a difference. In fact, it made no difference.

When Darwin himself was challenged to state his principles, he declared that while the subject of God was "beyond the scope of man's intellect," his moral obligation was clear: "man can do his duty."[78] That there was such a thing as duty which all right-thinking, right-feeling men were bound to respect was insisted upon by no one so fervently as the self-confessed agnostic. "I now believe in nothing," Leslie Stephen declared, "but I do not the less believe in morality." And those who could not agree with his *Science of Ethics* could happily endorse his intention: "I mean to live and die like a gentleman if possible."[79] Neither her professed disbelief nor the anomalies of her personal situation could shake George Eliot in her faith in "a binding belief or spiritual law, which is to lift us into willing obedience, and save us from the slavery of unrequited passion or impulse."[80] God might be inconceivable, immortality unbelievable, but duty was no less "peremptory and absolute."[81] And Frederic Harrison, arch-agnostic and high priest of English positivism, when asked by his son what a man who falls in love but cannot marry is to do, replied indignantly: "Do! Do what every gentleman does in such circumstances. Do what your religion teaches you. Do what morality prescribes as right." When his son persisted in wanting to know why love was proper only in marriage, Harrison could barely contain himself: "A loose man is a foul man. He is anti-social. He is a beast." He finally put an end to the matter: "It is not a subject that decent men do discuss."[82]

Neither science nor agnosticism could dampen such resolutely moral souls as these. Not science, because it could always be taken to confirm their moral prejudices. Prior to Darwin and independently of Spencer, Eliot was urging that truths of all varieties were revealed progressively by the unfolding of human history and development, and that the only principle governing them was the "undeviating law in the material and moral world . . . that invariability of sequence which is acknowledged to be the basis of physical science, but which is still perversely ignored in our social organization, our ethics, and our religion"; it was this undeviating law that enforced morality by speedily visiting retribution upon those who dared to defy it.[83] If anything, the agnostics were all the more solicitous of morality because of the need to redeem unbelief from the charge of immorality. Renan once explained: "I deliberately imposed upon myself the morals of a Protestant clergyman. A man should never take two liberties with popular prejudices at the same time. The free thinker should be very particular as to his morals."[84]

Indeed, so solicitous of morality were the Victorian agnostics that they were even willing to make concessions to religion in the interests of public morality. They were willing to suspend their own disbelief in order to bolster up other people's morals—not their own, for of their own they had no doubt. Thus Huxley agreed to the christening of his son and wrote polemics in favor of Bible reading and religious instruction in the schools. Hooker, who privately regarded theology as idle speculation, refused to permit heretical ideas to be discussed before women or children and approved of a national church on the grounds that it would inspire the young with respect for "knowledge, truth and pure living."[85] Leslie Stephen, shortly before his death, praised the evolutionary movement for having reawakened men to an appreciation of the social importance of the church, and for having discredited "the old 'negative' criticism which regarded creeds as simply false and churches as organised impostures."[86] Spencer, also in the ripeness of age, discovered that there was more to the question of religion than the truth or untruth of particular doctrines; against his will, as he said, he was forced to the conclusion that "the control exercised over men's conduct by theological beliefs and priestly agency, has been indispensable."[87] The most famous "apostate," of course, was Mill, whose last essay conceded that a familiarity, even an imagined familiarity, with a morally perfect being was bound to be a more effective guide to morality than any merely rational ideal.

There was little in ethical Darwinism to survive the exuberance of youth and first impressions. Upon maturity, the promise of a new dispensation evaporated, leaving the old homely truths in undisputed control. In practical morals and beliefs the post-Darwinian world was not so different from the pre-Darwinian one, or the late-Darwinian from the early one. Yet there were differences—not in what men did but how they thought, and not even so much in their principles as in the nuances of tone and feeling that lay behind those principles. The *status quo ante bellum* was never quite restored. Having learned to think of man as a creation of nature, the post-Darwinian could not entirely overcome a sense of presumptuousness in setting himself up against nature, in repudiating its values and denying its moral example. Similarly, having been encouraged to hope that nature had finally given up its secrets, that metaphysics, like ethics, had become the province of science, he could not conceal his humiliation at having to confess that the evolutionary process itself culminated in what Spencer called the "Unknowable" and Wallace the "Unknown Reality." His rediscovery of religion, or his invention of a quasi-religion, was not the exultant affirmation of a newly found truth but the abject admission of failure. It was an awakening from the Darwinian dream in which man had only to probe nature to elicit its metaphysical truths, and only to submit to nature to solve his moral problems.

CHAPTER 19

DARWINISM, POLITICS, AND SOCIETY

EVEN those who are entirely convinced of the validity of Darwin's scientific doctrines may be wary of their extension to political or social theory. They may feel about "social Darwinists" as Newman felt about secularists: "They persuade the world of what is false by urging upon it what is true."[1] More than most theories, Darwinism lent itself to such stratagems of persuasion, enjoying not only the prestige and authority attached to science but also the faculty of being readily translated into social terms. That this translation was necessarily rather free and loose was an added advantage, since it gave license to a variety of social gospels.

The duties of godfather to the new offspring pressed heavily upon Darwin, who never understood how this relationship had come about or what was expected of him in his new role. When a Manchester newspaper indignantly protested that the *Origin* had proved might to be right, thus justifying Napoleon and "every cheating tradesman," Darwin found it too ludicrous to take seriously.[2] Ten years later he wrote to one of the many German sociologists and economists who were busily construing the social implications of his theory, that it had not earlier occurred to him that his views on species were relevant to social questions. And when Wallace, himself an ardent social Darwinian, urged upon him the merits of Henry George, he replied that he had once read some books on political economy with "disastrous effect": he had come to distrust his own judgment and to doubt everyone else's.[3] Politically, he regarded himself as a "Liberal or Radical"[4] and professed great admiration for Mill and Gladstone. But he was so little concerned with politics that there is hardly any mention of either of these men in his *Life and Letters*, or of any contemporary affair except for an occasional allusion to the Franco-Prussian War. (Like many Liberals, including Mill, he favored the Germans, Bismarck appearing to be a lesser evil than Napoleon III.)

The one issue upon which he felt passionately was slavery. He was so ardent a supporter of the North in the American Civil War that he had no qualms at all about the shedding of blood: "In the long run, a million horrid deaths would be amply repaid in the cause of humanity" —upon which gory prospect he curiously apostrophized, "What wonderful times we live in!"[5] And when the *Times* persisted in its pro-Southern policy, he canceled his subscription, resigning himself to a boring after-lunch session with the *Daily News;* this was, as he pointed out, a far greater sacrifice for him than for his wife, who did not so much mind the *News*. Several months before the end of the war, when the *Times* (presumably as reported in the *News*) was predicting the joining of the Middle States with the South to eject the North, Darwin hoped that in that event the North would "marry Canada, and divorce England, and make a grand country, counter-balancing the devilish South."[6]

When the war ended with the victory of the North and without these matrimonial adventures, Darwin's moral indignation focused upon another disagreeable situation: the suppression by Governor Eyre, with great cruelty and vindictiveness, of a Negro revolt in Jamaica. Darwin promptly subscribed to the committee organized by John Stuart Mill for the prosecution of Eyre. When his son casually suggested that the funds had not been used entirely wisely, Darwin was so indignant that he brusquely rebuked him, thereupon passing a sleepless night until he could beg his forgiveness in the morning.

However passionate his feelings about slavery and the treatment of Negroes (the Jamaican rebels were not slaves), Darwin never pretended that they were related to his scientific views. Had he been tempted to do so, he would have been recalled to the truth by the fact that Hooker and Tyndall were as staunch in Eyre's support as he and Huxley were in his condemnation. Even Fitz-Roy, for whom all differences of opinion constituted a diabolical conspiracy, and who claimed to have had intimations of the Darwinian heresy as early as the voyage of the Beagle, did not claim that Darwin's offensive views on species were responsible for his equally offensive views on slavery. Evolutionists, it was obvious, might disagree on everything except evolution itself.

Yet there were deductions that might be drawn from the *Origin* in respect of race, and on the eve of the Civil War it was inevitable that they would be drawn. The only difficulty was that they could be made to favor either side. The most obvious deduction was the anti-racist one: the theory of evolution, by denying the separateness of varieties and species, also denied the separateness and thus the intrinsic inferiority or superiority of races. Slavery was therefore not an inviolable condition of nature but an ephemeral condition of history; by nature all men were brothers. When Asa Gray first read the *Origin*, he could not help wincing at the idea of being made kin to the Hottentot and Negro; a Northerner and abolitionist, however, he managed to suppress such unworthy thoughts. Others, starting from the unshakable conviction that Negroes and whites could not have come from a common stock, found this a decisive argument against the *Origin*. (It was also an argument against the Biblical theory of creation, but few men pursued the matter so far.) In strict logic, there is no reason why this reading of the *Origin* should have had these consequences: why an evolutionist could not have argued that history, if not nature, had made of the Negro an inferior race fit only for slavery; while the anti-evolutionist argued that whatever the facts of nature, morality and political expediency dictated a policy of abolition. Unfortunately, few men

had the firmness of mind to resist the authority of nature
and science, and to take their stand on purely moral or
political grounds.

It was not, however, necessary to confute the *Origin*
in order to justify the South. It was only necessary to
re-interpret it. For there were features of Darwin's theory
that could easily give comfort to the proponents of slavery
and racism. Although Darwin derived all races, like all
species, from a single historic ancestor, he by no means
denied the reality of separate races and species in the
present. He did not dissolve all species into an undistin-
guished mass of individuals; he did not even suggest, as
anti-racist theorists often do, that individuals constitute a
spectrum in which each differs from his neighbor so slightly
that only artificially, statistically, can varieties, races, or
species be distinguished. Indeed, the purpose of his doctrine was precisely to account for the reality of species,
to explain not only how species evolved but also how they
became stabilized and fixed in form, sometimes for very
long periods (sometimes—and it was one of Darwin's main
tasks to account for this—for all of recorded history). Nor
did he deny that under certain conditions it was desirable
to maintain, as far as possible, the purity of races. The
Origin did declare that crosses between varieties tended to
increase the number, size, and vigor of the offspring. But
this was true only in special cases: where, for example,
the crossed varieties had previously been exposed to fluctuating conditions and thus were especially hardy. Otherwise, such a cross might prove fatal to both varieties.

It was this argument against the crossing of races that
first impressed itself upon some of the readers of the *Origin*.
One month after its publication, on the occasion of John
Brown's raid at Harpers Ferry, the *Times* gave warning
that the abolitionists would turn the population of the
South into a "mixed race." The lesson of modern times, it
said, was that such a mixture of races "tends not to the
elevation of the black, but to the degradation of the white
man."[7] Reading this, a secretary at the American legation
in London observed: "This is bold doctrine for an English
journal and is one of the results of reflection on mixed

races, aided by light from Mr. Darwin's book, and his theory of 'Natural Selection.' "[8]

The subtitle of the *Origin* also made a convenient motto for racists: "The Preservation of Favoured Races in the Struggle for Life." Darwin, of course, took "races" to mean varieties or species; but it was no violation of his meaning to extend it to human races, these being as much subject to the struggle for existence and survival of the fittest as plant and animal varieties. Darwin himself, in spite of his aversion to slavery, was not averse to the idea that some races were more fit than others, and that this fitness was demonstrated in human history:

> I could show fight on natural selection having done and doing more for the progress of civilization than you seem inclined to admit. Remember what risk the nations of Europe ran, not so many centuries ago of being overwhelmed by the Turks, and how ridiculous such an idea now is! The more civilized so-called Caucasian races have beaten the Turkish hollow in the struggle for existence. Looking to the world at no very distant date, what an endless number of the lower races will have been eliminated by the higher civilized races throughout the world.[9]

From the "preservation of favoured races in the struggle for life," it was a short step to the preservation of favored individuals, classes, or nations—and from their preservation to their glorification. Social Darwinism has often been understood in this sense: as a philosophy exalting competition, power, and violence over convention, ethics, and religion. Thus it has become a portmanteau of nationalism, imperialism, militarism, and dictatorship, of the cults of the hero, the superman, and the master race. The hero or superman, most recently translated as *Führer*, is assumed to be the epitome of the fittest, the best specimen of his breed, the natural ruler who exercises his rule by right of might. As he is the instrument of providence to lead his nation to victory, so the nation is the instrument that will raise civilization to a more sublime state. And as he made his way by struggle and force, so the nation must make its way in the world by war and conquest. A German general has given the classical expression to this glorification of struggle:

> War is not merely a necessary element in the life of nations but an indispensable factor of culture, in which a truly civilized nation finds the highest expression of strength and vitality. . . . War gives a biologically just decision, since its decisions rest on the very nature of things. . . . It is not only a biological law, but a moral obligation, and, as such, an indispensable factor in civilization.[10]

For the general, it was the needs of war that came first, the imperialist adventures and nationalist experiments that followed. For others it was the reverse: the imperialist and nationalist aspirations brought war and militarism in their wake. There were even some who would have liked the virtues of war without the onus of militarism or nationalism; this was social Darwinism in its purest, most disinterested form. Sir Arthur Keith, the anthropologist, evolutionist, and biographer of Darwin, confessed that although he personally liked the thought of peace, he feared the results of such an experiment. At the end of five centuries of peace, he predicted, the world would look like "an orchard that has not known the pruning hook for many an autumn and has rioted in unchecked overgrowth for endless years." He was no champion of war, he protested, but he could not conceive of any substitute that would serve as well "for the real health of humanity and the building of stronger races."[11]

Recent expressions of this philosophy, such as *Mein Kampf*, are, unhappily, too familiar to require exposition here. And it is by an obvious process of analogy and deduction that they are said to derive from Darwinism. Nietzsche predicted that this would be the consequence if the Darwinian theory gained general acceptance:

> If the doctrines of sovereign Becoming, of the liquidity of all . . . species, of the lack of any cardinal distinction between man and animal—doctrines which

I consider true but deadly—are hurled into the people for another generation . . . then nobody should be surprised when . . . brotherhoods with the aim of robbery and exploitation of the non-brothers . . . will appear on the arena of the future.[12]

Perhaps the most common understanding of social Darwinism is as a philosophy of extreme individualism, of laissez-faire in economics and government. If this seems far removed from the nationalism and authoritarianism that also go by the name of "social Darwinism," it has no less right to that title.

It was not only the theory itself—the struggle for existence and survival of the fittest—that suggested the idea of laissez-faire, but also the genesis of that theory. From Malthus to Darwin and back to a Malthusian Darwinism: the system seemed to be self-sufficient and self-confirming. The theory of natural selection, it is said, could only have originated in England, because only laissez-faire England provided the atomistic, egotistic mentality necessary to its conception. Only there could Darwin have blandly assumed that the basic unit was the individual, the basic instinct self-interest, and the basic activity struggle. Spengler, describing the *Origin* as "the application of economics to biology," said that it reeked of the atmosphere of the English factory.[13] Nietzsche was similarly repelled by it: "Over the whole of English Darwinism there hovers something of the suffocating air of over-crowded England, something of the odour of humble people in need and in straits."[14] And commentators since then, although as often as not without any Nietzschean or Spenglerian bias, have repeated the refrain: natural selection arose and throve in England because it was a perfect expression of Victorian "greed-philosophy," of the capitalist ethic and Manchester economics.[15]

Social laissez-faire being an extension of economic laissez-faire, it was credited with the same Darwinian motives. In society, as in nature, there was presumed to be a "natural order" which, left alone, would insure the survival of the fittest. Any interference with that order, either to direct the organization of society or to protect special interests, would violate nature and enfeeble society. Darwin himself, bemoaning the future of humanity, once complained that "in our modern civilization, natural selection had no play, and the fittest did not survive."[16] It did not take the *Origin* to persuade men of the evils of social reform, of interfering with the natural competitive order. Malthus was persuaded of this, at the same time that he deprecated the idea of evolution. A theory of evolution did help, however, although it did not have to be Darwin's. Almost a decade before the *Origin*, Spencer protested against those misguided reformers who would shield men from the full rigors of the struggle for existence:

> The well-being of existing humanity, and the unfolding of it into this ultimate perfection, are both secured by that same beneficent, though severe discipline, to which the animate creation at large is subject; a discipline which is pitiless in the working out of good; a felicity-pursuing law which never swerves for the avoidance of partial and temporary suffering. The poverty of the incapable, the distresses that come upon the imprudent, the starvation of the idle, and those shoulderings aside of the weak by the strong, which leave so many "in shallows and in miseries," are the decrees of a large far-seeing benevolence.[17]

The leading American disciple of Spencer, William Graham Sumner, warned society that it faced the alternative of the survival of the fittest or the survival of the unfittest, the ill-conceived humanitarian efforts to alleviate poverty and preserve the weak having the effect of favoring the worst members of society and lowering the vitality of civilization. And John D. Rockefeller once regaled a Sunday-school audience with an account of how natural selection operated for the best interests of society as a whole. He compared the evolution of business with the evolution of a superior variety of rose: as the American Beauty could only have been produced "in the splendor and fragrance which bring cheer to its beholder," by sacrificing the buds which grow up around it, so the development of

a large business is "merely a survival of the fittest . . .
merely the working-out of a law of nature and a law of
God."[18]

Against this iron law of nature, there was no appeal. Certainly there was no appeal to such fictions as equality,
justice, or natural rights. As there were no such principles
in the jungle, so there were none in society. "There can be
no rights against Nature except to get out of her whatever
we can, which is only the fact of the struggle for existence
stated over again."[19] Society had to develop organically
out of its own inner compulsions, rather than artificially
out of the minds of reformers. There were no fixed species
in nature and, in the same way, no ideal values in society.
All was relative, historical, evolving. The only fixed point
of reference was liberty, the freedom to evolve in whatever way nature chose, the freedom of the fit to survive
and of the unfit to die.

What gave unity to these diverse and often contradictory
interpretations of social Darwinism was the idea of struggle, the unrelenting war to which nature and mankind
were eternally doomed. Yet it was possible to be diverted
from the competitive struggle of life today to the vision of
a non-competitive, pacific life in the future. Even Spencer
held out hope that the struggle for existence, so essential
to the present welfare of man and a necessary condition
for his evolution, would eventually be outmoded in consequence of that very evolution. As mankind adapted itself to the changing conditions of life, a new human nature
would develop. The military struggle would give way to
the milder industrial one, and society would find itself
encouraging the pacific, cooperative virtues in place of
the militant, competitive ones. Finally there would emerge
an ideal man, the epitome of perfection and happiness, in
complete harmony with his environment and unaffected
by struggle or competition. The American philosopher
John Fiske, predicted the same happy issue, the evolutionary processes of selection and adaptation serving ultimately
to bring about a condition in which the desires of each
individual would be in "proximate equilibrium" with the

desires of others and also with the means of satisfying them. When this happened, when selection and adaptation came to an end, and the ape and tiger in man were extinct, strife and sorrow would disappear, and the kingdom of this world would become the kingdom of Christ.[20]

It was this comforting reading of Darwinism, achieved by a judicious divorce of present and future, that made it possible for yet another creed to enter the amalgam of social Darwinism. Nothing, at first sight, could seem more remote from each other than the doctrines of laissez-faire and socialism. Yet socialists, too, have claimed descent from Darwin, and with some right.

When Marx read the *Origin*, he enthusiastically declared it to be "a basis in natural science for the class struggle in history."[21] Although his offer to dedicate *Das Kapital* to Darwin was refused Darwin politely acknowledged his gift of the book: "Though our studies have been so different, I believe that we both earnestly desire the extension of knowledge; and this, in the long run, is sure to add to the happiness of mankind."[22] If Darwin had not the least idea of what Marx was up to or what they might have in common, Marx knew precisely what he valued in Darwin. Recommending the *Origin* to Lassalle, he explained that "despite all deficiencies not only is the death-blow dealt here for the first time to teleology in the natural sciences, but their rational meaning is empirically examined."[23] The other reason for his interest in the *Origin* emerged in *Das Kapital*, where he complained of the abstract materialism of most natural science, "a materialism that excludes history and its process."[24] It was his hope that by focusing attention on change and development, the *Origin* would destroy both the old-fashioned supernaturalism and the equally old-fashioned materialism.

Yet there were obvious "deficencies" in the *Origin*, from the point of view of Marxism, which even Marx did not fully appreciate. For the socialist, the struggle was primarily between classes, with "solidarity" prevailing within each class. For Darwin the basic struggle took place within each species (the species being the counterpart to class). It was here, where Spencer and Marx diverged, that Darwinism may be said to favor Spencer. Another point of disagreement was Marx's assumption that the struggle for existence would be suspended or transcended to permit the emergence of the classless society. Here Marx and Spencer were at one, agreeing that the struggle would eventually become obsolete through the dialectic of change and adaptation. The struggle for existence that Darwin took to be a permanent condition of animal life, Marx saw as a condition only of particular epochs in human history. One such epoch was the bourgeois one, where it might fittingly be said that the social organization was that of the animals. Recalling Hegel's description of bourgeois society as a *geistiges Tierreich*, he and Engels amused themselves with the thought that what English liberals took to be the greatest social achievement of their country should have been revealed to be the natural state of the jungle.[25]

A more serious discrepancy between Darwinism and Marxism was expressed in Marx's contempt for "the crude English method of development."[26] Between evolution and revolution there would seem to be no possible accord. Yet even here Marxists, intent upon retaining the authority of Darwinism, found ground for reconciliation. Plekhanov, defending Marxism against those simple-minded souls who thought that the doctrine of evolution meant *natura non facit saltus* and that revolution and evolution were therefore mutually exclusive, quoted Hegel to show how the dialectic resolved this antinomy, as it did so many others. The relationship between evolution and revolution was identical with that between quantity and quality; as imperceptible increases of quantity could produce a sudden transformation of quality, so the imperceptible movement of evolution could, at the critical historical moment, erupt into revolution. Later Marxists found a more plausible way out of the difficulty in the concept of mutations, which seemed to them to re-introduce those leaps in nature that were analogous to revolutions in history.

Even without these later emendations, however, there is an important sense in which Marx and Darwin alike were evolutionists. There was truth in Engels' eulogy on Marx: "Just as Darwin discovered the law of evolution in organic nature, so Marx discovered the law of evolution in human history."[27] What they both celebrated was the internal rhythm and course of life, the one the life of nature, the other of society, that proceeded by fixed laws, undistracted by the will of God or men. There were no catastrophes in history as there were none in nature. There were no inexplicable acts, no violations of the natural order. God was as powerless as individual men to interfere with the internal, self-adjusting dialectic of change and development.

As their philosophical intent was similar, so was their practical effect. Against all those, socialists and others, who thought that men were moved by a basic harmony of interests, and that history advanced by making that harmony conscious and effective, Darwin and Marx insisted upon the basic fact of struggle and upon progress as its result. It was on this ground that Marxism had a legitimate claim to the title of "social Darwinism." And it was for this reason that some socialists, to Engels' horror, joined together Darwin, Spencer, and Marx as the trinity that would bring salvation to mankind.

While the "scientific socialists" sought to amalgamate Darwinism and Marxism into one irresistible system, other socialists hastened to dissociate themselves from the odious ideas of the struggle for existence and the survival of the fittest. Not struggle but cooperation was the spirit in which these socialists conceived their ideal—and not only their ideal but their means as well. They were happy to endorse Darwin's sentiment: "What a foolish idea seems to prevail in Germany on the connection between Socialism and Evolution, through Natural Selection."[28]

Natural selection, dubious enough in nature, seemed to them still more dubious in society. What, they asked, is the significance of such concepts as struggle for existence and survival of the fittest in a complex civilization where "existence" and "fittest" are not physiologically or biologically determined, but only socially determined? If society chooses to favor the rich, the well-born, or the beautiful, what is there about them to qualify them as the fittest. And what is there about the struggle for such values as these to make men submit to it as the right and proper mode of social conduct? As William James observed: "The entire modern deification of survival *per se,* survival returning to itself, survival naked and abstract, with the denial of any substantive excellence in *what* survives, except the capacity for more survival still, is surely the strangest intellectual stopping place ever proposed by one man to another."[29]

In fact, they argued, both the quality of the struggle and the values produced by it varied in time and place, some societies being so pacific in spirit as to discourage struggle itself. Thus not the strong and aggressive but the weak and diffident might find themselves favored. And this was not the anthropological rarity it was commonly supposed to be. From the meanest animal horde to the most exalted human society, there was evidence that cooperation was as much a feature of social behavior as competition or struggle. Kropotkin devoted an entire book to the evidence of "mutual aid" among animals, concluding that such aid was not only conspicuous among members of the same species but that it was itself an important factor in the evolution of species. Indeed, had the struggle for existence been as universal and unremitting as Darwin thought, it would have resulted in the degeneration rather than advance of the race: "survival values" being as arbitrary and artificial as society itself, the physically weak and infirm would have been as likely to survive as the strong and able. When Julian Huxley recently condemned wars on the ground that "intra-specific competition is often not merely useless but harmful to the species as a whole,"[30] he failed to add that it was intra-specific competition that was particularly encouraged in natural selection.

A more fundamental objection to Darwinism from the point of view of these socialists, although one which they themselves were not properly aware of, was their distrust of that impersonal, self-regulating process of adjustment implied in natural selection. As reformers, they could not be content to let the organism adapt itself, in its own good time, to its environment; if they could not themselves hasten this adaptation, at least they could busy themselves adapting the environment to the organism. In this sense, the very idea of reform is antithetical to Darwinism. The

Italian socialist Enrico Ferri, whose aim it was to base socialism upon "positive science," asserted that "socialism is naught but Darwinism economised, made definite, become an intellectual policy, applied to the conditions of human society"[31]—not realizing that in that process the essence of Darwinism was distorted beyond recognition. Even John Dewey, who prided himself on applying the evolutionary principle to ethics and politics as much as to metaphysics and epistemology, was guilty of this basic inversion of Darwinism. Having earlier objected to "evolutionary ethics" on the curious grounds that by banishing God from "the heart of things," it had also banished the ethical ideal from the life of man—"whatever exiles theology makes ethics and expatriate"[32]—he later accepted it for the equally curious reason that, in showing the origin and development of man in his environment, it showed intelligence how to fashion that environment for man's greater satisfaction. Not the unintended effects of unconscious struggle, which was the point of Darwinism, but the deliberate effect of intelligent control was Dewey's prescription for society.

Similarly, Francis Galton, Darwin's cousin and great champion, who made it his mission, as he thought, to give practical content to Darwin's theory, was by this very enterprise denying that theory. The science of eugenics, devoted to the improvement of the human stock, was designed "to further the ends of evolution more rapidly and with less distress than if events were left to their own course."[33] To these ends, he proposed such measures as keeping a register of "superior" families for purposes of interbreeding. Darwin's only comment on this proposal was that although "the object seems a grand one" and, indeed, was the "sole feasible" one, yet he feared it was utopian, since few people could be counted on to cooperate intelligently.[34] It did not seem to have occurred to him that it vitiated his essential principle, making survival independent of the natural struggle for existence. In Galton's scheme little was left to chance. The end was predetermined, and what struggle there was was carefully controlled so as to produce the desired result. Reformers everywhere have sought, like Galton, to "expedite the

changes that are necessary to adapt circumstances to race
and race to circumstances."[35] The point of Darwinism is
that such "expedition" was not only unnecessary but dangerous, for it was the processes of nature, unaided and un-
hindered, that were the sole means and warrant of evolution.

It was Walter Bagehot who sought in Darwinism, more
conscientiously than anyone else and without ulterior motive or prior prejudice, the principles of society; and it was
he who revealed more fundamentally, if unwittingly, its
inadequacy as a social theory. *Physics and Politics*, which
first appeared serially as essays in the *Fortnightly* in 1867—
well after the *Origin*, but before the *Descent*—was subtitled
"Thoughts on the Application of the Principles of 'Natural
Selection' and 'Inheritance' to Political Society." The subtitle has persuaded several generations of readers that the
book is an authoritative statement of social Darwinism, an
impression re-inforced by such unimpeachable Darwinian
sentiments as:

> In every particular state of the world, those nations
> which are strongest tend to prevail over the others;
> and in certain marked peculiarities the strongest tend
> to be the best.
> But why is one nation stronger than another? In the
> answer to that lies the key to the principal progress of
> early civilisation, and to some of the progress of all
> civilisation. The answer is that there are very many
> advantages—some small and some great—every one of
> which tends to make the nation which has it superior
> to the nation which has it not; that many of these
> advantages can be imparted to subjugated races, or
> imitated by competing races; and that, though some
> of these advantages may be perishable or inimitable,
> yet, on the whole, the energy of civilisation grows by
> the coalescence of strengths and by the competition
> of strengths.[36]

Yet, in spite of the apparent intention of the author and of
the generally accepted reading, Bagehot's work comes close
to being a travesty of Darwinism.

The first suspicious circumstance is that a work intended to do for the origin of society what Darwin did for the origin of species should so readily confess its inability to account for that origin in terms of natural selection. How the initial change from "no polity" to "polity" came about, Bagehot did not even venture to guess. The suggestion of Maine, that political society originated in the family, did not particularly recommend itself to him, although in the absence of any other, he was prepared to accept it as a working theory. What he was convinced of, however—and this was the main point of his book—was that the great need of primitive man and the essential prerequisite of progress was law: "rigid, definite, concise law."[37] It did not matter what the content of that law was, so long as it was binding. The quantity of government was more important than the quality, and while a good law might be better than a bad, any law was so infinitely preferable to none that the distinctions between good and bad were relatively unimportant. What was wanted was "a comprehensive rule binding men together, making them do much the same things, telling them what to expect of each other—fashioning them alike, and keeping them so."[38] The object was to create a "cake of custom," a "law of status."[39] For this purpose, a single absolute authority was necessary. Church and state, as they are now known, had to be one; there could be no division of power, no separation, let alone conflict, of authority, no differences of opinion. "That this regime forbids free thought is not an evil; or rather, though an evil, it is the necessary basis for the greatest good; it is necessary for making the mould of civilisation, and hardening the soft fibre of early man."[40]

It is difficult to conceive of any theory that would be more at variance with natural selection. The basic feature of natural selection—the struggle for existence out of which emerges the survival of the fittest—is precisely that which Bagehot was concerned to deny. In all the formative years of political society, according to his account, it was not competition between men and institutions that determined what was to survive, but rather the arbitrary imposition of a single law and authority. Competition was not only irrelevant to the early development of society; it was a positive evil, which had to be suppressed if society was to survive. Later, to be sure—and between, rather than within, political communities—natural selection played some part: that nation survived and dominated which was most coherent. Thus the cake of custom proved its worth in competition with other nations. But it is, after all, within, rather than between, nations that the development of society first took place and that its principles must be sought. Moreover, it is the formative period of society that is crucial in this theory—as it is the origin of species that was crucial for Darwin. And in this formative period natural selection had no part. Bagehot's invocation of an arbitrary law, arbitrarily imposed, is comparable to the invocation of an arbitrary act of creation to account for the origin of species.

His explanation of the development of national character is similarly inimical to Darwinism. All commentators on Bagehot quote his remark: "A national character is but the successful parish character."[41] But Bagehot never followed this up. Certainly he did not imply that it was in competition or struggle that the successful parish character proved itself. On the contrary, it was authority that he adduced as the leading agent in the formation of national character:

> I have only to show the efficacy of the tight early polity (so to speak) and the strict early law on the creation of corporate characters. These settled the predominant type, set up a sort of model, made a sort of *idol;* this was worshipped, copied, and observed, from all manner of mingled feelings, but most of all because it was the "thing to do," the then accepted form of human action. When once the predominant type was determined, the copying propensity of man did the rest.[42]

Men did not pick and choose among alternative types, selecting here and there the elements that gradually cohered into an accustomed way of life. On the contrary: they imitated what was set up before them as a model and worshiped it as an idol—an idol, preferably, whose features were fixed not only by custom but by law as well. In this imitative process there was no room for competing models. "Men imitate what is before their eyes, if it is before their eyes alone, but they do not imitate it if it is only one among many present things—one competitor among others, all of which are equal and some of which seem better."[43] To eliminate this undesirable competition, even the conflict between nations had to be suppressed. Foreign intercourse, the knowledge of alien idols, would have been a serious impediment to the fashioning of the national character. Ancient governments and classical philosophers, knowing this and wanting to "keep their type perfect,"[44] did not permit foreign influences to impinge upon it. They discouraged the mixture of races, they even discouraged commerce, lest the mingling of persons, ideas, and customs confuse and corrupt the people. Not through competition and mingling but through unity and isolation were nations formed. "All great nations have been prepared in privacy and in secret. They have been composed far away from all distraction."[45] Darwin too had assigned to isolation a role in the formation of species, but it was a secondary one, to account for the fixing, rather than evolving, of species. For Bagehot, isolation was the primary fact, accounting both for the evolution and the fixity of nations.

Even when nations finally emerged from their formative stage, when they were obliged to compete with each other, natural selection played only a minor role. The quality that determined their strength in war as in peace was their internal unity and coherence. And the most united and coherent nation was that with the tamest, the most docile people. Why a people cultivated for the virtues of tameness and docility should be expected to prevail over one cultivated for aggressiveness and adventurousness, Bagehot did not say. He did say, however, that having for so long eschewed conflict among themselves, they were reluctant to sanction even the mildest expression of civic differences. The bulk of his chapter on the "Use of Conflict" was devoted not to the role conflict actually plays in society but to the role it does not play:

The great difficulty which history records is not that of the first step, but that of the second step. What is most evident is not the difficulty of getting a fixed law, but getting out of a fixed law; not of cementing . . . a cake of custom, but of breaking the cake of custom; not of making the first preservative habit, but of breaking through it, and reaching something better.[46]

The cake of custom, having rendered men tame, sociable, and cooperative, also rendered them unfit for conflict and change.

Not only conflict but variation itself, Bagehot pointed out, is hateful to civilized men. It is hated equally by those whose interest it is to be conservative and by those who purport to be revolutionary. The revolutionary shares with the conservative a distrust of non-conformists—society's "monstrosities and anomalies"[47]—and differs from him only in wishing to replace the old despotism by a new. This is why progress in human history is so slow as to be denied by many philosophers. The advance of civilization depends upon the delicate balance of two elements: variability and legality. The amount of variation that can be tolerated at any one time is minute, and in its outward form it must bear as little the sign of variety as possible. "The doctrine of development means this—that in unavoidable changes men like the new doctrine which is most of a 'preservative addition' to their old doctrines."[48] Only within the narrowest bounds of a "selective conservatism" is progress possible. With both variation and conflict thus discouraged, natural selection has little room for play.

There was much truth in Bagehot's book, but little in the title or subtitle. In this respect, Darwin was as deceived as Bagehot. Pleased that Bagehot had undertaken to apply natural selection to political philosophy, Darwin quoted him approvingly in the *Descent*. This endorsement, however, serves less to confirm the principles of social Darwinism than to enhance the confusions and misapprehensions attending the subject.

In the spectrum of opinion that went under the name of social Darwinism almost every variety of belief was included. In Germany it was represented chiefly by democrats and socialists; in England by conservatives. It was appealed to by nationalists as an argument for a strong state, and by the proponents of laissez-faire as an argument for a weak state. It was condemned by some as an aristocratic doctrine designed to glorify power and greatness, and by others, like Nietzsche, as a middle-class doctrine appealing to the mediocre and submissive. Some socialists saw in it the scientific validation of their doctrine; others the negation of their moral and spiritual hopes. Militarists found in it the sanction of war and conquest, while pacifists saw the power of physical force transmuted into the power of intellectual and moral persuasion. Mill's doctrine was taken to be a sophisticated form of natural selection, in which the war of arms and might yielded to the war of words and ideas. Some complained because it exalted men to the level of supermen or gods; others because it degraded them to the status of animals. Political theorists read it as an assertion of the need for inequality in the social order corresponding to the inequality in nature, or alternatively as an egalitarian tract in which men as well as animals were in an undifferentiated state of equality. Bertrand Russell did not see how a resolute egalitarian could resist an argument in favor of "Votes to Oysters."[49]

Periodically, a work on the nineteenth century will come along to declare that Darwinism "placed all political and social problems in a new perspective."[50] The truth, however, is probably more with Herbert Spencer, who, having devoted his life to just this task of placing all political and social problems in the new evolutionary perspective, finally confessed: "The Doctrine of Evolution has not furnished guidance to the extent I had hoped. Most of the conclusions, drawn empirically, are such as right feelings, enlightened by cultivated intelligence, have already sufficed to establish."[51]

CHAPTER 20

THE CONSERVATIVE
REVOLUTION

THE *Origin* and *Descent* were the climax of Darwin's work, but they were not the end. In the eleven years remaining to him after the publication of the *Descent*, he continued to work, oblivious of the fact that anything else must be, for contemporaries as for history, anti-climactic. He himself, being as fond of little theories as of big ones, and not much troubled by the opinion of either contemporaries or history, continued in his customary routine of research and writing. If the titles of his later works—*The Expression of the Emotions in Man and Animals* (1872), *Insectivorous Plants* (1875), *The Effects of Cross and Self Fertilisation in the Vegetable Kingdom* (1876), *The Different Forms of Flowers on Plants of the Same Species* (1877), *The Power of Movement in Plants* (1880), and *The Formation of Vegetable Mould through the Action of Worms* (1881) —seem trivial compared with the momentous *Origin of Species* and *Descent of Man,* they were to him neither trivial nor any the less dear. And they testify not only to the fertility of his mind but also to the stability of his character, which remained unspoiled by acclaim and undistracted by the blandishments of success. There are not many cases of professionals having had so overwhelming a popular triumph, who did not succumb to the temptation of philosophizing and popularizing, of assuming the posture of an elder statesman. It is to Darwin's great credit that he was not for a moment deluded by his genius or beguiled by his ego, that he never exploited his fame or authority for extra-scientific purposes. While others used the prestige of Darwinism to promote their social or political views, Darwin himself forbore doing so—a forbearance made easier, to be sure, by indifference. He ended as he had started: a single-minded, hard-working scientist.

The Expression of the Emotions in Man and Animals had its beginnings as far back as 1838, in Darwin's first notebooks. The actual writing of it was begun in January 1871, two days after the last proofs of the *Descent* were sent off to the printers, and it had been so well thought out that the first draft of almost four hundred pages took only three months. The subject was the origin of the muscular contractions responsible for the characteristic emotional expressions. The research consisted of personal observation, particularly of his own children in their infancy and of the animals that came his way; correspondence with missionaries (to compare the expressions of savages with those of civilized men), supervisors of lunatic asylums (in the belief that the insane expressed themselves with fewer inhibitions than the sane), physicians, physiologists, friends, and whatever casual acquaintance could be drawn into the quest; the analysis of photographs, diagrams, and paintings (he deplored the fact that the great masters of painting and sculpture were too concerned with beauty to permit themselves to depict the more violent contortions); and the results of a questionnaire circulated among thirty-six selected correspondents to determine the exact correlation between particular muscular contractions and their emotional connotation.

The idea of such an inquiry, Darwin claimed, was meaningful only on the supposition of evolution, for so long as man and other animals were regarded as independent creations, there was no problem as to the "causes" of expression, their facial muscles being assumed to have been specially created to express their particular emotions. And his findings confirmed him in this supposition. He found that "movements which are serviceable in gratifying some desire, or in relieving some sensation, if often repeated, become so habitual that they are performed, whether or not of any service, whenever the same desire or sensation is felt, even in a very weak degree."[1] Thus expressions, like organs, were the product of the evolutionary experience of the race: having originated in some conscious and voluntary physical action, they have become habitual, hereditary, and instinctive.

The basic principle here was Lamarckian: not natural selection but adaptation and the inheritance of acquired characteristics were the governing factors. In specific details, however, Darwin was more eclectic, suggesting, for example, that the barking of dogs may have originated by imitation, "owing to dogs having lived so long in strict association with so loquacious an animal as man."[2] Other explanations of his were equally unsatisfactory. Confronted with the problem of the recognition of expressions—how it came about that particular expressions should be interpreted or responded to by the observer in particular and predictable ways—he proposed that just as the expressions themselves are instinctive, so the responses are. He was gratified to notice that his infant son responded to his smile with pleasure and to his nurse's feigned tears with grief, and he was only momentarily disconcerted when the child was amused by what he intended to be savage grimaces; he accounted for this by saying that he must have unwittingly smiled before grimacing. The amateurishness of these experiments and the lack of rigor in evaluating them was accentuated by the notes added by his son to the posthumous edition, in which reasonable objections were dismissed out of hand, and confirmation of the thesis was sought in George Eliot's novels.

Each of these later works was a footnote to the *Origin;* each was designed to show the utility of a particular structure or organism, thus qualifying it for a part in evolution. Feeling old and weak, and soberly reminding himself that "no man can tell when his intellectual powers begin to fail,"[3] he plodded on, with hardly a pause between books. One does not know whether to be more impressed by Darwin's perseverance or by his readers' loyalty. The volume on earthworms, his last work and an unpromising subject, one would think, for popular consumption, sold 8500 copies in its first three years. For Darwin it was the completion of a project started as early as 1837, when he read his first paper before the Geological Society. One can only admire the steadfastness of character that permitted him to return to the modest occupation of his youth after having enjoyed the laurels of maturity, and at the prescient self-knowledge which made him say, long before the *Origin:* "My life goes on like clockwork, and I am fixed on the spot where I shall end it."[4]

His mind and character remained the same—all the more remarkably so for the public distinctions that came his way. He was granted honorary membership in the scientific societies and academies of over a dozen countries. He received medals, prizes, honorary degrees, and an Order of Merit from the Prussian Government. He was besieged with requests for portraits and audiences from the great; he found the latter so tiring that he generally pleaded incapacity, as he did when the Emperor of Brazil proposed to visit him. Gladstone, however, could not be put off and subjected Darwin to a long harangue on Bulgarian atrocities; Darwin later marveled that such a great man should condescend to speak so naturally. These honors did not, to be sure, always proceed smoothly. His nomination to the French Academy was twice rejected before his election in 1878, and then it was only by a vote of twenty-six to fourteen—and not to the Zoological Section, as might have been expected, but to the Botanical Section; in the course of the debate he was charged with the typical vices of his nation: dilettantism, eclecticism, and empiricism. A year before, when Cambridge University broke its rule of never awarding honorary degrees to its own graduates, Huxley, replying on Darwin's behalf to the toast at the dinner following the ceremony (Darwin was dining quietly elsewhere with his wife), introduced a sour note in the festivities when he caustically congratulated the university upon the timing of the award: "Instead of offering her honours when they ran a chance of being crushed beneath the accumulated marks of approbation of the whole civilised world, the University has waited until the trophy was finished, and has crowned the edifice with the delicate wreath of academic appreciation."[5]

The books and monographs kept pace with the honors, and only occasionally was Darwin diverted from such occupations as the study of "The Parasitic Habits of Molothrus."[6] The most agreeable diversion was provided by the opportunity to help his friends: it was Darwin who was responsible for the granting of a government pension to Wallace, and for a subscription to permit Huxley to rest after a particularly serious illness. A less agreeable demand on his conscience was the duty to intervene in a public issue of great concern to science. This was the antivivisection campaign that had been building up in England until in 1875 it became the subject of bills in both houses of Parliament and of a Royal Commission inquiry. Darwin, who was so sensitive to cruelty that any intimation of it could bring on a bout of faintness and sleeplessness, was nevertheless convinced that physiology, one of the noblest and most valuable of sciences, could only progress by experiments on living animals. Testifying before the Royal Commission, he said that he could not endorse any of the proposals for its control: not the proposal to limit it to public laboratories, for this would unjustly restrict research to those living in a few large cities; nor the proposal to license private practitioners, for it was precisely the young and unknown, who might fail to receive licenses, who should be most encouraged; nor the proposal that animals be rendered insensible, for some experiments made this impractical. The only hope, he concluded, lay with the growth of humanitarian feeling. Although he was dissatisfied with the act that was finally passed, he was pleased with the report of the Royal Commission suggesting that cruelty was more commonly to be found among foreign scientists than among the English.

A far more disagreeable claim upon his attention was the one-sided feud instigated by Samuel Butler. In his youth Butler had been an enthusiastic admirer of Darwin, and even his two early works, *Darwin among the Machines* and *Erewhon,* which have been commonly taken to be satires on Darwinism, were, in fact, Butler's way of expressing appreciation by stretching an idea to its imaginative limits. In 1876, however, when he was working on his *Life and Habit,* he discovered in Lamarck the attractions of a theory that was at the same time voluntaristic and mechanistic. And perhaps because he could not think of ideas without personalizing them, or of differences of opinion as anything but conspiracies, he also thought he discovered a systematic effort on Darwin's part to depreciate the achievements of his predecessors. In *Life and Habit,* and a few years later in *Evolution Old and New,* he tried to redress this grievance by exposing Darwin as a fool and a knave who was wrong when he differed from his predecessors and guilty of plagiarism when he agreed with them; sometimes Butler managed to imply that even his errors were plagiarized, and this too was presumed to redound to Lamarck's credit and to Darwin's discredit.

With the publication in 1879 of a biography of Erasmus Darwin, Butler became still more venomous. The biography had originated in the form of an essay written by Ernst Krause for the special issue of the German evolutionary journal, *Kosmos,* commemorating Darwin's seventieth birthday. The essay, on Erasmus Darwin as a scientist, appealed to both Charles and his brother as suitable for English translation and publication—largely, apparently, because Charles wanted to take the occasion to append a life of his grandfather which would give the lie to the calumnies of Miss Seward, her unflattering biography being still the only one available. Later that year *Erasmus Darwin* appeared, incorporating the Krause article and a long biographical "preliminary notice" by Darwin, which was well over half the book.

What Darwin had not expected was that the book, rather than restoring the family reputation (the calumnies were less easily disproved than he had expected), would involve him in fresh scandal. By comparing the original essay with the English translation, Butler discovered changes that suggested to him unacknowledged borrowings from his own *Evolution Old and New,* combined with

an insidious attack upon himself—the latter in Krause's
statement that the attempt to revive the theories of Erasmus Darwin showed "a weakness of thought and a mental
anachronism which no one can envy."⁷ He also accused
Darwin of being a partner in this chicanery by making it
appear that Krause's essay had been reprinted without
change, thus concealing the fact that it had been altered
in response to Butler's book. Darwin replied by explaining
that the sentence in his notice mentioning Krause's revisions
had somehow been left out in the proofs. But he refused
to give Butler the satisfaction of issuing a public statement
to this effect, although his sons had urged him to do so
out of charity and even justice. Instead Darwin, for once
insensitive, retreated into a dignified silence, a policy advised by Huxley, who well knew that no retort is as effective, no conspiracy as deadly as ignoring an opponent. Butler's paranoia fed on this treatment: "If I am asked to lay
my hand on the theme which more than any other has
prevented my making way in my own generation, I should
say it was my quarrel with Charles Darwin and the dirty
tricks which he and his have never failed to play me when
they got a chance."⁸ That the public sided entirely with
Darwin aggravated the situation: "I attacked the foundations of morality in *Erewhon,* and nobody cared two straws.
I tore open the wounds of my Redeemer as he hung upon
the Cross in *The Fair Haven,* and people rather liked it.
But when I attacked Mr. Darwin they were up in arms in
a moment."⁹

Apart from such occasional vexations, of which the Butler affair was far the worst, Darwin's life went on, as he
had predicted, like clockwork. It was, in many ways, a
happier life than he had expected, even apart from his
extraordinary professional success. His children proved to
be not the helpless invalids he had imagined, and once the
two youngest boys, who had given most cause for anxiety,
married and left the parental home, where sickness was
assumed to be the natural human condition, their health
rapidly improved. Darwin's great pride and joy in them
was partly surprise that they had turned out so well. Wil-

liam, the eldest, was a successful banker who delighted his parents by marrying Sara Sedgwick, a very proper Bostonian, just as they were becoming sadly reconciled to his bachelorhood. George was a mathematician, and one of his father's last letters proudly announced his expectation of a professorship at Cambridge, a chair which he received the year after Darwin died. Francis, like Charles before him, had given up medicine and had instead devoted himself to his father's work; during the two years of his first marriage he moved only as far away as the village of Down, and after his wife's death in childbirth he returned home with his infant son Bernard, who became the pet of the family. Leonard went into the Army, where he did a variety of scientific jobs; and Horace, the youngest, was an engineer. Elizabeth remained unmarried, but Henrietta, the elder daughter, continued to be a vigorous and domineering invalid even after her marriage to Richard Litchfield, a barrister and one of the founders of the Working Men's College, who docilely submitted to her regimen.

Darwin's own health, surprisingly, also took a turn for the better during this last decade of his life. To be sure, not a day passed without some discomfort and fatigue, but the attacks had become milder and briefer. His occasional visits to Henrietta and Erasmus in London or William near Southampton, to the Hensleigh Wedgwoods in Surrey, to his friend and relation, Lord Farrer, at Abinger Hall, and even two trips as far afield as the lake country,' were no longer memorialized by letters complaining of unwellness but only by complaints at being separated from his work. Those who seek evidence of the psychological or psychosomatic nature of his illness may find it suggestive that the acute attacks subsided at about the time that the *Descent* was finished, and that he enjoyed better health when he was employed in more modest and uncontroversial occupations. They can also point to the fact that he did not permit his partial recovery to alter his habits of invalidism; he still avoided engagements, sent proxies to dinners given in his honor, and ventured no farther from home than his wife insisted upon. On the other hand, neither his physical improvement nor his continued invalidism is necessarily as suspicious as has been made out. It is not uncommon to find a quiescent old age following upon a chronic and virulent disorder. Nor is it unreasonable that a sickly man in his sixties, with no taste for social or public life, should prefer the habits of retirement cultivated earlier in life.

If sickness became less pressing, old age became more so, and the severity of his attacks relaxed only to be succeeded by a general sense of debility. Moreover, a new and disturbing symptom appeared. He began to complain of heart tremors, which the distinguished physician, Sir Andrew Clark, first dismissed as of little importance. (When the *Life* came to be written, Emma Darwin asked that this diagnosis not be mentioned.) The attacks became more frequent in the autumn of 1881 and again in the following spring, until finally he could no longer venture out on his beloved sand walk.

On the afternoon of Wednesday, April 19, 1882, at the age of seventy-three, Darwin died. He had been only intermittently conscious since the previous midnight, and his last hours were in keeping with the tenor of his life. He was peaceful and calm, assured his wife that he had no fear of death, and roused himself only to thank her and the children for being so good to him, and to apologize for inflicting his pain on them.

Darwin had expected to be buried in the churchyard at Down, near the grave of his brother Erasmus, who had died the previous year. Instead, a parliamentary petition requested his burial in Westminster Abbey, as being the wish of "a very large number of our countrymen of all classes and opinions."[10] The funeral took place in the Abbey on April 26, his grave being a few feet from that of Newton. The pallbearers included his old friends Huxley, Hooker, and Wallace (Lyell had died several years earlier), as well as more famous men of affairs, and the funeral was attended by representatives of many nations, universities, and learned societies and by a large number of relations, friends, and admirers. Even Herbert Spencer thought the occasion worthy enough to suspend his objections to religious ceremonies. The only incongruous note in these solemn proceedings was the action of his eldest son and chief mourner, William, who, ever vigilant to the threat of a draft, had covered his bald head with his gloves while the rest of the company sat, bareheaded and reverent.

Emma Darwin was equally unconforming. Preferring to mourn in private, she did not attend the funeral. In many ways, she was a stronger-minded, tougher person than Charles. When the family reproached itself for not appreciating the seriousness of Darwin's heart condition, she said it was better as it was, that she preferred her husband happily working to the last rather than being miserably idle in the hope of prolonging his life. She had no fear of death, either for herself or for others. Firm in the belief that life should not be purchased at the cost of too much pain, she was constantly expressing her gratitude for the death of old friends and relations. Her candor is sometimes startling, as when she wrote to her daughter of an ailing acquaintance: "You know what my feelings must be about the poor old man, but I am afraid he will recover."[11] Charles' death removed the center of her life; he had been the pivot around which every moment of her day had been organized. Yet somehow she managed to live on for fourteen years, in good cheer, filling up her day with writing amusingly informal and slightly incoherent letters, making the rounds of the poor in the village, knitting, reading a novel a day and more, and visiting and entertaining her children and grandchildren. (She bought a house in Cambridge so as to spend winters near most of the family.) She died in 1896, at the age of eighty-eight.

That the Dean of Westminster should have acquiesced so readily to the proposal of an Abbey burial testifies to the respectability enjoyed by both Darwin and his doctrine. If there were objections, they were voiced privately, and they came as much from those, like Spencer, who found the religious ceremonial offensive and unworthy of Darwin, as from those who found Darwin's ideas offensive and unworthy of religious sanction.

Everywhere the obituaries and, later, the reviews of the

Life and Letters, vied with each other in paying homage
to him. The *Times* put the stamp of scientific orthodoxy
on his theory, acclaiming him another Newton. At the
same time, the *Church Times* gave its religious sanction:
it recalled how men had first reviled the *Origin* for supposing nature to be autonomous or self-acting, and how a
little reflection had cleared it of the charges of atheism or
deism; if it had faults, they could only be of a scientific
order. Even *Punch,* which had always belabored the theory
in its usual sophomoric fashion, now joined in tribute: "Recorder of the long Descent of Man, and a most living
witness of his rise."[12] It was the *Athenaeum,* however,
which, among the usual hyperboles, pronounced the most
fitting eulogy upon Darwin in welcoming him to the "band
of gentle heroes" of which England was so proud.[13]

Darwin's friends were not entirely pleased to find themselves the custodians of the new orthodoxy. After the fury
of denunciation in the early years, there was something unseemly in the haste with which he was now canonized. It
was only three years after his death that a new series of
books devoted to "English Worthies" started publication,
with Darwin as its first title. Huxley complained that it
was a dull world, where all the propositions so roundly
anathematized only a quarter of a century before were now
being taught in the textbooks. He protested how queer it
was to hear the estimable Lord Salisbury, leader of the
Conservative Party and Chancellor of Oxford University,
assure the British Association that the theory of evolution
was now accepted with substantial unanimity. Recalling
the time when it was as unanimously denied, he apologized
for troubling the modern world "with such antiquarian
business."[14]

The new orthodoxy, however, was never quite so secure
as its proponents thought. In each generation a small
number of reputable scientists revived the "antiquarian"
controversy, reminding their colleagues of Huxley's warning about truths that begin as heresies and end as superstitions. Some of these dissidents also echoed Huxley's
early judgment that natural selection was not an established

theory but a tentative hypothesis, an extremely valuable and even probable hypothesis, but a hypothesis none the less. "It is not absolutely proven," Huxley had written in his review of the *Origin*, "that a group of animals, having all the characters exhibited by species in nature, has ever been originated by selection, whether artificial or natural."[15] This judgment came not only with the authority of Huxley but with that of Darwin himself:

> In fact the belief in Natural Selection must at present be grounded entirely on general considerations. . . . When we descend to details, we can prove that no one species has changed [i.e., we cannot prove that any one species has changed]; nor can we prove that the supposed changes are beneficial, which is the groundwork of the theory. Nor can we explain why some species have changed and others have not.[16]

At the end of the century, the same doubts were being repeated by scientists who were themselves committed to the theory, but who recognized that their commitment was more an act of faith than of demonstration—faith in the scientific enterprise in general, as they understood it, rather than a scientifically validated belief in a particular proposition. August Weismann, the geneticist and zoologist, was both perceptive and candid in describing the basis of his own evolutionary creed:

> Just as in this instance, so is it in every individual case of natural selection. We cannot demonstrate any of them. . . . We shall never be able to establish by observation the progress of natural selection . . . What is it then that nevertheless makes us believe in this progress as actual, and leads us to ascribe such extraordinary importance to it? Nothing but the power of logic; we must assume natural selection to be the principle of the explanation of the metamorphoses, because all other apparent principles of explanation fail us, and it is inconceivable that there could be yet another capable of explaining the adaptations of organisms, *without assuming the help of a principle of design*. . . . We accept it not because

we are able to demonstrate the process in detail, not even because we can with more or less ease imagine it, but simply *because we must, because it is the only possible explanation* that we can conceive.[17]

We must assume this [the theory of sexual selection] since otherwise secondary sexual characters remain inexplicable.[18]

Some years later, the biologist William Bateson reaffirmed the peculiar conjunction of doubt and faith that has been the heritage of the Darwinian:

Discussions of evolution came to an end primarily because it was obvious that no progress was being made. . . . Biological science has returned to its rightful place, investigation of the structure and properties of the concrete and visible world. We cannot see how the differentiation of species came about. Variation of many kinds, often considerable, we daily witness, but no origin of species . . . The particular and essential bit of the theory of evolution which is concerned with the origin and nature of *species* remains utterly mysterious. . . . I have put before you very frankly the considerations which have made us agnostic as to the actual mode and processes of evolution. . . . Let us then proclaim in precise and unmistakable language that our faith in evolution is unshaken. The difficulties which weigh upon the professional biologist need not trouble the layman.[19]

The many converging lines of evidence point so clearly to the central fact of the origin of the forms of life by an evolutionary process that we are compelled to accept this deduction, but as to almost all the essential features whether of cause or mode, by which specific diversity has become what we perceive it to be, we have to confess an ignorance nearly total. The transformation of masses of population by imperceptible steps guided by selection, is, as most of us now see, so inapplicable to the facts, whether of variation or of specificity, that we can only marvel both at the want of penetration displayed by the advocates of such a proposition, and at the forensic skill by which it was made to appear acceptable even for a time.[20]

More recently, so unimpeachable a witness as Bertrand Russell has said that "the particular mechanism of 'natural selection' is no longer regarded by biologists as adequate."[21] And a few years ago the professor of zoology at Cambridge posed the dilemma in its sharpest form:

> No amount of argument, or clever epigram, can disguise the inherent improbability of orthodox [Darwinian] theory; but most biologists feel it is better to think in terms of improbable events than not to think at all; there will always be a few who feel in their bones a sneaking sympathy with Samuel Butler's scepticism.[22]

It is no vulgar "act of faith" that is at issue here, no ignoble acquiescence in orthodoxy or submission to an establishment. What is at issue is the faith in science itself, or in what passes as the necessary logic of science. The theory of natural selection is in many respects almost the ideal scientific theory: it is eminently naturalistic, mechanical, objective, impersonal and economical. A maximum number of phenomena are accounted for in the simplest and most congenial way. "The desire for some such hypothesis," as the authors of a work on zoology put it,[23] is as powerful a factor in its perpetuation as it had been in its original acceptance. And when there is no alternative, or rather when the alternative is making do without any theory at all, the pull to Darwinism becomes very nearly irresistible. Science abhors gaps in its logical structure as it abhors leaps in nature—and for the same reason. Without the continuum of scientific theory, without the uniformity of nature, scientific knowledge, indeed science itself, feels jeopardized. Scientists cannot long—and a century is a long time as the history of modern science goes—live with the unknown, particularly when the unknown resides at the heart of their subject, and when it threatens to pass from the transient condition of the unknown into the permanent unknowable. Tyndall was once indiscreet enough to write: "The logical feebleness of science is not sufficiently borne in mind. It keeps down the weed of superstition, not by logic, but by slowly rendering the mental soil unfit for its cultivation."[24] The moral is that the mind must be so entirely habituated to the ideas of uniformity and continuity that even in the failure of fact and logic, the faith in science would remain intact.[25]

When it is not the "desire for some such hypothesis," or the feeling that "it is better to think in terms of improbable events than not to think at all," that draws men to Darwinism, it is often the confusion between the theories of evolution and natural selection. Every paleontological discovery that seems to have evolutionary significance is somehow taken as confirmation of the theory of natural selection, even when it has not the remotest bearing upon that theory. Similarly every evidence of natural selection, manifesting itself upon however small a scale, is taken as evidence of Darwinian evolution, of the "origin of species by means of natural selection." Thus the one item of empirical evidence that is cited again and again in reply to the skeptic's demand for proof is the famous experiment of H. B. D. Kettlewell in 1924, demonstrating that the smoke-blackened regions of industrial England favor the perpetuation of darker- rather than lighter-colored moths. But no one questions the operation of natural selection on this level, just as no one questions the evidence for the evolution of the horse. What is in question is the operation of natural selection in the evolution of one major species into another, and ultimately the operation of natural selection in the evolution of all species from the "one primordial form" posited by Darwin.

Yet even Darwin's most disgruntled critics could not deny the magnitude—enormity, they might have said—of his achievement. They might doubt the originality of his theory and deny its truth, but they had to concede that somehow it was revolutionary in effect. The anomaly increases as time goes on, as his predecessors seem to multiply in number, as their theories are discovered to approximate more and more closely to his, as the philosophical presuppositions of the theory are pushed further back in time, as critics find more and more flaws in the theory and the passage of years brings its vindication no nearer—and as, in spite of all this, history persists in dividing itself into a pre-Darwin and a post-Darwin epoch. That there was a Darwinian revolution, there is little doubt. But what kind of a revolution was it that was so generously prepared for beforehand and so strongly resisted afterward? How could men be so shocked by it, who also denied it all novelty, or so fearful of it, who found its errors so egregious?

Intellectual revolutions, it appears, are like political revolutions, only more so—political revolutions tending to vanish under the cold eye of the historian, until the new regime is only the legalization and continuation of a movement begun in the old, and intellectual revolutions often reducing themselves to little more than the synthesizing and popularizing of ideas long current. Yet, if the feelings of contemporaries and the judgment of history are not to be gainsaid, they are revolutions all the same. Such revolutions, intellectual or political, suggest the formula of the "conservative revolution." As the French Revolution extended, stabilized, and legalized the basic tendencies of the *ancien régime;* as Napoleon, bringing the Revolution to a halt, at the same time brought it to its fruition; so Darwin, dramatizing and bringing to a climax the ideas, sentiments, and conjectures of his age, may be thought of as the hero of a conservative revolution.

The Darwinian revolution was conservative, first, as a purely scientific affair. The observations on which it was based were largely familiar, the terms of the problem had been stated, even the crucial idea had been in circulation for half a century. And it influenced the practical work of the sciences less than might be supposed, each discipline assuming that its great import was in some other field. The job of the systematists was affected hardly at all, the "natural system" sought before Darwin's time being much the same as that posited by the theory of evolution. The final irony came when Darwin himself, at the end of his life, established and endowed the *Index Kewensis,* a published catalogue of all the species of flowering plants. This was not so much, as one botanist has suggested, the act of the iconoclast paying a tribute of remorse to a damaged idol,[26] as the act of the innocent who had never intended any injury to that particular idol. In many ways, as has been seen, Darwin was personally more conservative—"non-modern," as his son put it[27]—in his habits of mind and work than is commonly thought. He was much closer to the older school of naturalists, ranging widely but often superficially over many subjects, than to the newer generation of specialized, formal, and rigorous scientists.

Many of Darwin's friends must have felt as Huxley did when he upbraided himself for not having thought of so simple an idea himself. And many of his enemies must have agreed with Butler: "Buffon planted, Erasmus Darwin and Lamarck watered, but it was Mr. Darwin who said 'That fruit is ripe,' and shook it into his lap."[28] For what they were experiencing was not the shock of discovery but rather the shock of recognition. They were so quickly converted because there was little to be converted to. And those who chose not to be converted were also as quick in their responses; they knew instantly where to direct their attacks because they had long been mobilized for just such a threat. It is this shock of recognition that explains the reaction of Darwin's closest friends, those like Hooker and Huxley who were well acquainted with his thesis long before its publication, but who did not feel truly converted until they were confronted with the physical fact of the printed book—until, that is, they were able to recognize what it was that they already knew.

A more important sense in which Darwin was conservative, even old-fashioned, was his lack of self-consciousness as a scientist confronting his subject, his unquestioned faith in an objective universe in which both he and his subject occupied fixed and independent positions. He never doubted that he was a passive, disinterested observer accurately recording the laws revealed in nature. In this faith in the possibility of an objective science he was reverting to a tradition that even in his own time had begun to be questioned. Kant had already challenged philosophy to show how the scientist, himself an active agent, a participant rather than observer in the order of nature, could formulate laws that were presumed to be independent and objective. Darwin was never troubled by this dilemma. He was content to seek man's origin in nature, to assimilate man into nature, without trying to assimilate man's knowledge into the processes of nature.

It was for this reason that Darwinism did not turn out to be the implacable enemy of religion that was first suspected. For Darwinism shared with religion the belief in an objective knowledge of nature. If religion's belief was based on revelation and Darwinism on science, with good will the two could be—as indeed they were—shown to coincide. The true challenge to orthodox religion came with the denial of the possibility of all objective knowledge, with the skepticism of a Kierkegaard who refused to religion and science alike the claims of truth, and forswore the reason of religion equally with the reasons of science. Compared with this radical assertion of an arbitrary, willful faith inaccessible to all reason, the dispute between Darwin and his religious critics was little more than the friendly bickering of old friends. Pre-Kant and pre-Kierkegaard, Darwinism appears as the citadel of tradition.

This double role of Darwinism—at once conservative and revolutionary—may be deduced from some remarks of C. S. Lewis[29] on the dating of modernity. Strictly speaking, he said, the line dividing the modern world from the pre-modern should be drawn at the end of the seventeenth century, with the general acceptance of Copernicus and Descartes and the foundation in England of the Royal Society. From a wider cultural point of view, however, it had to be drawn in the middle of the nineteenth century, when the full effects of the revolution finally permeated all areas of human activity and consciousness. Until then, the prevailing tone continued to be ethical, rhetorical, and juristic, and Johnson's voice could still be heard: The knowledge of nature and science was not "the great or frequent business of the human mind." Science became the business of man when man became the business of science: when Watt applied himself to the invention of the steam engine and Darwin to the ancestry of mankind. Only then were the technology and mechanism implicit in the earlier revolution finally unleashed and the old humanistic order swept away.

For any particular generation, it is only the latest revolution that is judged worthy of the title; all others, in comparison, seem to fall short of it. In the light of recent intellectual events, Darwinism may appear to be hardly a revolution at all. That it was, however, revolutionary as well as conservative may be seen in the actual experiences of contemporaries. The test is in the "crisis of faith" which is so abundantly testified to in the memoirs of the time and so commonly attributed to the *Origin*.

Examined closely, almost all these religious crises appear to have been well on their way, and sometimes fully accomplished, quite independently of the *Origin*. Some of the older agnostics were even notably unenthusiastic about the *Origin*, cherishing their own sources of disbelief. John Stuart Mill greeted it as an interesting hypothesis of dubious truth. George Eliot, who had been nurtured in the hardier school of German Biblical criticism, found it unimpressive and uncongenial. Tennyson denied that his anguish bore any relation to Darwin's. Harrison, enumerating the influences that could lead a man from High Anglicanism to High Comteanism, put Darwin last in a long list of philosophers, theologians, poets, and scientists. William Hale White, alias Mark Rutherford, owed his "deliverance" to Wordsworth's "Lyrical Ballads."

Even the younger men managed to find their way to agnosticism without the help of Darwin. Both Huxley and Spencer lost their faith long before they discovered the theory of evolution, and apparently out of a temperamental repugnance to the idea of the supernatural. John Morley abandoned his Methodism while occupying Wesley's rooms at Oxford several months before the publication of the *Origin*. Walter Pater lost his faith soon after the book's appearance—by an assiduous reading of the local Oxford theologians; while William Lecky achieved the same effect at Dublin University, also as a result of his theological studies. Meredith's naturalism, deriving from the Greek stoics, made him more sympathetic to Lamarck than to

Darwin. At the same time, Swinburne and his young friends were finding their inspiration and their notoriety in the Marquis de Sade.

Even Leslie Stephen, who is generally taken as the archetype of the young man who fell from grace as a result of the *Origin,* and who made the memorable remark that as Montaigne had taken his consolations out of Lucretius, so he had to take his out of Darwin, was not at all the dramatic case of conversion he has been made out to be. Not until 1862 did it occur to him that the story of the flood was not literally true, and long afterward he continued to hold his college fellowship and the clerical orders that went with it. When he finally resigned these, it was not because he experienced a sudden loss of faith but because he had gradually discovered that he had never had any faith. Nor was it any particular book or doctrine that brought him to this realization. Cambridge was in a glacial epoch, when ideas lurked for so long below the surface of consciousness that by the time they broke through, they were no longer a heresy but a new conformity:

> We were in one of the periods at which a crust of conventional dogma has formed, like the paleocrystic ice of the polar sea, upon the surface of opinion. The accepted formulas are being complacently repeated in all good faith by the respectable authorities. And yet new currents are everywhere moving beneath, and the superincumbent layer of official dogma is no longer conformable to the substratum of genuine belief. Then a sudden cataclysm begins to break up the crust, and to sweep away the temporary bridging of the abyss which superficial observers had mistaken for solid earth. The alarm caused by the collapse of the ancient dogmas may perhaps be exaggerated. In time we come to see that the change is mainly in the open manifestations of the old, rather than in the intrusion of the really new modes of thought; and somehow or other as the new doctrines lose their strangeness we are sagacious enough to discover that we always believed them in substance. However that may

be, old-fashioned people had to bear some severe shocks.[30]

The *Origin* was the cataclysm that broke up the crust of conventional opinion. It expressed and dramatized what many had obscurely felt. More than this: it legitimized what they felt. Coming from so unexceptional a source, with all the authority of science and without the taint of ulterior ideology, it became the receptacle of great hopes and great fears. Those who were already partial to the mode of thought it represented—which could mean anything from a mild naturalism or deism to a belligerent atheism—often fastened upon it as the symbol and warrant of their belief; if they later loosely spoke of it as the cause of their conversion, the error is understandable, the leap from justification to cause being all too easily effected. Similarly, those who had already committed themselves to the other side, finding naturalism uncongenial or unpersuasive, tended to look upon the *Origin* as the incarnation of all that was hateful and fearful. There were, to be sure, some who experienced a genuine crisis of faith upon reading it, as there were also some who came to it with an open mind and left unconverted; if the former have been more publicized, it may be because the loss of faith is a more dramatic affair than the retention of faith. For most men, however, the *Origin* was not an isolated event with isolated consequences. It did not revolutionize their beliefs so much as give public recognition to a revolution that had already occurred. It was belief made manifest, revolution legitimized.

Whether these new beliefs are truer than the old is irrelevant to the fact of an intellectual revolution, just as the question of the justice of new institutions is irrelevant to the fact of a political revolution. But if it is important for later generations not to deny the fact of revolution because they cannot concede its truth or justice, it is no less important not to concede truth or justice merely because they cannot deny the fact of revolution.

NOTES

Life and Letters alone refers to Darwin's *Life and Letters*. References to the *Origin* are to the reprint of the first edition, unless otherwise specified. Titles are abbreviated, the complete forms appearing in the bibliography.

CHAPTER 1

Pages 2-30

1. Judd, "Darwin and Geology," in Seward (ed.), *Darwin and Modern Science*, p. 337.

2. Dr. Stukeley, "Account of the Almost Entire Sceleton of a Large Animal," *Philosophical Transactions*, April–May 1719, cited in Darwin's introduction to Krause, *Eramus Darwin*, p. 4.

3. Anna Seward, *Memoirs of the Life of Dr. Darwin*.

4. Darwin's introduction, Krause, p. 61.

5. Another explanation of Erasmus' abstention has recently suggested itself, too late to be incorporated into the text. This appears in a recently published letter from Erasmus to his son Robert. Robert, then twenty-six and probably contemplating marriage, was apparently worried about the possibility of insanity inherited from his mother and grandfather, and Erasmus had written to allay his fears. Mr. Howard (Robert's maternal grandfather), he assured him, was not in the least insane, although he was a drunkard "both in public and private" and died from gout and debility of digestion. Robert's mother (Erasmus' first wife) was a more complicated and serious case. She suffered severe pains beneath the liver, followed by violent convulsions; these were sometimes relieved by great doses of opium and wine which induced intoxication. At other times she experienced a temporary delirium lasting for half an hour—"what by some," Erasmus remarked, "might be termed insanity" but which he took to be a form of epilepsy. She resorted to more and more liquor until she died of inflammation of the liver. Erasmus gave it as his opinion that "all the drunken diseases

are hereditary in some degree," that "epilepsy and insanity are produced originally by drinking," but that one sober generation would cure the evils which one drunken one had created. He concluded that so long as Robert remained temperate, he had nothing to fear. He himself, he added, feeling the weakness of old age, had started to drink two glasses of home-made wine with water at dinner instead of water alone. (This letter appears in Darwin's *Autobiography* edited by his granddaughter, Nora Barlow, pp. 223–25.)

6. Erasmus Darwin, *Botanic Garden,* preface.

7. Krause, p. 208.

8. Darwin's introduction, *ibid.,* pp. 93, 97.

9. This particular form of infection and death seems to have made a great impression upon Charles, for it appears several times in his writings. One of his early notebooks suggests the experiment: "Puncture one animal with recent dead body of other, and see if same effect as with man" (*Darwin and the Voyage of the Beagle,* p. 265). And in the *Journal of the Voyage of the Beagle* he speculated upon the possibility that the "effluvium" of one set of men might be poisonous when inhaled by others: "Mysterious as this circumstance appears to be, it is not more surprising than that the body of one's fellow-creature, directly after death, and before putrefaction has commenced, should often be of so deleterious a quality, that the mere puncture from an instrument used in its dissection should prove fatal" (1st ed., p. 522).

10. Darwin's introduction, Krause, p. 76.

11. *Ibid.,* p. 85.

12. Sir Joseph Hooker, Darwin's good friend, had a different version of this event: "I understood from Darwin that his father had not only scientific proclivities, but ambition, and that he presented to the Royal Society a communication on some optical subject, which, being rejected, disgusted him, and led to his stifling his own early scientific tendencies and scoffing at those of others" (Hooker to Huxley, May 2, 1888: Hooker, *Life and Letters,* II, 306).

13. Autobiography, *Life and Letters,* I, 20.

14. *Ibid.*

15. These recollections of Emma Darwin were confided to her son Francis and recorded by him soon after the death of Charles Darwin (Cambridge Mss.). An expurgated version may be found in Emma Darwin, *Letters,* I, 40.

16. *Life and Letters,* I, 11.

17. Catherine Darwin to Fanny Wedgwood, Dec. 1830: Emma Darwin, *Letters,* I, 227–28.

18. *Ibid.,* I, 140.

19. Unpublished passage of autobiography (Cambridge Mss.). In Emma Darwin's *Letters,* this has been toned down to: "Doctor Darwin used to say she was the only woman he ever

knew who thought for herself in matters of religion" (I, 164).

20. Cambridge Mss.

21. *Ibid.*

22. Smiles, *Josiah Wedgwood*, pp. 12–13.

23. *Ibid.*, p. 298.

24. Darwin's introduction, Krause, p. 45.

25. Autobiography, *Life and Letters*, I, 27.

26. Kempf, *Psychoanalytic Review*, V (1918), 155–58.

27. Darwin's recollections are contained in a fragment of an autobiographical memoir dealing with his childhood and written in 1838 (*More Letters*, I, 1–5); an unpublished journal recording the main events in his life—births of children, holidays, illnesses, publications, etc. (Cambridge Mss.); and the autobiography, "Recollections of the Development of My Mind and Character," most of which was written in 1876. The original manuscript of the autobiography is among the Cambridge Mss., and an expurgated version is in the *Life and Letters*, I, 11–20, 21–22, 26–107, and passim. (Since this was written, the original has been published in Nora Barlow's edition of the *Autobiography*.)

28. Autobiography, *Life and Letters*, I, 32.

29. Autobiography, Cambridge Mss. The deletion without ellipses of several sentences in the published version (*Life and Letters*, I, 27) makes it appear that Catherine's greater facility in learning emerged at the day school to which Charles was later sent. In fact, Catherine probably did not attend that school. The original version, with the unpublished part in italics, reads: "In the spring of this same year I was sent to a day school in Shrewsbury, where I stayed a year. *Before going to school I was educated by my sister Caroline, but I doubt whether this plan answered.* I have been told that I was much slower in learning than my younger sister Catherine, and I believe that I was in many ways a naughty boy. *Caroline was extremely kind . . .*" (The quotation continues as in the text above.)

30. A parenthetical "perhaps" does not much soften the sentiment. Fanny Allen to Elizabeth Wedgwood, Feb. 9, 1866: Emma Darwin, *Letters*, II, 184.

31. Carlyle, *Reminiscences*, I, 208.

32. Autobiography, Cambridge Mss.

33. *Life and Letters*, I, 21.

34. Emma Darwin, *Letters*, I, 7.

35. Autobiography, *Life and Letters*, I, 44.

36. *Ibid.*, p. 28.

37. *Ibid.*

38. *Ibid.*

39. *More Letters*, I, 4.

40. Dr. Samuel Butler, *Life and Letters*, I, 195.

41. Autobiography, *Life and Letters*, I, 31.

42. Dec. 18, 1818: Dr. Butler, *Life and Letters*, I, 163.
43. *Ibid.*, p. 157.
44. Dr. Butler to Dr. Darwin, Feb. 4, 1819: *ibid.*, p. 164.
45. These memories appear in the Recollections of George Darwin. He could not remember whether he had had them from his father or his uncle; a marginal note in the manuscript, probably written by Francis Darwin, ascribed them to Erasmus (Cambridge Mss.).
46. Dr. Butler, *Life and Letters*, I, 211.
47. *Ibid.*, pp. 196–97.
48. Autobiography, *Life and Letters*, I, 32.
49. *Ibid.*
50. *Ibid.*, p. 35.
51. *Ibid.*
52. George Darwin's Recollections, Cambridge Mss.
53. Autobiography, *Life and Letters*, I, 32.
54. *Ibid.*, p. 37.
55. *Ibid.*, p. 36.
56. Darwin to Hooker, April 18, 1847: *ibid.*, p. 355.
57. West, *Darwin*, p. 60.
58. Huxley, Obituary notice of Darwin, *Darwiniana*, p. 262.
59. Edinburgh notebooks, Cambridge Mss.
60. Autobiography, *Life and Letters*, I, 43.
61. Charles to Susan Darwin, Jan. 29, 1826: Cambridge Mss.
62. Darwin to Henslow, July 2, 1848: Cambridge Mss.
63. George Darwin's Recollections, Cambridge Mss.
64. Autobiography, *Life and Letters*, I, 41.
65. *Ibid.*
66. *Ibid.*, p. 43.
67. *Ibid.*, pp. 43–44.
68. *Ibid.*, p. 45.

CHAPTER 2

Pages 31–49

1. Psychoanalysis reads much more significance into Charles's acquiescence: "It is permissible to infer, therefore, that Darwin's *consecration* of himself as a naturalist for the welfare of humanity, besides gratifying and beautifully sublimating his mother-attachment, also gratified his father's desire that he should religiously consecrate himself to the welfare of humanity, which is remarkably like the mechanism of the sacrifice of the devoted son, Christ, if we consider other facts" (Kempf, *Psychoanalytic Review*, V, 172; italics in the original). It may be doubted whether any consideration of other facts, such as Robert Darwin's total indifference to religion and the welfare of humanity, could discourage this analyst.

2. Autobiography, *Life and Letters*, I, 45.
3. Autobiography, Cambridge Mss.
4. *Life and Letters*, I, 171.
5. Autobiography, *ibid.*, I, 45.
6. Winstanley, *Early Victorian Cambridge*, p. 397.
7. Autobiography, *Life and Letters*, I, 48.
8. Cambridge Mss.
9. Autobiography, *Life and Letters*, I, 46.
10. Darwin to Henslow, July 2, 1848: Cambridge Mss.
11. Autobiography, *Life and Letters*, I, 47.
12. *Ibid.*, p. 48.
13. "Science," *Oxford Dictionary*.
14. Stimson, *Scientists and Amateurs*, p. 128.
15. Jenyns, *Memoir of Henslow*, p. 13.
16. *Ibid.*, p. 14.
17. *Ibid.*, p. 35.
18. Sedgwick, *Life and Letters*, I, 160–61.
19. *Ibid.*, p. 161.
20. Stimson, p. 6.
21. *Ibid.*, p. 191.
22. Autobiography, *Life and Letters*, I, 52.
23. *Ibid.*, p. 187.
24. *Ibid.*, p. 55.
25. The analogy of the house burnt down by fire reappeared in the *Origin of Species* (6th ed.), p. 412.
26. *Life and Letters*, II, 348.
27. Herschel, *Preliminary Discourse*, pp. 14–15.
28. *Ibid.*, p. 25.
29. *Ibid.*, p. 360.
30. Darwin to Hooker, Feb. 10, 1845: *Life and Letters*, I, 336.
31. Von Hagen, *South America Called Them*, p. 180.
32. Darwin to Hooker, Aug. 6, 1881: *Life and Letters*, III, 247.
33. Autobiography, *ibid.*, I, 55.
34. *Ibid.*, p. 51.

CHAPTER 3

Pages 52–81

1. Autobiography, *Life and Letters*, I, 61.
2. Peacock to Darwin, Aug. 1831: *ibid.*, p. 194.
3. Henslow to Darwin, Aug. 24, 1831: *ibid.*, p. 192.
4. The story as it always appears in biographies of Darwin, including the *Life and Letters*, is that Jenyns had instantly accepted the offer and had already packed his clothes, when a sober reconsideration suggested to him scruples about leaving

his parish. This account originated with Darwin himself in a letter to his sister Susan (Sept. 4, 1831: *ibid.*, p. 200), and apparently always remained his understanding of it. Yet a letter from Jenyns (then Blomefield) to Francis Darwin contains an entirely different version of the affair: "I was not at all inclined to it, though Henslow was very anxious I should consent. I took a day to consider of it, at the end of which I quite determined against going" (May 1, 1888: Cambridge Mss.).

5. Henslow to Darwin, Aug. 24, 1831: *Life and Letters,* I, 192.

6. *Ibid.,* p. 193.

7. Autobiography, *ibid.,* p. 59.

8. Josiah Wedgwood to Robert Darwin, Aug. 31, 1831: *ibid.,* p. 198.

9. FitzRoy, *Proceedings of the Second Expedition,* p. 18.

10. Oct. 25, 1831: Darwin, *Diary of the Voyage,* p. 5.

11. Autobiography, *Life and Letters,* I, 59.

12. Darwin to Henslow, Nov. 15, 1831: *ibid.,* p. 215.

13. *Ibid.*

14. Sept. 14, 1831: *Darwin and the Voyage,* p. 50.

15. Dec. 17, 1831: *Diary,* p. 15.

16. Dec. 27, 1831: *ibid.,* p. 18.

17. Darwin to FitzRoy, Oct. 17, 1831: *Life and Letters,* I, 214.

18. Dec. 30, 1831: *Diary,* p. 19.

19. Darwin was in error in putting his age at twenty-three (Charles to Susan Darwin, Sept. 4, 1831: *Life and Letters,* I, 200).

20. One of the more trivial examples of censorship suffered by Darwin's autobiography was his mention of FitzRoy's striking resemblance to a Count d'Albanie, "an illegitimate descendant" of Charles II; the word "illegitimate" was deleted from the published version.

21. Charles to Susan Darwin, Sept. 6, 1831: *Life and Letters,* I, 207.

22. Charles to Caroline Darwin, Apr. 25, 1832: *Darwin and the Voyage,* p. 65.

23. Darwin to Lyell, Aug. 9, 1838: Cambridge Mss.

24. FitzRoy, p. 61.

25. *Journal* (revised ed.), p. 499.

26. March 12, 1832: *Diary,* p. 42.

27. July 3, 1832: *ibid.,* p. 77.

28. FitzRoy, pp. 657–58.

29. *Darwin and the Voyage,* p. 37.

30. George Darwin's Recollections, Cambridge Mss.

31. Hamond to Francis Darwin, Sept. 19, 1882: Cambridge Mss.

32. The pamphlet was written in June 1836 and appeared in the *South African Christian Recorder* in September.

33. Darwin to Fox, Nov. 1832: Cambridge Mss.

34. May 23, 1833: *ibid.*

35. Most biographers, carried away by the zeal of hindsight, tend to hasten the development of Darwin's religious views, as they do his evolutionary views. Thus one recent writer says of the period of the voyage: "Subtle changes in his ideas about species had been mysteriously accompanied by subtle changes in his ideas about Christianity" (Irvine, *Apes, Angels, and Victorians,* p. 51). No amount of subtlety and mystery can conceal the total lack of evidence on this score.

36. FitzRoy, p. 4.

37. *Ibid.,* p. 12.

38. Jan. 20, 1832: *Diary,* p. 130.

39. Jan. 23, 1832: *ibid.,* p. 132.

40. FitzRoy, p. 327.

41. Feb. 24, 1834: *Diary,* p. 213.

42. FitzRoy, p. 18.

43. Later gold was discovered there, but by that time it was also discovered that, contrary also to Darwin's findings, the land was fertile, so that the mining industry was never developed to the same extent as sheep grazing. In view of Darwin's other preoccupations, these minor blunders have understandably been ignored.

44. Feb. 10, 1832: *Darwin and the Voyage,* p. 59.

45. March 5, 1832: *Diary,* p. 42.

46. Charles to Catherine Darwin, April 6, 1834: *Darwin and the Voyage,* p. 96.

47. Jan. 16, 1832: *Diary,* p. 25.

48. Jan. 17, 1832: *ibid.*

49. Jan. 19, 1832: *ibid.,* p. 26.

50. Apr. 6, 1832: *ibid.,* p. 49.

51. Oct. 8, 1832: *ibid.,* p. 106.

52. Oct. 21, 1833: *ibid.,* p. 190.

53. May 3, 1835: *ibid.,* p. 308.

54. Feb. 13, 1834: *ibid.,* p. 211.

55. Dec. 29, 1834: *ibid.,* p. 261.

56. Feb. 20, 1835: *ibid.,* p. 277.

57. June 10, 1835: *ibid.,* p. 318.

58. Sept. 17, 1835: *ibid.,* p. 334.

59. Sept. 26–27, 1835: *ibid.,* p. 337.

60. Charles to Caroline Darwin, July 1835: *Darwin and the Voyage,* p. 125.

61. Aug. 1–6, 1836: *Diary,* pp. 416–17.

62. *Ibid.,* p. 426.

63. *Ibid.,* p. 430.

64. Autobiography, *Life and Letters,* I, 64.

65. They were read at a meeting of the Cambridge Philosophical Society on Nov. 16, 1835, and printed for private circulation the next month. Extracts were also published in the

Entomological Magazine, April 1836, pp. 457–60, and in the *Proceedings of the Geological Society*, II, 210–12.

66. Nov. 1835: Dr. Butler, *Life and Letters*, II, 116. Butler forwarded this letter or perhaps quoted it to Dr. Darwin; and Susan transmitted it to Charles in her letter of Nov. 22, 1835 (Cambridge Mss.). It may be that Sedgwick did not actually call upon Dr. Darwin, as Charles thought, but that the report of his praise came by way of Butler.

67. Strangely enough, Francis Darwin, in compiling the *Life and Letters*, did not seem to know of these early reprints of Darwin's volume. They do not appear in the bibliography, and in the text there is a comment to the effect that it did not become well known "until it was separately published in 1845" (I, 323). There is a reference, however, to a German translation of 1844.

68. The *Diary* has been published by one of Darwin's descendants, Nora Barlow, in an exact transcription, including all of the youthful Darwin's idiosyncrasies of spelling, punctuation, and syntax, and even passages deleted by Darwin himself at the time of writing. In quoting from this, I have corrected such peculiarities of spelling and punctuation as might distract from the content.

CHAPTER 4

Pages 82–106

1. 1832 notebook: *Darwin and the Voyage*, p. 156.
2. Darwin to Fox, May 1832: *Life and Letters*, I, 233–34.
3. Feb. 2, 1832: *Diary*, p. 32.
4. Autobiography, *Life and Letters*, I, 62.
5. For an excellent discussion of the scientific controversies in this pre-Darwin period, see Eiseley, *Darwin's Century*, which appeared after this was completed.
6. Lyell, *Principles* (6th ed.), I, 112.
7. Lyell, *Life, Letters and Journals*, I, 253.
8. Buckland, *Life and Correspondence*, p. 82.
9. Sedgwick, *Address, Feb. 1830*, p. 21.
10. *British Critic*, IX (1831), 206.
11. Sedgwick to Canon Wodehouse, 1836: Sedgwick, *Life and Letters*, I, 452.
12. Geikie, *Founders of Geology*, p. 404.
13. Geikie, *Life of Murchison*, I, 103.
14. Sedgwick, *Address, Feb. 1831*, pp. 22–23.
15. Huxley, "Science and Pseudo-Science," *Essays upon some Controverted Questions*, pp. 272–73.
16. Lyell, *Principles* (1st ed.), I, 74.
17. *Ibid.*, p. 88.
18. *Ibid.*, pp. 76, 78.

19. *Ibid.*, p. 80.

20. *Life and Letters*, I, 189.

21. An example of the unreliability of the wisdom-of-hindsight approach to the history of ideas is the strained interpretation placed upon Darwin's remark by an otherwise responsible commentator: "May we not read in this passage an indication that the self-taught geologist [Darwin] had, even at this early stage, begun to feel a distrust for the prevalent catastrophism, and that his mind was becoming a field in which the seeds which Lyell was afterwards to sow would 'fall on good ground?'" (Judd, "Darwin and Geology," in Seward [ed.], *Darwin and Modern Science*, pp. 344–45).

22. Feb. 2, 1832: *Diary*, p. 32. His autobiographical recollection, that it was St. Jago that first converted him to Lyell, is confirmed in his letter to Henslow of about this time, in which he remarked that the geology of St. Jago would be of particular interest to Lyell—presumably in confirming his thesis (May 18, 1832: *Life and Letters*, I, 62).

23. March 4, 1835: *Diary*, p. 280.

24. April 18, 1835: *More Letters*, I, 21.

25. July 19, 1835: *Darwin and the Voyage*, p. 245.

26. *Journal* (1st ed.), pp. 376, 380–81. In the revised edition of the *Journal* this is put in the past tense: ". . . which one *was* accustomed to attribute . . ." (p. 308).

27. *Journal* (1st ed.), p. 386.

28. Huxley, "On the Reception of the 'Origin of Species,'" Darwin, *Life and Letters*, II, 190.

29. Huxley, Obituary notice of Darwin, *Darwiniana*, p. 268.

30. Darwin to Fox, July 1835: *Life and Letters*, I, 263.

31. Darwin to Horner, Aug. 29, 1844: *More Letters*, II, 117.

32. Autobiography, *Life and Letters*, I, 65.

33. *Ibid.*, p. 149.

34. *Ibid.*, p. 70.

35. *Ibid.*, p. 65.

36. Darwin, *Structure and Distribution of Coral Reefs* (1st ed.) pp. 93–94.

37. *Ibid.*, p. 95.

38. *Life and Letters*, I, 325.

39. *Ibid.*, p. 324.

40. Geikie, *Darwin as Geologist*, p. 36.

41. Darwin to K. Semper, Oct. 2, 1879: *Life and Letters*, III, 182.

42. Darwin to A. Agassiz, May 5, 1881: *ibid.*, p. 184.

43. Aug. 9, 1838: *ibid.*, p. 293.

44. Darwin to Lyell, Sept. 6, 1861: *More Letters*, II, 188.

45. *Life and Letters*, I, 69.

46. *Ibid.*, p. 103.

47. *Diary*, p. 430; *Journal*, p. 608.

CHAPTER 5

Pages 107–123

1. *Origin*, p. 21.
2. *Life and Letters*, I, 82.
3. Sept. 22, 1832: *Diary*, pp. 102–3.
4. Oct. 8, 1832: *ibid.*, pp. 105–6.
5. Nov. 25, 1833: *ibid.*, p. 196.
6. Charles to Caroline Darwin, Aug. 9, 1834: *Darwin and the Voyage*, p. 105.
7. Charles to Caroline Darwin, Sept. 20, 1833: *ibid.*, p. 91.
8. Darwin to Henslow, Nov. 24, 1832; July 24, 1834; April 18, 1835: *More Letters*, I, 12, 18, 23.
9. Charles to Caroline Darwin, March 30, 1833: *Darwin and the Voyage*, p. 82.
10. *Journal*, pp. 209–10.
11. Huxley, Obituary notice of Darwin, *Darwiniana*, p. 271. Darwin agreed: "Another of my occupations was collecting animals of all classes, briefly describing and roughly dissecting many of the marine ones; but from not being able to draw, and from not having sufficient anatomical knowledge, a great pile of Ms. which I made during the voyage has proved almost useless" (Autobiography, *Life and Letters*, I, 62). And to Henslow he wrote in 1836: "I knew no more about the plants which I had collected than the man in the moon" (*ibid.*, p. 276).
12. Darwin to FitzRoy, [Aug.] 1834: Cambridge Mss.
13. June 20, 1835: *Diary*, p. 322.
14. Lyell, *Principles* (1st ed.), I, 73.
15. *Journal*, p. 208.
16. *Ibid.*, p. 209.
17. *Origin*, p. 389.
18. *Life and Letters*, I, 82. At another point in the autobiography, Darwin wrote: "Nor must I pass over the singular relations of the animals and plants inhabiting the several islands of the Galapagos archipelago, and of all of them to the inhabitants of South America" (*ibid.*, p. 65). In the original manuscript of the autobiography, this sentence appears as a later insertion. It is apparent that, with the passage of time, the events of the trip fixed themselves in his mind more and more in the shape of a formula. This may account for the fact that the same phrases repeat themselves with suspicious exactness in his recollections, an exactness testifying not so much to the trustworthiness of his memory as to the rigid pattern imposed by memory on the events.
19. Mar. 22, 1833: *Darwin and the Voyage*, p. 179.
20. 1834 notebook: *ibid.*, p. 219.

21. Charles to Caroline Darwin, Dec. 27, 1835: *ibid.*, p. 129.
22. Jan. 1836: Cambridge Mss.
23. Sept. 16–17, 21, 1835: *Diary*, pp. 334–35.
24. Sept. 26–27: *ibid.*, p. 337.
25. There is one other allusion to the Galapagos, among the ornithological notes, that would be relevant here, if the editor, Nora Barlow, were right in ascribing it to the period of the voyage:

> When I recollect the fact, that from the form of the body, shape of scales and general size, the Spaniards can at once pronounce from which island any tortoise may have been brought:—when I see these islands in sight of each other and possessed of but a scanty stock of animals, tenanted by these birds but slightly differing in structure and filling the same place in Nature, I must suspect they are only varieties. The only fact of a similar kind of which I am aware is the constant asserted difference between the wolf-like fox of East and West Falkland Islands. If there is the slightest foundation for these remarks, the zoology of archipelagoes will be well worth examining; for such facts would undermine the stability of species.

The editor, assuming these notes to be contemporaneous with the voyage, cited this passage as "placing beyond doubt the date when these ideas crystallized" (*Darwin and the Voyage*, p. 246). Unfortunately, however, it appears not in one of the small pocket books where he made his original entries but rather in the large unbound folio pages obviously composed later. These folio pages constitute an annotated catalogue of ornithological specimens. Because of occasional references to places and dates (generally month and year, sometimes year alone), they might at first glance be taken to be a summary of his findings composed, like his diary, several weeks or months after the events to which they refer. This was apparently the assumption of Nora Barlow. A closer examination, however, shows that this was not so. Neither the consecutive numbering on the specimens nor the chronology of the voyage is strictly adhered to. Thus on page 42 there are premature references to the Galapagos and to specimens numbered 3297, 3298, etc., while page 52 reverts to specimens numbered 2004, and not until page 71 is the proper section on the Galapagos reached. Nor can it be argued that the pagination is at fault: the inking of the page numbers corresponds exactly with that on the sheet itself. The only plausible explanation for the departure from consecutive and chronological order (a matter not of weeks or months but of years) is that the whole was composed after Darwin's return from the voyage, probably after the specimens were identified and in preparation for the ornithological volume of the *Zoology of the Voyage*. This dating is further suggested by a note in his

"Fragment of Autobiography" (Cambridge Mss.), under the date of June 1838: "Preparing first part of Birds . . . little species theory." The conjunction of these two events—the work on birds for which this catalogue was a preparation and the reflections on species—is significant, as is the resemblance of the passage to others written about this time.

26. *Journal*, pp. 454, 474.

27. *Ibid.*, p. 474.

28. March 31, 1844: Cambridge Mss.

29. *Journal*, p. 466.

30. Lyell, *Principles* (1st ed.), II, 66.

31. *Journal*, p. 460.

32. *Ibid.*, p. 466.

33. The biographers' habit of finding significance in the mere statement of problems rather than in their answers has been carried to such extremes that the very mention of the words "origin of species" inspires them to portentous comments. Even the notorious Professor Jameson of Edinburgh, for whom Darwin had such contempt and who very nearly turned him away from the study of geology, has been resurrected as an "influence":

> Professor Jameson, too . . . cannot fail to have influenced Darwin somewhat; and we find that the first lecture of the concluding portion of Jameson's zoological course, dealing with "The Philosophy of Zoology," had the suggestive title of "The Origin of the Species of Animals." Thus we must acknowledge that already at Edinburgh Darwin was fairly started in the paths of zoological inquiry, and the northern university must be admitted to share with Cambridge, the distinction of being the foster-parent of this giant-child (Bettany, *Life of Darwin*, p. 23).

34. Sept. 26–27, 1835: *Diary*, p. 337.

35. Aug. 5, 1834: *ibid.*, p. 236.

36. *Journal*, p. 53.

37. *Ibid.*, pp. 399–400.

38. *Ibid.*, p. 469.

39. *Ibid.*, p. 211.

40. *Ibid.*, p. 212.

41. *Diary*, p. 383.

42. *E.g.*, Nora Barlow, referring to the period spent on the Galapagos: "Though a revolution was taking place in his views on the immutability of species . . ." (*Darwin and the Voyage*, p. 245); or Irvine's reference to "subtle changes in his ideas about species . . ." (*Apes, Angels and Victorians*, p. 51).

43. Darwin to Otto Zacharias, 1877: *More Letters*, I, 367.

CHAPTER 6

Pages 126–146

1. Feb. 12, 1836: Cambridge Mss.

2. Years later Darwin wrote to the naturalist, J. Jenner Weir: "It is most curious the number of persons of the name of Jenner who have had a strong taste for Natural History. It is a pity you cannot trace your connection with the great Jenner, for a duke might be proud of his blood" (April 18, 1868: *More Letters*, II, 75).

3. So Erasmus reported it to Charles (Autobiography, *Life and Letters*, I, 75). Lord Farrer, who was also present at the party, recalled Buckle's words as: "Mrs. Wedgwood, what a very inferior man Charles Darwin is to his book" (Cambridge Mss.).

4. Autobiography, *Life and Letters*, I, 77.

5. *Ibid.*, p. 160.

6. Darwin to Henslow, Oct. 14, 1837: *ibid.*, p. 287.

7. "What made him ill was the recognition that, as a family man, work was impossible for him in London. . . . Darwin by his psychoneurosis secretly and passionately nourished his genius" (Hubble, *Lancet*, Jan. 30, 1943, p. 132).

8. Darwin to Lyell, June 1841: *Life and Letters*, I, 272.

9. June 1, 1844: Cambridge Mss.

10. This was Henrietta's recollection. Emma Darwin later denied the fact of the loss of memory and asked Francis to ignore the entire episode in the *Life and Letters* (Henrietta Litchfield to Francis Darwin, March 18, 1887: Cambridge Mss.).

11. March 31, 1845: Cambridge Mss.

12. Hubble, *Lancet*, Dec. 26, 1953, p. 1354.

13. Barlow, *Listener*, Aug. 23, 1956.

14. R. Good, letter in *Lancet*, Jan. 9, 1954, p. 106. Great significance is also attached to the fact that in his autobiography Darwin misdated his father's death, writing Nov. 13, 1847, in place of Nov. 13, 1848.

15. Hubble, *Lancet*, Jan. 30, 1943, p. 133.

16. Gould, *Biographic Clinics*. The other sufferers from eye-strain (that they strained their eyes is deduced from the fact that they read and wrote a lot) include De Quincey, Carlyle and Mrs. Carlyle, Huxley, Browning, George Eliot, Lewes, Wagner, Parkman, Spencer, Whittier, Margaret Fuller, and Nietzsche.

17. Alvarez, *Nervousness, Indigestion, and Pain*, pp. 240–43; Alvarez, *Neuroses*, p. 216.

18. Mme. Sismondi to Emma Wedgwood, Nov. 23, 1838: Emma Darwin, *Letters* (1904 ed.), I, 422–23. This was suppressed in the regular edition of the letters.

19. Emma Wedgwood to Mme. Sismondi, Nov. 15, 1838: *ibid.* (1915 ed.), II, 5.

20. *Ibid.*, I, 277; II, 1.

21. Emma to Mme. Sismondi, Nov. 15, 1838: *ibid.*, I, 7.

22. Jan. 20, 1839: *More Letters*, II, 29.

23. Sept. 1871: Emma Darwin, *Letters* (1904 ed.), II, 249.

24. There were ten children, seven of whom survived into adulthood: William, b. 1839; Anne, 1841–51; Mary, b. 1842 (died in infancy); Henrietta ("Etty"), b. 1843; George, b. 1845; Elizabeth, b. 1847; Francis, b. 1848; Leonard, b. 1850; Horace, b. 1851; Charles, 1856–58.

25. At his death, his estate had been augmented by half of his brother's fortune (Eramus predeceased him by less than a year). He had also been named the heir of a Mr. Anthony Rich. Rich had written to him in 1878 introducing himself and declaring his intention of bequeathing his fortune to him in appreciation of his work. It was a comic situation, with Darwin insisting that he was wealthy enough, and Rich protesting that with all those mouths to feed, he could surely use more money. As it happened, Darwin died before his would-be benefactor, and Rich's estate ultimately went to Darwin's children. The affair was made even more ridiculous by Darwin's feeling that the source of the money was not quite proper. "The property," he described it to his brother, "is not of a nice kind, namely a share in houses in Cornhill, which will bring in rather above £1100 annually" (Dec. 12, 1878: Cambridge Mss.).

26. *Life and Letters*, III, 179. The chapter on "The Man of Business" in Keith's *Darwin Revalued* contains a record of his financial affairs based on his original account books.

27. Dec. 25, 1844: Cambridge Mss. Flinders was the author of *Voyage to Terra Australis in 1801–03, in H.M.S. Investigator* (London, 1814), which contained a valuable botanical appendix by Robert Brown.

28. Charles to Susan Darwin, Sept. 3, 1845: Emma Darwin, *Letters*, II, 97. His notes written when he was contemplating marriage reveal the same obsessive anxieties about money:

> If marry—means limited—feel duty to work for money. . . . If I were moderately rich I would live in London, with pretty big house and do as B—but could I act thus with children and poor? No. Then . . . live in country near London; better; but great obstacles to science and poverty. Then Cambridge, better, but fish out of water, not being Professor and poverty. Then Cambridge Professorship—and make best of it—do duty as such and work at spare times. My destiny will be Camb. Prof. or poor man; outskirts of London—some small square etc.—and work as well as I can. . . . The expense and anxiety of children. . . . Less money for books etc. If many children

> forced to gain one's bread. (But then it is very bad for one's
> health to work too much.)

After more speculations in the same vein, he concluded that his
fate was to suffer a "horrid poverty"—unless, it parenthetically
occurred to him, his wife happened to have money (*Autobiography*, ed. Nora Barlow, pp. 231–34; I have slightly altered
the punctuation).

29. Darwin to Fox, March 7, 1852: *Life and Letters*, I, 382.

30. Erasmus to Charles Darwin, Sept. 25, 1873: Cambridge
Mss.

31. *Life and Letters*, I, 121.

32. Today the village is known as Downe, although Darwin's
house preserves the old spelling of Down House.

33. Kempf, *Psychoanalytic Review*, p. 172.

34. Darwin to Fox, March 28, 1843: *Life and Letters*, I, 321.

35. *Ibid.*

36. *Ibid.*, III, 178.

37. Autobiography, *ibid.*, I, 101.

38. Francis Darwin's Recollections, Cambridge Mss.

39. Autobiography, *Life and Letters*, I, 101–2.

40. Reminiscences of Francis Darwin, *ibid.*, p. 117.

41. *Ibid.*, p. 147.

42. *Origin*, p. 481.

43. *Darwin and the Voyage*, p. 151.

CHAPTER 7

Pages 147–167

1. The words "evolve" and "evolution" do not actually appear
in Darwin's early writings, including the first few editions of
the *Origin*. Although Lyell had used "evolution" in its present
sense in his *Principles*, and Spencer more prominently in his
essay on "The Development Hypothesis" in 1852, it was not
then commonly used, and entered the popular and scientific
vocabularies only later. "Change," "variation," "transformation,"
"transmutation," and "mutability" were the accepted terms of
the doctrine, with "chain of being," "tree of life," or "organization of life" to connote the evolutionary hierarchy. "Evolve"
and "evolution" appear in this discussion only when they convey
the same meaning as Darwin's less familiar terms.

2. *Life and Letters*, I, 276.

3. The geologist J. W. Judd (in Seward [ed.], *Darwin and
Modern Science*, pp. 350–53) and Francis Darwin (in the introduction to *The Foundations of the Origin*, pp. x–xii) date
Darwin's conversion from his first reading of Lyell's *Principles*,
on the grounds that he must have realized that "mutability was

the logical conclusion of Lyell's doctrine, though this was not acknowledged by Lyell himself." The difficult problems of Lyell's implications and intentions will be discussed in the following chapter. It need only be said here that there is no evidence that Darwin took Lyell at other than face value, and there is much evidence that he did not become an evolutionist until after his return from the voyage.

4. Some of these quotations appear in *Life and Letters*, II, 5–10.

5. *Origin*, p. 21.

6. Autobiography, *Life and Letters*, I, 83.

7. Quoted by Nordenskiöld, *History of Biology*, p. 462.

8. *Life and Letters*, I, 104. The published version of this passage is even more ambiguous:

> On the other hand, I am not very sceptical—a frame of mind which I believe to be injurious to the progress of science. *A good deal of scepticism in a scientific man is advisable to avoid much loss of time,* for I have met with not a few men, who, I feel sure, have often thus been deterred from experiment or observations, which would have proved directly or indirectly serviceable. In illustration . . . [There follows an example in which scepticism was desirable.]

In the original, the italicized sentence appears as an addendum, and an obviously misplaced caret has led the editor to print it in this sequence, instead of in its proper place, which is directly before "in illustration." It is evident that Darwin intended to say that an excess of skepticism was injurious to science in deterring men from experiments, etc., although some skepticism was necessary to avoid much loss of time. (Darwin's sons have commented on his strange habit of containing two contradictory or conflicting thoughts in his mind at the same time and trying to express them both simultaneously, or introducing his qualifications before his main statement.)

9. Darwin to Fawcett, Sept. 18, 1861: *More Letters*, I, 195.

10. *Life and Letters*, I, 149.

11. Darwin to Hooker, March 19, 1845: Cambridge Mss.

12. Darwin to Hooker, 1855: *Life and Letters*, II, 55.

13. Darwin to Lyell, Dec. 12, 1859: *ibid.*, II, 241.

14. Dewey, *The Influence of Darwin on Philosophy*, p. 13.

15. Sept. 13, 1838: *Life and Letters*, I, 298.

16. Cambridge Mss.

17. *Ibid.*

18. Autobiography, *Life and Letters*, I, 69.

19. *Ibid.*, p. 83.

20. Bonar, *Malthus and His Work*, p. 407.

21. Malthus, *Essay on the Principle of Population*, I, 5–6. This is a reprint of the seventh edition, which, except for the

addition of appendices, is substantially the same as the second to the sixth, any one of which Darwin may have read.

22. *Ibid.*, p. 29.

23. *Ibid.*, p. 60.

24. Loren Eiseley has pointed out that Lyell had used the expression in his *Principles*, having borrowed it from de Candolle (*Darwin's Century*, pp. 101–2).

25. Malthus, *Essay*, II, 7.

26. *Ibid.*

27. *Ibid.*

28. *Ibid.*, p. 10.

29. June 18, 1862: Marx-Engels, *Briefwechsel*, p. 77.

30. For such a critique, see my article on Malthus in *Encounter*, V (1955), 53–60.

31. Darwin to Lyell, June 6, 1860: *Life and Letters*, II, 317.

32. Wallace, *My Life*, I, 232.

33. Hooker to Asa Gray, 1867 or 1868: Hooker, *Life and Letters*, II, 43.

34. Wallace, *My Life*, I, 362.

35. Polanyi, "From Copernicus to Einstein," *Encounter*, V (1955), p. 60.

36. James, "Great Men and their Environment," *Selected Papers on Philosophy*, p. 193.

37. Jones, *Sigmund Freud*, I, 348.

CHAPTER 8

Pages 168–194

1. Newton, *Opticks*, p. 349.

2. It was de Maillet's work, *Telliamed*, that Huxley said was "surely deserving more respectful consideration than it usually receives" (*Darwiniana*, p. 208).

3. Darwin to I. A. Crawley, Feb. 12, 1879: Cambridge Mss.

4. Hagberg, *Linnaeus*, p. 197, quoting Linnaeus' *Dissertation on Peloris*, 1744.

5. Jackson, *Linnaeus*, p. 361.

6. Buffon, *Natural History*, IX, 167.

7. *Ibid.*, p. 176.

8. *Origin* (6th ed.), p. 9. The "Historical Sketch" first appeared in the first German edition (1860) and in the third English edition (1861).

9. *Life and Letters*, III, 45. The letter is tentatively dated July 12, 1865, but it seems more likely, from the chronology given in his diary, that it was written the following year.

10. *Origin* (6th ed.), p. 10.

11. Autobiography, *Life and Letters*, I, 38.

12. He defended this method in the Apology to *Botanic*

Garden: "Extravagant theories, however, in those parts of philosophy where our knowledge is yet imperfect are not without their uses; as they encourage the execution of laborious experiments, or the investigation of ingenious deductions, to confirm or refute them."

13. Erasmus Darwin, *Zoonomia*, II, 237.

14. *Ibid.*, pp. 238–39.

15. *Ibid.*, p. 239.

16. *Ibid.*, p. 240.

17. *Ibid.*, p. 244.

18. *Botanic Garden*, p. 8.

19. *Phytologia*, p. 556.

20. *Ibid.*, p. 557.

21. Autobiography, *Life and Letters*, I, 38.

22. Lamarck, *Zoological Philosophy*, p. 114.

23. *Ibid.*, p. 37.

24. *Ibid.*, pp. 109–10.

25. *Ibid.*, p. 54.

26. *Ibid.*, p. 46.

27. *Ibid.*, p. 114.

28. *Hydrogéologie*, quoted by Geikie, *Founders of Geology*, pp. 356–57.

29. Cambridge Mss.

30. *Ibid.*

31. Darwin to Hooker, 1844: *Life and Letters*, II, 29, 39.

32. *Ibid.*, p. 23.

33. *Origin*, p. 407.

34. Oct. 11, 1859: *Life and Letters*, II, 215.

35. Darwin to Huxley, June 9, 1859 (?): *More Letters*, I, 125.

36. *Origin* (6th ed.), p. 10.

37. More, *The Dogma of Evolution*, p. 178.

38. Cuvier, *Essay on the Theory of the Earth*, p. 112.

39. Nordenskiöld, *History of Biology*, p. 338.

40. Goethe, *Conversations with Eckermann*, p. 481.

41. Wells, *Two Essays*, p. 439.

42. Cambridge Mss.

43. Lyell, *Principles* (1st ed.), I, 155.

44. *Ibid.*, p. 156.

45. *Ibid.*, II, 65.

46. *Ibid.*, p. 21.

47. Lamarck, *Zoological Philosophy*, p. 43.

48. Lyell, *Principles* (1st ed.), II, 179.

49. Lyell to G. Mantell, March 2, 1827: Lyell, *Life, Letters and Journals*, I, 168.

50. Lyell to George Poulett Scrope, June 14, 1830: *ibid.*, pp. 270–71.

51. Lyell to Sedgwick, Jan. 20, 1838: *ibid.*, II, 36–37.

52. Lyell to Herschel, June 1, 1836: *ibid.*, I, 467.

53. *Ibid.*, pp. 468–69.

54. Lyell to G. Mantell, Dec. 29, 1827: *ibid.*, p. 174. General Campbell was commander of the Burmese campaign and later Governor of Burma.

55. Lyell to his sister, Feb. 26, 1830: *ibid.*, p. 263.

56. Lyell to Dr. Fleming, Feb. 3, 1830: *ibid.*, p. 260. Another letter by Lyell may be taken as an example of the esoteric technique at its most extreme, the technique of saying one thing while implying its opposite. Lyell was ostensibly praising his correspondent for having taken, in a public lecture, "the very highest ground in doing homage to the martyrs for truth." With tongue in cheek, he pretended to deprecate the "low tone of morality" by which truth is suppressed, while clearly admiring and counseling it:

> The extent to which the concealment of nearly all the newly discovered truths in every branch, moral and physical, is defended, if opposed to the popular notion, is one of the worst vices of the times, against which the shafts of satire should be aimed.
>
> Speaking of martyrs, do you remember the low tone of morality with which, in a letter cited in Sir S. Romilly's Life, Mirabeau, with all his characteristic force and eloquence, apologizes for Fontenelle. "He well knew," he says, "that philosophers do not multiply like fanatics under the axe of the executioner or in the dungeons of the Inquisition," etc., and went on to show how slowly and cautiously "truth should be unveiled."

(Lyell to William Grove, Feb. 14, 1832: *ibid.*, II, 277.)

57. For a different interpretation of Lyell's behavior, see the very interesting discussion in Eiseley, *Darwin's Century*, Chapter 4.

CHAPTER 9

Pages 195–215

1. Autobiography, *Life and Letters*, I, 83–4.

2. There are several references to a sketch written in 1839. A note in Darwin's hand in the bound copy of the 1844 manuscript states: "This was sketched in 1839." On the basis of this note, Lyell and Hooker, in their preface to the joint paper of Darwin and Wallace read before the Linnean Society in 1858, spoke of a first sketch of 1839; and a letter written shortly after by Darwin to Wallace referred to a sketch "written in 1839, now just twenty years ago" (Jan. 25, 1859: *Life and Letters*, II, 146).

On the other hand, not only Darwin's autobiography, written

of course many years later, but also his diary, written at the time, specified 1842 as the date of the first sketch and recalled the exact chronology and circumstances of the event: "1842. May 18th went to Maer. June 15th to Shrewsbury, and on 18th to Capel Curig . . . During my stay at Maer and Shrewsbury, (five years after commencement) wrote pencil sketch of my species theory." This is borne out by the original manuscript—not the bound copy—of the 1844 sketch, on which Darwin had written: "This was written and enlarged from a sketch in 37 pages in pencil (the latter written in summer of 1842 at Maer and Shrewsbury) in beginning of 1844, and finished it [sic] in July, and finally corrected the copy by Mr. Fletcher in the last week of September." The wrapper around another manuscript, found by his son long after Darwin's death, has a note, also in Darwin's hand, describing it as "First Pencil Sketch of Species Theory. Written at Maer and Shrewsbury during May and June 1842", the pencil sketch to which this refers is no longer in the packet, although a single penciled page bears the heading: "Maer, May 1842, useless." In addition, a letter to Lyell of June 18, 1858 (Life and Letters, II, 116), and the first edition of the Origin gave 1842 as the date of the first sketch. The problem is complicated by the fact that the dates appearing on his manuscripts were often added long after the actual time of their composition, so that what appears to be contemporary is not necessarily so.

The weight of the evidence, therefore, rests with the diary entry, which is the only certain original source and which also, in tone and circumstantial details, has the ring of authenticity. Shocked by Wallace's letter and naturally seeking to defend his priority, Darwin later seized upon the date of 1839 when the theory had taken definite form in his mind; in his autobiography 1839 appears as the date "when the theory was clearly conceived" (Life and Letters, I, 88). He probably added this date to the copy of the sketch he turned over to Lyell and Hooker, who then published it in the Linnean Society report. (Much of the evidence in favor of 1842 as the date of the first sketch is given by Francis Darwin in his introduction to the published edition of this sketch, Foundations of the Origin, xv–xvii.)

3. July 5, 1844: Life and Letters, II, 16.
4. Ibid., p. 17.
5. Jan. 11, 1844: ibid., p. 23.
6. Foundations of the Origin, p. 6.
7. Ibid., p. 13.
8. Ibid., p. 52.
9. Life and Letters, I, 84.
10. Foundations of the Origin, p. 77. Francis Darwin agreed that the lack of a more explicit statement did not seem to be a serious flaw (Life and Letters, II, 15–16).
11. Foundations of the Origin, p. 7.

12. The chronology given in the *Life and Letters* (I, 327) is inaccurate. According to that, the *Geology of South America* occupied Darwin from July 1844 to April 1845, and the second edition of the *Journal* from October 1845 to October 1846. In fact, the geology volume took from July 1844 to October 1846 (only the first part had been finished by April 1845), and the *Journal* from April to August 1845. This agrees with the chronology in his unpublished diary (Cambridge Mss.) and in *More Letters*, I, xix–xx.

13. Darwin to Hooker, 1845: *Life and Letters*, I, 346.

14. Huxley to Francis Darwin: *ibid.*, p. 347.

15. Francis Darwin to Huxley, Dec. 24, 1885: Huxley Mss., vol. xiii.

16. *Life and Letters*, I, 346.

17. Darwin to Hooker, May 10, 1848: *More Letters*, I, 65.

18. Darwin to Hooker, Sept. 25, 1853: *Life and Letters*, II, 40.

19. *Ibid.*

20. April 9, 1849: *Life and Letters*, I, 376.

21. Darwin to Fox, Jan. 1841: Cambridge Mss.

22. Darwin to Hooker, Feb. 23, 1844: *Life and Letters*, II, 25.

23. Darwin to Hooker, July 1844: *ibid.*, p. 30.

24. Darwin to Fox, March 19, 1855: *ibid.*, p. 46.

25. Darwin to Fox, Nov. 13, 1858: Cambridge Mss.

26. An unpublished portion of Darwin's autobiography exposes this weakness: "He was very fond of society, especially of eminent men and of persons high in rank; and this over-estimation of a man's position in the world seemed to me his chief foible. He used to discuss with Lady Lyell as a most serious question, whether or not they should accept some particular invitation."

Hooker complained to Darwin that Lyell had once approached Mrs. Hooker to ask whether her husband had written a review of his own book. Not content with this "monstrous and revolting" insinuation, as the indignant Hooker reported, "by way of mending the matter, when my wife told him I was not, he added that he did not think it very well done!" (Dec. 7, 1856: Cambridge Mss.).

27. Lyell, *Principles* (6th ed.), I, 287.

28. Darwin to Hooker, 1844: *Life and Letters*, II, 29.

29. July 13, 1856: *More Letters*, I, 96.

30. Lyell to Hooker, July 25, 1856: Lyell, *Life, Letters and Journals*, II, 214–15.

31. Hooker, *Life and Letters*, I, 41.

32. Darwin to Hooker, Sept. 25, 1853: Darwin, *Life and Letters*, II, 40.

33. Hooker's notes: *ibid.*, p. 27.

34. Hooker, *Life and Letters*, I, 491.

35. Jan. 28, 1859: *ibid.*, p. 501.

36. *More Letters*, I, 39.

37. Hooker, *Life and Letters*, I, 497.

38. Lyell to C. J. F. Bunbury, Nov. 13, 1854: Lyell, *Life, Letters and Journals*, II, 199.

39. Hooker to Harvey, Jan. 1, 1859: Hooker, *Life and Letters*, I, 481–82.

40. *Ibid.*

41. Huxley, *Life and Letters*, I, 11.

42. *Ibid.*, p. 129.

43. Huxley Mss., vol. V; also copy in Cambridge Mss.

44. Autobiography, Cambridge Mss.

45. *Life and Letters*, II, 196–97. The biographer who wrote: "To those of us who began biological work after the idea of evolution had been impressed upon anatomical work, it is very difficult to follow Huxley's papers [of the 1850's] without reading into them evolutionary ideas" (Mitchell, *Huxley*, p. 60), was suffering from the familiar occupational fallacy of the historian, the habit of reading the future into the past.

46. *Life and Letters*, II, 195.

47. *Ibid.*, p. 196.

48. Lyell to Horner, Jan. 19, 1842: Lyell, *Life, Letters and Journals*, II, 63.

49. A partial list of Darwin's main correspondents before the publication of the *Origin*—the list becomes immensely extended after the publication—includes, apart from those already mentioned: the zoologists Leonard Jenyns (later the Rev. Blomefield), Edward Blyth, John Gould, Thomas Bell, G. R. Waterhouse, and Thomas Wollaston; the botanists John Henslow, J. F. Royle, Robert Brown, John Lindley, Dean Herbert, H. C. Watson, Hugh Falconer, Sir Charles Bunbury, and George Bentham; the geologists Hugh Strickland, S. P. Woodward, Edward Forbes, and Sir A. C. Ramsay; the anatomist-paleontologist Richard Owen; the American geologist-zoologists Louis and Alexander Agassiz and James Dana; and scores of practical breeders and agriculturalists of whom the most prominent was the poultry breeder W. B. Tegetmeier.

50. Darwin to Gray, Sept. 5, 1857: Cambridge Mss.

51. A curious variation on the usual priority claim came when Darwin found himself obliged to quote someone else on a problem which he had earlier and independently worked out for himself. In the *Origin* he graciously acknowledged the priority of Forbes on a particular problem of the distribution of alpine plants, although in his personal letters and autobiography he could not resist mentioning the privately circulated paper in which he had earlier made the same point.

CHAPTER 10

Pages 216–240

1. Disraeli, *Tancred*, I, 225–26.
2. *Vestiges of the Natural History of Creation*, p. 148.
3. *Ibid.*, p. 157.
4. *Ibid.*
5. *Ibid.*, p. 231.
6. Bosanquet, *Vestiges: Its Argument Examined and Exposed*, p. 3.
7. *Edinburgh Review*, LXXXII (1845), pp. 2–3.
8. *Ibid.*, pp. 3–4.
9. *Life and Letters*, II, 189.
10. Huxley, "Science and Pseudo-Science," *Controverted Questions*, pp. 281–83.
11. *More Letters*, I, 114.
12. *Origin*, p. 3.
13. *Vestiges*, p. 238.
14. *Ibid.*, p. 374.
15. Dec. 30, 1844: Cambridge Mss.
16. *Life and Letters*, I, 333.
17. *Ibid.*, II, 188.
18. *Vestiges*, pp. 224–25.
19. *Origin* (6th ed.), p. 14.
20. Sept. 2, 1854: *More Letters*, I, 75.
21. Dec. 30, 1844: Cambridge Mss.
22. Darwin to Hooker, 1845: *Life and Letters*, II, 39.
23. Spencer, *Autobiography*, I, 335, 400.
24. Francis Darwin's Recollections, Cambridge Mss. The exact wording is Francis Darwin's, not Spencer's.
25. Spencer, *Autobiography*, I, 403.
26. *Ibid.*, p. 505.
27. *Ibid.*, p. 384.
28. Spencer, *Essays Scientific, Political, and Speculative*, I, 6.
29. Spencer, "Theory of Population," p. 34.
30. *Autobiography*, I, 472.
31. James, *Memories and Studies*, p. 126.
32. Darwin to E. Ray Lankeaster, March 15, 1870: *Life and Letters*, III, 120.
33. Darwin to Hooker, Dec. 10, 1866: *ibid.*, p. 56.
34. Darwin to Lyell, Feb. 25, 1860: Darwin-Lyell Mss., American Philosophical Society Library.
35. Autobiography, Cambridge Mss.
36. Wordsworth, "The Tables Turned."
37. Wordsworth, "A Poet's Epitaph."

38. John Wain, quoting Cyril Connolly, *Twentieth Century*, May 1953, p. 383.

39. Tennyson, "In Memoriam," invocation.

40. T. S. Eliot, *Essays Ancient and Modern*, p. 190.

41. Johnson, *The Alien Vision of Victorian Poetry*, xvi.

42. Hallam Tennyson, *Lord Tennyson*, p. 251.

43. *Ibid.*, p. 253.

44. Eliot, p. 184.

45. Morley, *Recollections*, I, 14–15.

46. Carlyle to Emerson, Aug. 5, 1844: *Correspondence of Carlyle and Emerson*, II, 66.

47. Eliot, p. 187.

48. "In Memoriam," lv–lvi.

49. *Ibid.*, xcvi.

50. *Ibid.*, cxviii.

51. Huxley to Tyndall, Oct. 15, 1892: Huxley, *Life and Letters*, III, 270.

52. "In Memoriam," cxxviii.

53. Aug. 17, 1868: Hallam Tennyson, *Lord Tennyson*, p. 464.

54. *Ibid.*, p. 270.

55. *Ibid.*, p. 262.

56. *Ibid.*, p. 186.

57. Samuels, *Young Henry Adams*, p. 128.

58. Geikie, *Life of Murchison*, II, 333.

59. *Ibid.*

60. Gillispie, *Genesis and Geology*, p. xi. I am much indebted to this excellent work on the relations between religion and science in the first half of the nineteenth century.

61. Clark, *Victorian Mountaineers*, pp. 35–36.

62. Sedgwick, *Discourse on the Studies of the University*.

63. Miller, *Testimony of the Rocks*, p. 75.

64. Gillispie, p. 50.

65. *Eclectic Review*, I, 672.

66. Gosse, *Father and Son*, p. 118.

67. There is very little that the highly personal God of fundamentalism cannot be—indeed, has not been—credited with. In an *éloge* of Cuvier, one French scientist declared: "It would almost appear as if the Almighty had placed this vast and wonderful collection of fossil remains for the express purpose of being discovered and descanted upon by our great country-man!" (Basil Hall to Leonard Horner, Sept. 7, 1833: Lyell, *Life, Letters and Journals*, II, 466).

68. Drachman, *Studies in the Literature of Natural Science*, p. 48.

69. Ruskin T. Acland, 1851: Cook, *Life of Ruskin*, II, 19–20.

70. Carlyle, *Life of Sterling*, p. 9.

71. Jevons, *Letters and Journals*, p. 23.

72. Autobiography, *Life and Letters*, I, 87.

CHAPTER 11

Pages 242–254

1. June 18, 1858: *Life and Letters*, II, 116.
2. June 25, 26, 1858: *ibid.*, pp. 117–18.
3. Wallace, *My Life*, I, 359.
4. *Ibid.*, p. 255.
5. *Annals and Magazine of Natural History*, 2nd series, xvi (1855), 186, 191.
6. May 1, 1857: *Life and Letters*, II, 95–96.
7. Twelve years before, he now specified; although earlier in the autobiography he had said that he had read Malthus while teaching at Leicester in 1844 (*My Life*, I, 232, 362).
8. *Ibid.*, pp. 362–63.
9. Wallace, *Letters and Reminiscences*, I, 114–16.
10. At the celebration held in 1908 by the Linnean Society commemorating the Darwin-Wallace papers, Hooker said that Darwin had been elected a Fellow of the Society only a month before the presentation of his paper. In fact, the records of the society show that he had been proposed on Dec. 20, 1853, elected on March 7, 1854, and admitted on May 2, 1854.
11. Gage, *History of the Linnean Society*, p. 56.
12. *Life and Letters*, II, 126.
13. March 28, 1859: *ibid.*, p. 152.
14. Paston, *At John Murray's*, p. 170.
15. *Ibid.*, p. 172.
16. Haynes, *Cornhill Magazine*, XLI (1916), 233.
17. 1867: *Life and Letters*, III, 60.
18. May 3, 1859: *More Letters*, I, 121.
19. Darwin's biographers often speak as if the book was sold out in the bookstores on the first day of its sale. In fact, it was "sold out" only so far as the publisher was concerned—that is, it was fully subscribed by the dealers.
20. Cruse, *Victorians and Their Books*, pp. 329–30.

CHAPTER 12

Pages 255–267

1. Darwin to Hooker, Oct. 23, 1859: *Life and Letters*, II, 175–76.
2. Darwin to Lyell, Sept. 30, 1859: *ibid.*, p. 169; Sept. 2, 1859: *ibid.*, p. 165.
3. Lyell to C. J. F. Bunbury, Nov. 13, 1854: Lyell, *Life*,

Letters and Journals, II, 199. Lyell attributed the phrase "ugly facts" to Hooker.

4. *Life and Letters,* II, 166.

5. *Origin,* p. 264.

6. Darwin to Lyell, Oct. 20, 1859: *Life and Letters,* II, 174. The words are Darwin's.

7. Sept. 2, 1859: *ibid.,* p. 165.

8. Oct. 11, 1859: *ibid.,* p. 208.

9. Dec. 5, 1859: Darwin-Lyell Mss., American Philosophical Society Library.

10. Darwin to Gray, July 22, 1860: *Life and Letters,* II, 326.

11. Lyell to Hooker, Dec. 19, 1859: Lyell, *Life, Letters and Journals,* II, 327–28.

12. Darwin to Lyell, Oct. 25, 1859: *Life and Letters,* II, 176.

13. Lyell had spoken of writing such a book as early as 1856 or 1857.

14. Lyell, *Geological Evidences of the Antiquity of Man,* pp. 412, 469.

15. *Ibid.,* p. 506.

16. Lyell to Darwin, March 15, 1863: Lyell, *Life, Letters and Journals,* II, 365.

17. Feb. 24, 1863: *Life and Letters,* III, 9.

18. Lyell to Darwin, March 11, 1863: Lyell, *Life, Letters and Journals,* II, 363–64.

19. Darwin to Hooker, May 15, 1863: *More Letters,* I, 241.

20. Lyell, *Antiquity of Man* (2nd ed.), p. 469.

21. *Ibid.,* p. 412.

22. Lyell to Hooker, March 9, 1863: Lyell, *Life, Letters and Journals,* II, 362.

23. Lyell to George Ticknor, Jan. 9, 1860: *ibid.,* p. 329.

24. Lyell to Haeckel, Nov. 23, 1868: *ibid.,* pp. 436–37.

25. Darwin to Wallace, Jan. 25, 1859: *Life and Letters,* II, 146.

26. Dec. 1859: Hooker, *Life and Letters,* I, 511.

27. From Darwin's *Life and Letters* it might appear that Hooker's essay followed, rather than preceded, the *Origin.* On October 23 Darwin first congratulated Hooker on its completion (II, 175); on November 19 or 20 he asked whether it had yet been published (p. 225); on November 21 Hooker replied that "those lazy printers have not finished my luckless Essay" (p. 228); and it was not until December 22 that Darwin received a copy of it (p. 246). Yet Hooker's *Life and Letters* refers to the publication of the essay "nearly a month earlier" than the *Origin* (I, 504); and Hooker himself, while denying the significance of its priority, did not deny its fact (p. 502). The explanation is simply that the entire *Flora Tasmaniae,* including the introductory essay, was published almost a month before the

Origin, but that the separate reprint of the essay did not appear until later that year.

28. Fawcett, *Macmillan's Magazine,* III (1860), 92.

29. Hooker to Harvey, Hooker, *Life and Letters,* I, 516–17.

30. Huxley, Review of the *Origin* in *Westminster Review, Darwiniana,* p. 52.

31. Ernle, *Quarterly Review,* CCXXXIX (1923), 224.

32. *Life and Letters,* II, 198.

33. Huxley, Review in the *Times, Darwiniana,* pp. 19–21.

34. Huxley, Review in the *Westminster Review, Darwiniana,* pp. 74–75, 78.

35. Hooker to George Bentham, July 17, 1859: Hooker, *Life and Letters,* I, 485.

36. Huxley to Lyell, June 25, 1859: Huxley, *Life and Letters,* I, 250.

37. March 22, 1861: *ibid.,* p. 276.

38. Darwin to Hooker, March 3, 1860: *Life and Letters,* II, 293. The table listed, under geologists: Lyell, Ramsay, Jukes, and H. D. Rogers; under zoologists and paleontologists, Huxley, J. Lubbock, L. Jenyns ("to large extent"), and Searles Wood; under physiologists, W. B. Carpenter and Sir H. Holland ("to large extent"); and under botanists, Hooker, H. C. Watson, Asa Gray ("to some extent"), Dr. Boott ("to large extent"), and Thwaites.

39. Carpenter, *National Review,* X (1860), 214.

40. Gray, "Natural Selection not Inconsistent with Natural Theology" (1860), *Darwiniana,* pp. 91–92.

41. Darwin to Hooker, May 11, 1859: *Life and Letters,* II, 158.

42. Gray to Booth, Jan. 16, 1860: Lady Lyell's Album, Mss. American Philosophical Society Library.

43. Gray, *Darwiniana,* p. 11.

44. The title of the pamphlet based on his *Atlantic Monthly* articles.

45. Darwin to Gray, Nov. 26, 1860: *Life and Letters,* II, 354.

CHAPTER 13

Pages 268–286

1. Sedgwick, *Life and Letters,* II, 290.

2. Darwin, *Life and Letters,* II, 248–50. The letter is here dated Dec. 24, 1859. This is apparently an error for Nov. 24, as Francis Darwin pointed out (*More Letters,* I, 136), and as is apparent from the date of Darwin's reply, which clearly appears on the original manuscript as Nov. 26 (British Museum Mss. Eg. 3020A).

3. British Museum Mss. Eg. 3020A.

4. *Ibid.*

5. Darwin to W. H. Miller: *More Letters*, I, 123; to Henslow, May 14, 1860: *ibid.*, p. 150.

6. Feb. 19, 1860: British Museum Mss. Eg. 3020A.

7. March 6, 9, 1860: *ibid.* Other letters from the Archbishop of Dublin included among these manuscripts are dated Feb. 13 and 25, March 16, and April 13. Sedgwick's part of the correspondence has not been preserved, but its substance may be deduced from the Archbishop's.

8. Owen, *Life*, II, 95.

9. *Spectator*, March 24, 1860, pp. 285–86. The letter was reprinted on April 7 in a slightly revised and expanded form. The editor explained that the author had not had an opportunity to correct the proofs of the first version.

10. When Clark's son, many years later, denied that his father had supported Sedgwick, a footnote was added to Darwin's account of the episode suggesting that Darwin must have misunderstood Henslow's report of the meeting (*Life and Letters*, II, 308). However, the subsequent publication of Henslow's letter (the letter had been addressed to Hooker, with instructions to forward it to Darwin) leaves little doubt that Clark did, in fact, endorse Sedgwick's remarks. The suggestion of Hooker's biographer, that Clark's son must have intended to dissociate his father from the attack on Darwin's morals as distinct from that on his science, seems to have been less a serious attempt to resolve the discrepancy than a gesture of respect for the son's filial piety.

11. May 10, 1860: Hooker, *Life and Letters*, I, 513.

12. The reference was to Baden Powell, who had cited Darwin in his essay, "On the Study of the Evidences of Christianity," in the recently published *Essays and Reviews*.

13. Darwin to Gray, July 22, 1860: *Life and Letters*, II, 327.

14. Henslow to Hooker, May 10, 1860: Hooker, *Life and Letters*, I, 513.

15. Jenyns, *Memoir of Henslow*, p. 213. This is Jenyns' account of Henslow's views.

16. Wilberforce's authorship was not made public until 1874, when the review was reprinted in his *Essays Contributed to the 'Quarterly Review.'*

17. "Wilberforce," *Dictionary of National Biography*, LXI, 207.

18. Wilberforce, *Essays*, I, 59.

19. *Ibid.*, p. 70.

20. *Ibid.*, p. 103.

21. *Ibid.*, p. 93.

22. *Ibid.*, p. 94.

23. *Ibid.*, p. 98.

24. *Ibid.*, p. 100. It is curious that he should have singled out

Oken for abuse, since Owen's philosophy was so largely and explicitly based on Oken's.

25. Oct. 22, 23, 1886: Hooker, *Life and Letters*, II, 301–2.

26. Darwin to Hooker, July 1860: Darwin, *Life and Letters*, II, 324; to Brodie Innes: *ibid.*, p. 325.

27. *Journal of the Proceedings of the Linnean Society*, 1857, p. 20.

28. Only once did he let the secret escape: when he complained to a friend that the editor of the *Edinburgh Review* had mutilated his manuscript (Darwin to Hooker, April 23, 1861: *More Letters*, I, 185). Even his grandson-biographer, writing of this episode thirty-five years later, after the death of both protagonists, preserved this fiction of anonymity, being content to quote without comment a letter from Sedgwick diffidently suggesting Owen as the author of the review (Owen, *Life*, II, 96). Because this anonymity was so carefully observed, others found it necessary to be discreet about any definite attribution to Owen. When Huxley, in his chapter on the reception of the *Origin* in the *Life and Letters*, wrote: ". . . the author of an article in the *Edinburgh Review* who is so familiar with the works of Sir R. Owen, that he had even caught his peculiarities of style," Francis Darwin requested that he omit the sentence. He gave as his reasons, first, his mother's objections to it; second, his intention to print Darwin's criticism of the article, which would make such an open identification embarrassing; and third, the information that had been given to him that although Owen supplied the material for the article, it had actually been written by a Carter Blake (Huxley Papers, vol. xiii).

29. Darwin to Henslow, May 8, 1860: *More Letters*, I, 149.

30. *Edinburgh Review*, CXI (1860), 516.

31. *Ibid.*, pp. 500–3.

32. *Ibid.*, p. 511.

33. *Ibid.*, p. 532.

34. Owen, *Life*, II, 40.

35. Darwin to Jeffries Wyman, Oct. 3, 1860: Dupree, *Isis*, XLII (1951), 107.

36. Darwin referred to this at least twice: March 26, 1862: *More Letters*, I, 200; Dec. 5, 1863: *ibid.*, II, 338.

37. *Athenaeum*, XXXIV (1859), 660.

38. *Annals and Magazine of Natural History*, V (1860), 138.

39. Harvey, "An Inquiry," p. 2.

40. *North British Review*, XXXII (1860), 476.

41. *Ibid.*, p. 457.

42. *Saturday Review*, VIII (1859), 775–76; IX (1860), 573.

43. *Dublin Review*, XLVIII (1860), 51.

44. *Saturday Review*, VIII (1859), 521.

45. *North British Review*, XXXII (1860), 486.

1. *Life and Letters,* II, 186.

2. June 25, 1860: Darwin-Lyell Mss., American Philosophical Society Library.

3. Darwin to Huxley, July 3, 1860: *Life and Letters,* II, 324.

4. Andrade, *Sir Isaac Newton,* p. 69.

5. Darwin to J. D. Dana, July 30, 1860: *More Letters,* I, 159–60.

6. It must be emphasized that these were not direct quotations but rather paraphrases. Unfortunately, the only contemporary account of the meeting, apart from some letters written by those present, was in the *Athenaeum,* July 7 and 14, 1860, and this was incomplete.

7. Huxley to Francis Darwin, June 27, 1891: Huxley, *Life and Letters,* I, 187–88.

8. Hooker to Darwin, July 2, 1860: Hooker, *Life and Letters,* I, 525–26. Only one critic has pointed out that the Yankee donkey was, in fact, not an American but an Englishman who had come to the United States only in his twenties, and that the accent which so amused Oxford was probably as much North of England as American (Eiseley, *Yale Review,* 1955, p. 634). A more serious error is the impression left by almost all biographers, when they have not stated it as a fact, that the "metaphysical fulminations" (Irvine, *Apes, Angels, and Victorians,* p. 5) which Draper was droning out in his quaint American accent were directed against Darwin. In fact, if Draper did put the question, as reported by one witness and thereafter quoted in most accounts of the episode—"Air we a fortuitous concourse of atoms?" ("A Grandmother's Tales," *Macmillan's Magazine.* LXXVIII [1898], 433)—he intended it in criticism of Darwin's opponents.

9. Hooker to Darwin, July 2, 1860: Hooker, *Life and Letters,* I, 526. The letter is reprinted in Darwin, *Life and Letters,* II, 231, but with the word "ugliness" omitted.

10. Huxley to Francis Darwin, June 27, 1891: Huxley, *Life and Letters,* I, 188.

11. This is the account of one member of the audience, John Richard Green, the eminent historian, then an undergraduate at Oxford (Green to W. B. Dawkins, July 3, 1860: Green, *Letters,* p. 45). Since there is no verbatim account of the debate, there has been much dispute about Huxley's exact words. Another member of the audience, recalling the event many years later, thought that Green's account lacked the simplicity and incisiveness of Huxley's style. Others took exception to the

word "equivocal"; Huxley himself was certain he had not used it. Nevertheless, the substance of the remark is not in doubt, and, questioned about it many years later, Huxley was satisfied with its essential accuracy. Some of the more vulgar renditions were almost certainly invented: for example, the version in the biography of Wilberforce by his son, in which Huxley is made to say, "I would rather be descended from an ape than a bishop" (Wilberforce, *Life*, II, 45).

12. G. J. Storey to Francis Darwin, May 17, 1895: Cambridge Mss. Most accounts agree that FitzRoy spoke at the same meeting as Wilberforce and Huxley (e.g., Tuckwell, *Reminiscences of Oxford*, pp. 52–53), but Darwin, who was not himself present, referred to the incident of "poor FitzRoy with the Bible" as taking place at the Geographical Section of the Association (Darwin to Henslow, July 16, 1860: Cambridge Mss.).

13. July 2, 1860: Hooker, *Life and Letters*, I, 526.

14. Huxley, *Life and Letters*, I, 184.

15. *Ibid.*, p. 183.

16. *Ibid.*, p. 189.

17. *Ibid.*, p. 188.

18. *Ibid.*, p. 184.

19. *Illustrated London News*, XXXV (1859), 633.

20. Hooker to Henslow, May 1860: Hooker, *Life and Letters*, I, 514.

21. Darwin to Lyell, August 11, 1860: Darwin, *Life and Letters*, II, 331.

22. *Origin*, p. 408.

23. *Ibid.*

24. Geikie, *A Long Life's Work*, p. 72.

25. Shaler, *Autobiography*, p. 128.

26. Pearson, *Charles Darwin*, p. 23.

27. Weismann, *The Evolution Theory*, p. 28.

28. Darwin to Gray, Dec. 21, 1859: *Life and Letters*, II, 245.

29. Bryce, *Proceedings of the American Philosophical Society*, XLVIII (1909), x.

30. Morley, *Critical Miscellanies*, III, 126.

31. Buckle, *History of Civilization in England*, I, 806.

32. Buckle to Mrs. Woodhead, Jan. 17, 1860: Buckle, *Life and Writings*, II, 28.

33. Darwin to Hooker, Feb. 23, 1858: *Life and Letters*, II, 110.

34. 1862: *ibid.*, p. 386.

35. Fawcett, *Macmillan's Magazine*, III (1860), 84.

36. July 16, 1861: *More Letters*, I, 189. See also Mill, *System of Logic*, p. 328.

37. Mill to Alexander Bain, April 11, 1860: Mill, *Letters*, I, 236.

38. Mill, "Theism," *Three Essays on Religion*, pp. 172–74.

39. Journal, Nov. 23–24, 1859: Eliot, *Life*, II, 109–10.

40. Eliot to Mrs. Bodichon, Dec. 5, 1859: *ibid.*, p. 113.

41. This is the quotation as it appeared in the *Origin* (2nd ed., 1860), p. 481. Either Kingsley in a subsequent letter, perhaps that in which he gave permission to use this quotation, altered its wording or Darwin himself altered it for greater clarity. In the *Life and Letters* (Kingsley to Darwin, Nov. 18, 1859: II, 288), it reads:

> I have gradually learnt to see that it is just as noble a conception of Diety, to believe that He created primal forms capable of self development into all forms needful *pro tempore* and *pro loco*, as to believe that He required a fresh act of intervention to supply the *lacunas* which He Himself had made.

42. Bree, *Species not Transmutable*, p. 244.

43. Darwin to Henslow, Oct. 26, 1860: *More Letters*, I, 174.

44. Nov. 18, 1859: *Life and Letters*, II, 287.

45. Kingsley, *Water Babies*, p. 183.

46. *Essays and Reviews*, p. 139.

47. Simpson, *Landmarks in the Struggle between Science and Religion*, p. 177.

48. Manning, "On the Subjects Proper to the Academia," *Essays on Religion and Literature*, p. 51.

49. Jan. 2, 1860: *Life and Letters*, II, 261.

50. Carlyle to Blunt: Wilson and MacArthur, *Carlyle in Old Age*, p. 328.

51. Wilson, *Carlyle to Threescore and Ten*, p. 517.

52. Jane Carlyle to Mrs. Russell, Jan. 28, 1860: Jane Carlyle, *Letters and Memorials*, II, 119–20.

53. *Times*, Jan. 17, 1877.

54. Eramus Darwin to Charles Darwin, Jan. 27, 1877: Cambridge Mss.

55. Schopenhauer, *World as Will and Idea*, I, 356; Copleston, *Schopenhauer*, p. 193; Murray, *Science and Scientists in the Nineteenth Century*, p. 207.

56. Geikie, *Long Life's Work*, p. 129.

57. Zahm, *Evolution and Dogma*, p. xvii.

58. Darwin to Gray, July 3, 1860: Cambridge Mss.

59. Kingsley, *Letters and Memories*, II, 171.

60. Sedgwick to Dr. Livingstone, March 16, 1865: Sedgwick, *Life and Letters*, II, 411.

61. Sept. 12, 1868: Huxley, *Life and Letters*, I, 297.

62. July 28, 1868: *More Letters*, I, 304.

63. Wallace, *Letters and Reminiscences*, I, 233.

64. Marchand, *The Athenaeum*, p. 92.

65. *Origin* (6th ed.) p. 522.

66. Froude, *The Earl of Beaconsfield*, p. 176.

67. *Punch*, April 8, 1871, p. 145.

1. *Origin*, p. 26.

2. The term "survival of the fittest" was first introduced in the fifth edition (1869) with the remark: "The expression often used by Mr. Herbert Spencer of the Survival of the Fittest is more accurate [than natural selection], and is sometimes equally convenient" (p. 72).

3. *Origin*, p. 72.

4. This was made even more emphatic in the *Descent of Man*. See below, Chapter 17, pp. 363–66.

5. *Origin*, p. 65.

6. Dobzhansky, *Evolution, Genetics and Man*, p. 112.

7. *Origin*, p. 173.

8. *Ibid.*, p. 52.

9. *Ibid.* (6th ed.), pp. 233–34.

10. Darwin to Lyell, Dec. 12, 1859: *Life and Letters*, II, 241.

11. Bateson, "Heredity and Variation in Modern Lights," in Seward (ed.), *Darwin and Modern Science*, p. 99.

12. *Origin* (6th ed.), pp. 232, 243.

13. *Ibid.*, p. 105.

14. Darwin to Galton, Nov. 7, 1875: *More Letters*, I, 306.

15. Butler, *Life and Habit*, p. 263.

16. *Origin* (6th ed.), pp. 94–95.

17. Darwin to Lyell, Sept. 28, 1860: *Life and Letters*, II, 346; Darwin to Wallace, July 5, 1866: *ibid.*, III, 46.

18. *Origin*, p. 11. In the sixth edition, the laws of inheritance, from being "quite" unknown, had become "for the most part" unknown.

19. *Foundations of the Origin*, p. 2.

20. *Ibid.*, p. 85.

21. Fisher, *Genetical Theory of Natural Selection*, p. 5.

22. *More Letters*, I, 103.

23. *Origin*, p. 139.

24. *Ibid.* (6th ed.), p. 169.

25. *Variation of Animals and Plants under Domestication*, II, 357.

26. Darwin to Huxley, May 27, 1865 (?): *Life and Letters*, III, 44; Darwin to Bentham, April 22, 1868: *More Letters*, II, 371.

27. *Variation*, I, 403–5.

28. Zirkle, *Proceedings of the American Philosophical Society*, LXXXIV (1941), 74.

29. Fisher, quoted by J. Huxley (ed.), *Evolution as a Process*, p. 5.

30. J. Huxley, *ibid.*

31. *Ibid.*

32. *Origin,* p. 82. The sixth edition deleted the word "infinitesimally."

33. *Ibid.,* chap. IX.

34. *Ibid.,* p. 257.

35. *Ibid.,* p. 255.

36. *Ibid.,* p. 262.

37. Huxley, "Paleontology and the Doctrine of Evolution" (address delivered in 1870, quoting his opinion of 1862), *Critiques and Addresses,* pp. 182–83.

38. *Ibid.,* p. 200.

39. *Origin,* p. 245.

40. *Ibid.* (6th ed.), p. 340.

41. *Life and Letters,* I, 103.

42. Whewell, *Astronomy and General Physics,* pp. xvii–xviii.

43. *Variation,* I, 8–9.

44. In later editions this appeared as "Difficulties of the Theory."

45. *Origin,* p. 409.

46. Darwin to Bentham, May 22, 1863: *Life and Letters,* II, 24–25.

47. *Origin,* p. 175.

48. Polanyi, *Personal Knowledge,* p. 47.

49. Even one who accepts the theory of natural selection may be repelled by Darwin's own presentation and defense of it. C. D. Darlington, professor of botany at Oxford, in his recent lecture on "Darwin's Place in History," is extremely harsh on the subject of Darwin's "slippery" and "wriggly" tactics: "a flexible strategy which is not to be reconciled with even average intellectual integrity"; "his intellectual opportunism, his willingness or at least ability to confound theoretical issues and conceal the confusion from his indiscriminating readers by loose writing or, shall we say, by making a mystery wherever it was wanted" (pp. 53, 55, 56).

CHAPTER 16

Pages 337–352

1. *Origin* (6th ed.), pp. 58–59.

2. Darwin to Gray, April 3, 1860: *Life and Letters,* II, 296.

3. *Origin,* pp. 160–61.

4. Weismann, "The Selection Theory," in Seward (ed.), *Darwin and Modern Science,* p. 25.

5. *Origin* (6th ed.), p. 214.

6. Helmholtz, *Popular Lectures on Scientific Subjects,* p. 201.

7. Bateson, in Seward (ed.), *Darwin and Modern Science*, p. 100.

8. Medawar, *Twentieth Century*, Sept. 1955, pp. 236–45.

9. *Origin* (6th ed.), p. 213.

10. *Ibid.* (1st ed.), p. 195.

11. *Ibid.*, p. 193.

12. *Ibid.* (6th ed.), p. 136.

13. *Spectator*, March 24, 1860, p. 285.

14. Hutton, *Aspects of Religious and Scientific Thought*, pp. 45–53.

15. Bree, *An Exposition of Fallacies in the Hypothesis of Mr. Darwin.*

16. See above, Chapter 7, pp. 157–58.

17. Nordenskiöld, *History of Biology*, p. 471.

18. *Life and Letters*, II, 201.

19. Bergson, *Creative Evolution*, p. 43.

20. *Ibid.*, pp. 72–73.

21. The first edition contained only Whewell and Bacon; Butler made his appearance in the second edition.

22. *Origin*, p. 172.

23. Paley, *Natural Theology*, p. 8.

24. *Origin*, p. 87.

25. *Ibid.*, p. 189. In the sixth edition the word "odious" was omitted, apparently in an effort to minimize the anthropomorphism (p. 276).

26. *Ibid.*, p. 77. In later editions he added the phrase, "in the eyes of the female bird," after "ornamental" (6th ed., p. 103); but instead of improving matters, this only served to call attention to the blatant anthropomorphism of the sentiment by presuming that Darwin could judge what the female bird would regard as ornamental.

27. July 13, 1856: *More Letters*, I, 94.

28. *Origin*, p. 68.

29. *Ibid.* (6th ed.), pp. 528–29. The only serious change between this and the passage in the first edition is the addition here of the phrase, "by the Creator."

30. Gray, *Darwiniana*, p. 57.

31. Darwin to Gray, May 8, 1868: *Life and Letters*, III, 85.

32. 1866: Hooker, *Life and Letters*, II, 106.

33. *Ibid.*

34. Lyell, *Principles* (10th ed.), II, 492.

35. June 5, 1874: *Life and Letters*, III, 189.

36. Darwin, *Descent of Man*, I, 152–53.

37. *Life and Letters*, II, 197.

38. Newton, *Macmillan's Magazine*, LVII (1888), 244.

39. Hooker to Darwin, Nov. 21, 1859: Darwin, *Life and Letters*, II, 228.

40. Wallace to Sims, March 15, 1861: Wallace, *Letters and Reminiscences*, I, 77.

41. *Spectator*, March 24, 1860, p. 285.

CHAPTER 17

Pages 353–378

1. See above, Chapter 12, p. 259.

2. *Ibid.*, Chapter 7, p. 151.

3. *Origin*, pp. 410, 414.

4. Darwin to Lyell, Jan. 10, 1860: *Life and Letters*, II, 265–66. (Darwin's italics.)

5. *Ibid.*, III, 91, 98.

6. Of Haeckel's book, Darwin said: "If this work had appeared before my essay had been written, I should probably never have completed it" (*Descent*, p. 4; all quotes, unless otherwise indicated, are from the first edition). Yet it had, in fact, appeared before most of the *Descent* was written. Like most authors, Darwin was not as easily dissuaded from writing his book as he would have us think.

7. Darwin to Hooker, Jan. 1871: *Life and Letters*, III, 131.

8. Paston, *At John Murray's*, p. 232.

9. Hooker to Darwin, March 26, 1871: Hooker, *Life and Letters*, II, 125.

10. *Ibid.*

11. *Life and Letters*, III, 133.

12. Darwin to Innes, May 29, 1871: *ibid.*, p. 140.

13. Darwin to Dr. Dohrn, Feb. 3, 1872: *ibid.*, p. 133.

14. *Edinburgh Review*, CXXXIV (1871), 195.

15. *Saturday Review*, XXXI (1871), 276.

16. *Times*, April 8, 1871, p. 5.

17. Morley, *On Compromise*, p. 17.

18. *Athenaeum*, March 4, 1871, p. 275.

19. *Edinburgh Review*, CXXXIV, 196.

20. *Spectator*, March 11, 1871, p. 288.

21. *Dublin Review*, new series, XVII (1871), 3.

22. Huxley, "Mr. Darwin's Critics" (1871), *Darwiniana*, p. 149.

23. *Contemporary Review*, XVII (1871), 274–75.

24. *Spectator*, March 18, 1871, p. 320.

25. *Macmillan's Magazine*, XXIV (1871), 51.

26. *Saturday Review*, XXXI (1871), 277.

27. Darwin to Wallace, Jan. 30, 1871: *Life and Letters*, III, 135–36.

28. July 9, 1871: *ibid.*, p. 145.

29. Darwin to Hooker, Sept. 16, 1871: *More Letters*, I, 333.

30. Wright, *Darwinism*, p. 23.

31. July 12, 1871: Wallace, *Letters and Reminiscences*, I, 265–66.

32. Huxley, *Darwiniana*, p. 183.

33. Mivart, *On the Genesis of Species*, p. 239.

34. *Quarterly Review*, CXXXI (1871), 47–90.

35. Cambridge Mss.

36. *Descent*, II, 389–90.

37. *Ibid.*, p. 385.

38. *Ibid.*, p. 384.

39. Wallace, "The Development of Human Races under the Law of Natural Selection" (1864), *Contributions to the Theory of Natural Selection*, pp. 303–31.

40. *Descent*, II, 370.

41. *Ibid.*, pp. 381–82.

42. *Ibid.*, p. 383.

43. *Ibid.*, p. 337.

44. *Ibid.* (2nd ed.), p. 137.

45. *Ibid.* (1st ed.), II, 384.

46. In the first edition he had doubted whether these long-continued habits of mutilation could be inherited, but by the second edition he was conceding its possibility (1st ed., II, 380; 2nd ed., p. 920).

47. *Ibid.* (1st ed.), I, 152–53.

48. *Ibid.* (2nd ed.), p. 92.

49. *Ibid.* (1st ed.), I, 153.

50. *Ibid.*, p. 154.

51. *Ibid.*, II, 388.

52. *Ibid.*, p. 387.

53. *Ibid.* (2nd ed.), p. 93.

54. *Ibid.* (1st ed.), I, 154.

55. *Ibid.*, p. 43.

56. *Ibid.*, p. 46.

57. *Ibid.*, p. 53.

58. *Ibid.*, p. 54.

59. *Ibid.*, p. 57.

60. *Ibid.*

61. *Ibid.*, p. 60.

62. *Ibid.*, p. 71.

63. *Ibid.*, p. 97.

64. *Ibid.*, p. 96.

65. *Ibid.*, p. 65.

66. *Ibid.*, p. 68.

67. *Ibid.*

68. *Ibid.*, p. 99.

69. *Ibid.*, p. 71.

70. Eiseley, *The Immense Journey*, p. 91.

71. Perhaps the best critique of Piltdown man was published long before its exposure; it is an analysis of the evidence and reasoning upon which eminent scientists were prevailed to accept

the findings. See Miller, *Annual Report of the Smithsonian Institution,* 1928, pp. 413–65.

CHAPTER 18

Pages 380–411

1. "Darwinism" may, of course, be used to refer to the scientific doctrine itself. In this context, however, it is meant to apply particularly to the social, political, moral, and religious theories based on the scientific doctrine.

2. Cambridge Mss. The published version omits the last half of this second sentence.

3. 1839: Emma Darwin, *Letters,* II, 174.

4. 1861: *ibid.,* p. 175.

5. Darwin to W. D. Fox, Sept. 1841: Cambridge Mss.

6. Emma to Henrietta Litchfield, 1870: Emma Darwin, *Letters,* II, 196.

7. Darwin to F. E. Abbott, Nov. 16, 1871: *Life and Letters,* I, 306.

8. Cambridge Mss. (The friend was Lyell.)

9. Keith, *Darwin Revalued,* p. 234.

10. Cambridge Mss. Henrietta went so far as to speak of legal proceedings to stop its publication (Darwin, *Autobiography,* ed. Nora Barlow, p. 12).

11. Cambridge Mss. In view of this episode, it is disquieting to recall that Henrietta, before her marriage, had acted as Darwin's secretary and editor. One wonders whether in the course of the rephrasing or restyling for which Darwin so profusely thanked her, she managed to insinuate her prejudices, as William seemed to think she would have done given the opportunity.

12. *Life and Letters,* I, 307–13.

13. F. Julia Wedgwood to Francis Darwin, Oct. 3, 1884: Cambridge Mss.

14. See above, Chapter 1, p. 12. In one of her early letters Emma Darwin had suggested that Charles had been unduly influenced by his brother Erasmus in the matter of disbelief (1839: Emma Darwin, *Letters,* II, 173).

15. This passage was preceded by the note, not in Darwin's hand: "Written in 1879. Copied out April 22, 1881."

16. Darwin to Gray, May 22, 1860: *Life and Letters,* II, 312.

17. F. Julia Wedgwood to Francis Darwin, Oct. 3, 1884: Cambridge Mss.

18. Aveling, *New Century Review,* I (1897), 323.

19. *Life and Letters,* I, 317.

20. Oct. 21, 22, 24, 1873: Cambridge Mss.

21. Huxley, "Mr. Darwin's Critics" (1871), *Darwiniana,* p. 147.

22. Strauss, *The Old Faith and the New*, p. 205.

23. Bradford, *Darwin*, pp. 245–47.

24. Quoted by Sherrington, *Man on His Nature*, p. 297.

25. Romanes, *A Candid Examination of Theism*, pp. 51–52.

26. Tennyson, *Lord Tennyson*, pp. 86–87.

27. *Life and Letters*, III, 274.

28. Gray to G. F. Wright, Aug. 14, 1875: Gray, *Letters*, II, 656.

29. Dewey, *Influence of Darwin*, p. 12.

30. Paley, *Natural Theology*, p. 12.

31. Chalmers, *On the Power, Wisdom and Goodness of God*, I, ix.

32. *Principal Speeches and Addresses of the Prince Consort*, pp. 111–12.

33. *Essays and Reviews* (12th ed.), p. 154.

34. *Ibid.*, p. 141.

35. Quoted by Gillispie, *Genesis and Geology*, pp. 182–83.

36. Temple, "Present Relations of Science to Religion," p. 13.

37. Beecher, *Evolution and Religion*, I, 115.

38. Drummond, *Ascent of Man*, p. 426.

39. Abbott, *Theology of an Evolutionist*, p. 15.

40. *National Review*, XI (1860), 484.

41. *Life and Letters*, II, 237.

42. Gladstone to Jevons, May 10, 1874: Magnus, *Gladstone*, p. 233.

43. Schneider, *Journal of the History of Ideas*, VI (1945), 6–9.

44. *Athenaeum*, XXXIV (1859), 268.

45. Dorlodot, *Darwinism and Catholic Thought*, p. 6.

46. Russell, *Month*, Jan. 1956, pp. 41, 45.

47. Adams, *Education*, pp. 225–26.

48. Huxley, "The Interpreters of Genesis and the Interpreters of Nature" (1885), *Essays upon some Controverted Questions*, p. 89.

49. Coleridge, *Biographia Literaria*, p. 63.

50. Dewar and Shelton, *Is Evolution Proved?*, p. 5.

51. Davey, *Darwin, Carlyle and Dickens*, p. 26.

52. Julian Huxley, *Religion Without Revelation*.

53. Haeckel, *Last Words on Evolution*, p. 112.

54. Agassiz, *Methods of Study in Natural History*, p. iv.

55. Hutton, "The Materialists' Stronghold" (1874), *Aspects of Religious and Scientific Thought*, p. 48.

56. Hutton, "Dr. Abbott on Natural and Supernatural" (1879), *ibid.*, pp. 173–74.

57. Butler, *Analogy of Religion*, preface.

58. Carlyle, "Characteristics" (1831), *Critical and Miscellaneous Essays*, III, 310.

59. Froude, *Carlyle: His Life in London*, I, 290–91.

60. Gaskell, *Life of Brontë*, II, 204.

61. Spencer, *Data of Ethics*, p. 324.

62. Hirst, *Early Life and Letters of Morley*, I, 317.

63. Paley, *Natural Theology*, p. 490.

64. Rashdall, *Philosophy and Religion*, p. 60.

65. Arnold, "In Harmony with Nature."

66. Swinburne, *Blake*, p. 158.

67. Huxley to Kingsley, Sept. 23, 1860: Huxley, *Life and Letters*, I, 316–17.

68. Huxley to Kingsley, May 5, 1863: *ibid.*, p. 348.

69. Huxley, *Evolution and Ethics*, p. 146.

70. *Ibid.*, pp. 196, 199–200.

71. Huxley, *Science and Christian Tradition*, pp. 256–57.

72. *Evolution and Ethics*, p. 58.

73. *Ibid.*, pp. 81, 83.

74. Huxley, *Life and Letters*, III, 293.

75. *Evolution and Ethics*, p. 82.

76. *Science and Christian Tradition*, pp. 215, 307.

77. Huxley, *Life and Letters*, III, 220–21.

78. April 2, 1873: Darwin, *Life and Letters*, I, 307.

79. Stephen, *Life and Letters*, pp. 144–45.

80. Eliot to Cross, Oct. 20, 1873: Eliot, *Life*, III, 179.

81. Myers, *Essays: Modern*, p. 269.

82. Harrison, *Frederic Harrison: Thoughts and Memories*, pp. 127–29.

83. 1851: Eliot, *Life*, I, 192. Yet she once admitted: "Life, though good to men on the whole, is a doubtful good to many, and to some not a good at all. To my thought it is a source of constant mental distortion to make a denial of this a part of religion" (Morley, *Recollections*, I, 58). Huxley would have pointed out that if Christianity was condemned by this fact, so was nature.

84. Renan, *Recollections of My Youth*, pp. 215–16.

85. Hooker to Huxley, Oct. 8, 1893: Hooker, *Life and Letters*, II, 67.

86. Stephen, "Evolution and Religious Conceptions," in *Nineteenth Century: A Review of Progress*, p. 380.

87. Spencer, *Autobiography*, II, 467.

CHAPTER 19

Pages 412–431

1. Newman, *Idea of a University*, p. 78.

2. Darwin to Lyell, Jan. 1860: *Life and Letters*, II, 262.

3. July 12, 1881: Wallace, *My Life*, II, 4.

4. *Life and Letters*, III, 178.

5. Darwin to Gray, June 5, 1861: *ibid.*, II, 374.

6. Darwin to Gray, Sept. 13, 1864: Cambridge Mss.

7. *Times*, Dec. 28, 1859.

8. Moran, *Journal, 1857–65,* I, 619.

9. Darwin to W. Graham, July 3, 1881: *Life and Letters,* I, 316.

10. Hofstadter, *Social Darwinism in American Thought,* p. 197.

11. Keith, preface to Machin, *Darwin's Theory Applied to Mankind,* p. viii.

12. Kaufmann, *Nietzsche,* p. 141.

13. Spengler, *Decline of the West,* I, 373, 369.

14. Nietzsche, *Joyful Wisdom,* p. 290.

15. Peirce, *Collected Papers,* VI, 199; Zirkle, *Proceedings of the American Philosophical Society,* LXXXIV (1941), 72; Sandow, *Quarterly Review of Biology,* XIII (1938), 315–36; Cowles, *Isis,* XXVI (1937), 341–48; Cruikshank, *Dickens and Early Victorian England,* p. 213.

16. Wallace, *Studies Scientific and Social,* I, 509. (The words are Wallace's.) On another occasion, however, he suggested that greater evil might result from trying to suppress the sentiments of benevolence and sympathy than by permitting them to exist, even at the risk of perpetuating the weak and sick (Darwin to G. A. Gaskell, Nov. 15, 1878: *More Letters,* II, 50).

17. Spencer, *Social Statics,* pp. 322–23.

18. Hofstadter, *Social Darwinism in American Thought,* p. 45.

19. *Ibid.,* p. 59, quoting William Graham Sumner.

20. Fiske, *Destiny of Man,* pp. 103, 118–19; *Outlines of Cosmic Philosophy,* II, 228.

21. Marx to Lassalle, Jan. 16, 1861: Marx, *Correspondence,* p. 125.

22. Aveling, *New Century Review,* I (1897), 243.

23. Marx to Lassalle, Jan. 16, 1861: Marx, *Correspondence,* p. 125.

24. Marx, *Capital,* I, 367.

25. Marx to Engels, June 18, 1862: Marx, *Correspondance* (French ed.), VII, 119; Engels, "Dialectics of Nature," in Marx, *Selected Works,* I, 172–73.

26. Marx to Lassalle, Jan. 16, 1861: Marx, *Correspondence* (English ed.), p. 125.

27. Marx, *Selected Works,* I, 16.

28. Darwin to Dr. Scherzer, Dec. 26, 1879: *Life and Letters,* III, 237.

29. Quoted by Hofstadter, *Social Darwinism in American Thought,* p. 201.

30. Julian Huxley, *Evolutionary Ethics,* pp. 44–45.

31. Ferri, *Socialism and Positive Science,* vii–viii.

32. Quoted by Quillian, "Evolution and Moral Theory in America," in Persons (ed.), *Evolutionary Thought in America,* p. 402.

33. Galton, *Inquiries into Human Faculty and its Development,* p. 1.

34. Darwin to Galton, Jan. 4, 1873: *More Letters*, II, 43.

35. Galton, *Inquiries*, p. 218.

36. Bagehot, *Physics and Politics*, pp. 43, 49–50.

37. *Ibid.*, p. 21.

38. *Ibid.*, p. 25.

39. *Ibid.*, pp. 27, 29.

40. *Ibid.*, p. 27.

41. *Ibid.*, p. 37.

42. *Ibid.*, pp. 37–38. (Bagehot's italics.)

43. *Ibid.*, p. 39.

44. *Ibid.*

45. *Ibid.*, p. 214.

46. *Ibid.*, p. 53.

47. *Ibid.*, p. 60.

48. *Ibid.*, p. 104.

49. Russell, *History of Western Philosophy*, p. 727.

50. Bowle, *Politics and Opinion in the Nineteenth Century*, p. 474.

51. Spencer, *Principles of Ethics*, II, v.

CHAPTER 20

Pages 432–452

1. *Expression of the Emotions in Man and Animals*, p. 369.

2. *Ibid.*, p. 378.

3. Darwin to Haeckel, 1872: *Life and Letters*, III, 171.

4. *Ibid.*, I, 318.

5. *More Letters*, I, 371.

6. *Nature*, XXV (1881–82), 51, 529.

7. Krause, *Erasmus Darwin*, p. 216.

8. Butler, *Notebooks*, p. 263.

9. *Ibid.*, p. 167.

10. *Life and Letters*, III, 360.

11. Emma Darwin to Henrietta, Feb. 1881: Emma Darwin, *Letters*, II, 244.

12. *Punch*, April 29, 1882, p. 203.

13. *Athenaeum*, Nov. 26, 1887, p. 715.

14. Huxley, *Life and Letters*, III, 321.

15. Huxley, *Darwiniana*, p. 74.

16. Darwin to Bentham, May 22, 1863: *Life and Letters*, III, 25.

17. Weismann, *Contemporary Review*, LXIV (1893), 323–36.

18. Weismann, "The Selection Theory," in Seward (ed.), *Darwin and Modern Science*, p. 49.

19. *William Bateson: Naturalist*, pp. 390–98.

20. Bateson, *Problems of Genetics*, p. 248.

21. Russell, *Scientific Outlook*, pp. 43–44.

22. J. Gray, *Nature*, CLXXIII (1954), 227.

23. Robson and Richards, *Variation of Animals in Nature*, p. 186.

24. Quoted by Mivart, *Genesis of Species*, p. 269.

25. A reviewer of the first edition of the present book made the point neatly: "It must have been at a fairly early age that I decided that Mr. Darwin had a better explanation of my existence than God. I forgot whether his idea, once implanted, spread like a cancerous growth to oust the other belief, or whether it filled a vacuum left by loss of belief. However it happened, it seemed satisfyingly right, and it still does. Dr. Gertrude Himmelfarb in *Darwin and the Darwinian Revolution* knocks holes in his data and his logic, but even if she took the bottom out of him altogether (which she does not, nor set out to do) I should still find him satisfying even if not right." (G. F. Seddon, *Manchester Guardian*, July 17, 1959.)

26. Darlington, *The Place of Botany in the Life of a University*, p. 17.

27. *Life and Letters*, I, 155. See also Darlington's characterization of Darwin's scientific "outlook" as "conservative" and "old-fashioned." (*Darwin's Place in History*, pp. 60–61, 63.)

28. Butler, *Luck or Cunning*, p. 291.

29. Lewis, *De Descriptione Temporum*, pp. 10–11.

30. Stephen, *Some Early Impressions*, pp. 52–53.

BIBLIOGRAPHY

This bibliography contains only works cited in the footnotes. It does not include the contemporary periodical literature which is too extensive to be individually itemized and which is adequately referred to in the footnotes.

Abbott, Lyman. *Theology of an Evolutionist*. London, 1897.

Adams, Henry. *Education*. Modern Library edition.

Agassiz, Louis. *Methods of Study in Natural History*. Boston, 1863.

Albert, Prince. *The Principal Speeches and Addresses of His Royal Highness*. London, 1862.

Alvarez, Walter C. *Nervousness, Indigestion, and Pain*. London, 1943.

——. *Neuroses*. Philadelphia, 1951.

Andrade, E. N. da C. *Sir Isaac Newton*. London, 1954.

Aveling, Edward B. "Charles Darwin and Karl Marx," *New Century Review*, I (1897).

Bagehot, Walter. *Physics and Politics*. Revised ed., London.

Barlow, Erasmus Darwin. "The Dangers of Health," *Listener*, Aug. 23, 1956.

Bateson, William. *Problems of Genetics*. London, 1913.

——. *William Bateson: Naturalist*. Cambridge, 1928.

Beecher, Henry Ward. *Evolution and Religion*. 2 vols., London, 1885.

Bergson, Henri. *Creative Evolution*. Tr. A. Mitchell. London, 1911.

Bettany, G. T. *Life of Charles Darwin*. London, 1887.

Bonar, James. *Malthus and His Work*. London, 1924.

Bosanquet, S. R. *Vestiges of the Natural History of Creation: Its Argument Examined and Exposed*. 2nd ed., London, 1845.

Bowle, John. *Politics and Opinion in the Nineteenth Century*. London, 1954.

Bradford, Gamaliel. *Darwin*. Boston, 1926.

Bree, C. R. *An Exposition of Fallacies in the Hypothesis of Mr. Darwin*. London, 1872.

——. *Species Not Transmutable, nor the Result of Secondary Causes*. London, 1860.

Bryce, James. "Personal Reminiscences of Charles Darwin and of the Reception of the 'Origin of Species,'" *Proceedings of the American Philosophical Society*, XLVIII (1909).

Buckland, William. *Life and Correspondence*. Ed. Mrs. Gordon. London, 1894.

Buckle, Henry Thomas. *History of Civilization in England*. 2 vols., London, 1857–61.

——. *Life and Writings*. Ed. Alfred H. Huth. 2 vols., New York, 1880.

Buffon, Georges Louis Leclerc. *Natural History*. Tr. Barr. 10 vols., London, 1807.

Butler, Joseph. *The Analogy of Religion*. London, 1791.

Butler, Dr. Samuel. *Life and Letters*. 2 vols., London, 1896.

Butler, Samuel. *Life and Habit*. London, 1878.

——. *Luck or Cunning*. London, 1887.

——. *Notebooks*. Ed. Geoffrey Keynes and Brian Hill. New York, 1951.

Carlyle, Jane Welsh. *Letters and Memorials*. 2 vols., New York, 1883.

Carlyle, Thomas. *Correspondence of Thomas Carlyle and Ralph Waldo Emerson*. 2 vols., London, 1883.

——. *Critical and Miscellaneous Essays*. 5 vols., London, 1840.

——. *Life of John Sterling*. 2nd ed., London, 1852.

——. *Reminiscences*. Ed. J. A. Froude. 2 vols., London, 1881.

Chalmers, Thomas. *On the Power, Wisdom and Goodness of God*. 2 vols., London, 1833.

Chambers, Robert. *Vestiges of the Natural History of Creation*. 3rd ed., London, 1845.

Clark, Ronald. *The Victorian Mountaineers*. London, 1953.

Coleridge, Samuel Taylor. *Biographia Literaria*. Everyman ed.

Cook, E. T. *Life of John Ruskin*. 2 vols., London, 1911.

Copleston, Frederick. *Arthur Schopenhauer*. London, 1946.

Cowles, Thomas. "Malthus, Darwin and Bagehot: A Study in the Transference of a Concept," *Isis*, XXVI (1937).

Cruikshank, R. J. *Charles Dickens and Early Victorian England*. New York, 1949.

Cruse, Amy. *The Victorians and Their Books*. London, 1935.

Cuvier, Georges. *Essay on the Theory of the Earth*. Ed. Jameson. 5th ed., London, 1827.

Darlington, C. D. *The Place of Botany in the Life of a University*. Oxford, 1954.

——. *Darwin's Place in History*. New York, 1961.

Darwin, Charles. Manuscripts: American Philosophical Society Library; British Museum, Eg. 3020A; Cambridge University Library.

——. *Autobiography*. Ed. Nora Barlow. London, 1958.

Darwin, Charles. *Charles Darwin and the Voyage of the Beagle.* Ed. Nora Barlow. London, 1945.

——. *Descent of Man.* 1st ed., 2 vols., London, 1871; 2nd ed., 1 vol. reprint, London, 1901.

——. *Diary of the Voyage of H.M.S. Beagle.* Ed. Nora Barlow. Cambridge, 1933.

——. *The Expression of the Emotions in Man and Animals.* 2nd ed., London, 1904.

——. *The Foundations of the Origin of Species.* Ed. Francis Darwin. Cambridge, 1909.

——. *Journal and Remarks.* Vol. III of *Narrative of the Surveying Voyages of H.M.S. Adventure and Beagle between 1826 and 1836.* London, 1839.

Revised ed. under title *Journal of Researches into the Natural History and Geology of the Countries Visited during the Voyage of H.M.S. Beagle Round the World.* London, 1845.

——. *Life and Letters.* Ed. Francis Darwin. 3 vols., London, 1887.

——. *More Letters.* Ed. Francis Darwin and A. C. Seward. 2 vols., London, 1903.

——. *On the Origin of Species.* Reprint of 1st ed., London, 1950; reprint of 6th ed., New York, 1909.

——. *The Structure and Distribution of Coral Reefs.* Part 1 of *The Geology of the Voyage of the Beagle.* London, 1842; 2nd ed., ed. J. W. Judd, London, 1900.

——. *The Variation of Animals and Plants under Domestication.* 2 vols., London, 1868.

Darwin, Emma. *A Century of Family Letters, 1792–1896.* Ed. Henrietta Litchfield. 2 vols., Cambridge, 1904 (private ed.); 2 vols., London, 1915 (regular ed.).

Darwin, Erasmus. *Botanic Garden.* London, 1791.

——. *Phytologia.* London, 1800.

——. *Zoonomia.* 3rd ed., 4 vols., London, 1801.

Davey, Samuel. *Darwin, Carlyle and Dickens.* London, 1876.

Dewar, Douglas, and Shelton, H. S. *Is Evolution Proved?* London, 1947.

Dewey, John. *The Influence of Darwin on Philosophy.* New York, 1910.

Disraeli, Benjamin. *Tancred.* 3 vols., London, 1847.

Dobzhansky, Theodosius. *Evolution, Genetics and Man.* New York, 1955.

Dorlodot, Henry. *Darwinism and Catholic Thought.* London, 1922.

Drachman, Julian M. *Studies in the Literature of Natural Science.* New York, 1930.

Draper, John William. *History of the Conflict between Religion and Science.* London, 1927.

——. "The Intellectual Development of Europe Considered

with Reference to the Views of Mr. Darwin," *Report of the British Association for the Advancement of Science*, 1860.

Drummond, Henry. *Ascent of Man*. London, 1894.

Dupree, A. Hunter. "Some Letters from Charles Darwin to Jeffries Wyman," *Isis*, XLII (1951).

Eiseley, Loren. *Darwin's Century*. New York, 1958.

———. *The Immense Journey*. New York, 1957.

———. Review of Irvine's *Apes, Angels, and Victorians, Yale Review*, 1955.

Eliot, George. *Life*. Ed. J. W. Cross. 3 vols., "Illustrated Cabinet Ed.," Boston.

Eliot, T. S. *Essays Ancient and Modern*. London, 1936.

Ernle, Lord. "Victorian Memoirs and Memories," *Quarterly Review*, CCXXXIX (1923).

Essays and Reviews. 1st ed., London, 1860; 12th ed., London, 1869.

Fawcett, Henry. "A Popular Exposition of Mr. Darwin on the Origin of Species," *Macmillan's Magazine*, III (1860).

Ferri, Enrico. *Socialism and Positive Science*. Tr. Edith C. Harvey. London, 1905.

Fisher, R. A. *The Genetical Theory of Natural Selection*. Oxford, 1930.

Fiske, John. *Destiny of Man*. Boston, 1884.

———. *Outlines of Cosmic Philosophy*. 2 vols., London, 1874.

FitzRoy, Robert. *Proceedings of the Second Expedition, 1831–1836*. Vol. II of *Narrative of the Surveying Voyages of H.M.S. Adventure and Beagle between 1826 and 1836*. London, 1839.

Froude, James Anthony. *The Earl of Beaconsfield*. Everyman ed.

———. *Thomas Carlyle: His Life in London*. 2 vols., London, 1902.

Gage, A. T. *History of the Linnean Society of London*. London, 1938.

Galton, Francis. *Inquiries into Human Faculty and its Development*. Everyman ed.

Gaskell, E. C. *Life of Charlotte Brontë*. 2nd ed., 2 vols., London, 1857.

Geikie, Archibald. *Charles Darwin as Geologist*. Cambridge, 1909.

———. *The Founders of Geology*. London, 1905.

———. *Life of Sir Roderick I. Murchison*. 2 vols., London, 1875.

———. *A Long Life's Work*. London, 1924.

Gillispie, Charles C. *Genesis and Geology*. Cambridge (Mass.), 1951.

Goethe, Johann Wolfgang von. *Conversations with Eckermann*. Tr. J. Oxenford. London, 1874.

Gosse, Edmund. *Father and Son*. London, 1907.

Gould, George M. *Biographic Clinics*. 6 vols., Philadelphia, 1903–9.

Gray, Asa. *Darwiniana*. New York, 1876.

———. *Letters*. Ed. James L. Gray. 2 vols., London, 1893.

Gray, J. "The Case for Natural Selection," *Nature*, CLXXIII (1954).

Green, John Richard. *Letters*. Ed. Leslie Stephen. New York, 1901.

Haeckel, Ernst. *History of Creation*. Tr. E. Ray Lankester. 2 vols.; London, 1876.

———. *Last Words on Evolution*. Tr. J. McCabe. London, 1906.

Hagberg, Knut. *Carl Linnaeus*. Tr. Alan Blair. London, 1952.

Hagen, Victor W. von. *South America Called Them*. London, 1949.

Harrison, Austin. *Frederic Harrison: Thoughts and Memories*. London, 1926.

Harvey, William H. "An Inquiry into the Probable Origin of the Human Animal on the Principles of Mr. Darwin's Theory of Natural Selection, and in Opposition to the Lamarckian Notion of a Monkey Parentage." Privately printed pamphlet, 1860.

Haynes, E. S. P. "Master George Pollock," *Cornhill Magazine*, new series, XLI (1916).

Helmholtz, Hermann von. *Popular Lectures on Scientific Subjects*. Tr. E. Atkinson. London, 1893.

Herschel, J. F. W. *A Preliminary Discourse on the Study of Natural Philosophy*. 2nd ed., London, 1842.

Himmelfarb, Gertrude. "Malthus," *Encounter*, V (1955).

Hirst, F. W. *Early Life and Letters of John Morley*. 2 vols., London, 1927.

Hofstadter, Richard. *Social Darwinism in American Thought*. 2nd ed., Boston, 1955.

Hooker, Joseph Dalton. *Life and Letters*. Ed. Leonard Huxley. 2 vols., London, 1918.

Hubble, Douglas. "Charles Darwin and Psychotherapy," *Lancet*, Jan. 30, 1943.

———. "Life of the Shawl," *Lancet*, Dec. 26, 1953.

Humboldt, Alexander von. *Personal Narrative of Travels to the Equinoctial Regions of the New Continent during the Years 1799 to 1804*. Tr. Helen M. Williams. 6 vols., London, 1818.

Hutton, Richard Holt. *Aspects of Religious and Scientific Thought*. Ed. E. M. Roscoe. London, 1899.

Huxley, Julian, Hardy, A. C., and Ford, E. B. (eds.). *Evolution as a Process*. London, 1954.

Huxley, Julian. *Evolutionary Ethics*. Oxford, 1943.

———. *Religion Without Revelation*. London, 1927.

Huxley, Thomas Henry. Manuscripts: Library of the Imperial College of Science and Technology, London.

———. *Critiques and Addresses*. London, 1883.

———. *Darwiniana*. New York, 1893.

———. *Essays upon some Controverted Questions*. London, 1892.

————. *Evolution and Ethics and other Essays.* New York, 1898.

————. *Life and Letters.* Ed. Leonard Huxley. 3 vols., London, 1903.

————. *Science and the Christian Tradition.* New York, 1898.

Irvine, William. *Apes, Angels, and Victorians.* New York, 1955.

Jackson, B. D. *Linnaeus.* London, 1923.

James, William. *Memories and Studies.* London, 1911.

————. *Selected Papers on Philosophy.* Everyman ed.

Jenyns, Leonard. *Memoir of the Reverend John Stevens Henslow.* London, 1862.

Jevons, W. Stanley. *Letters and Journals.* Ed. Mrs. Jevons. London, 1886.

Johnson, E. D. H. *The Alien Vision of Victorian Poetry.* Princeton, 1952.

Jones, Ernest. *Sigmund Freud.* 3 vols., London, 1953–57.

Kaufmann, Walter A. *Nietzsche.* Princeton, 1950.

Keith, Arthur. *Darwin Revalued.* London, 1955.

Kempf, Edward J. "Charles Darwin: The Affective Sources of His Inspiration and Anxiety Neurosis," *Psychoanalytic Review* (Washington), V (1918).

Kingsley, Charles. *Letters and Memories of His Life.* Ed. Mrs. Kingsley. 2 vols., London, 1877.

————. *Water Babies.* London, 1954.

Krause, Ernst. *Erasmus Darwin.* London, 1879.

Kropotkin, Peter. *Mutual Aid.* London, 1903.

Lamarck, Jean Baptiste. *Zoological Philosophy.* Tr. Hugh Elliot. London, 1914.

Lewis, C. S. *De Descriptione Temporum.* Cambridge, 1955.

Lovejoy, Arthur O. *The Great Chain of Being.* Cambridge (Mass.), 1936.

Lyell, Charles. *Geological Evidences of the Antiquity of Man.* London, 1863; 2nd ed. under title *Antiquity of Man*, London, 1863.

————. *Life, Letters and Journals.* Ed. Mrs. Lyell. 2 vols., London, 1881.

————. *Principles of Geology.* 1st ed., 3 vols., London, 1830–33; 6th ed., 3 vols., London, 1840; 9th ed., 1 vol., London, 1853; 10th ed., 2 vols., London, 1867.

Machin, Alfred. *Darwin's Theory Applied to Mankind.* London, 1937.

Magnus, Philip. *Gladstone.* London, 1954.

Malthus, T. R. *An Essay on the Principle of Population.* Everyman ed.

Manning, Henry Edward. *Essays on Religion and Literature.* London, 1865.

Marchand, Leslie A. *The Athenaeum.* Chapel Hill, 1941.

Marx, Karl. *Briefwechsel.* Part 3, vol. 3 of Marx-Engels *Gesamtausgabe*, Berlin, 1930.

————. *Capital.* Tr. S. Moore and E. Aveling. London, 1908.

Marx, Karl. *Correspondance.* Paris, 1933.
———. *Correspondence.* London, 1934.
———. *Selected Works.* 2 vols., London, 1942.
Medawar, P. B. "The Imperfections of Man," *Twentieth Century,* Sept. 1955.
Mill, John Stuart. *Letters.* Ed. Hugh S. R. Elliot. 2 vols., London, 1910.
———. *System of Logic.* London, 1949.
———. *Three Essays on Religion.* London, 1874.
Miller, Gerrit S. "The Controversy over Human 'Missing Links,'" *Annual Report of the Smithsonian Institute, 1928.*
Miller, Hugh. *Testimony of the Rocks.* Edinburgh, 1857.
Mitchell, P. Chalmers. *Thomas Henry Huxley.* New York, 1900.
Mivart, St. George. *On the Genesis of Species.* London, 1871.
Moran, Benjamin. *Journal, 1857–65.* Ed. S. A. Wallace and F. E. Gillespie. Chicago, 1948.
More, L. T. *The Dogma of Evolution.* Princeton, 1925.
Morley, John. *Critical Miscellanies.* 3 vols., London, 1886.
———. *On Compromise.* London, 1886.
———. *Recollections.* 2 vols., New York, 1917.
Murray, Robert H. *Science and Scientists in the Nineteenth Century.* London, 1925.
Myers, F. W. H. *Essays: Modern.* London, 1883.
Newman, John Henry. *The Idea of a University.* London, 1873.
Newton, Alfred. "Early Days of Darwinism," *Macmillan's Magazine,* LVII (1888).
Newton, Isaac. *Opticks.* 3rd ed., London, 1721.
Nietzsche, Friedrich. *Complete Works.* Ed. Oscar Levy. 18 vols., New York, 1910.
Nordenskiöld, Erik. *History of Biology.* Tr. Leonard B. Eyre. New York, 1928.
Owen, Richard. *Life.* Ed. by his grandson. 2 vols., London, 1894.
Paley, William. *Natural Theology.* 6th ed., London, 1803.
Paston, George (pseud. for E. M. Symonds). *At John Murray's.* London, 1932.
Pearson, Karl. *Charles Darwin.* London, 1923.
Peirce, Charles Sanders. *Collected Papers.* Ed. Charles Hartshorne and Paul Weiss. 6 vols., Cambridge (Mass.), 1935.
Persons, Stow (ed.). *Evolutionary Thought in America.* New Haven, 1950.
Pius XII. *Humani Generis.* Tr. Ronald A. Knox, London, 1955.
Polanyi, Michael. "From Copernicus to Einstein," *Encounter,* V (1955).
———. *Personal Knowledge.* London, 1958.
Rashdall, Hastings. *Philosophy and Religion.* London, 1909.
Renan, Ernest. *Recollections of My Youth.* London, 1897.
Robson, G. C., and Richards, O. W. *The Variation of Animals in Nature.* London, 1936.

Romanes, George. *A Candid Examination of Theism*. London, 1878.

Russell, Bertrand. *History of Western Philosophy*. New York, 1945.

———. *The Scientific Outlook*. London, 1931.

Russell, John L. "The Theory of Evolution: The Present State of the Evidence," *Month*, Jan. 1956.

Samuels, Ernest. *The Young Henry Adams*. Cambridge (Mass.), 1948.

Sandow, Alexander. "Social Factors in the Origin of Darwinism," *Quarterly Review of Biology*, XIII (1938).

Schneider, Herbert W. "The Influence of Darwin and Spencer on American Philosophical Theology," *Journal of the History of Ideas*, VI (1945).

Schopenhauer, Arthur. *World as Will and Idea*. Tr. R. B. Haldane and J. Kemp. 3 vols., London, 1883.

Sedgwick, Adam. *Address Delivered to the Geological Society, Feb. 19, 1830*. London, 1831.

———. *Address Delivered to the Geological Society, Feb. 18, 1831*. London, 1831.

———. *A Discourse on the Studies of the University*. 2nd ed., Cambridge, 1834.

———. *Life and Letters*. Ed. J. W. Clark and T. M. Hughes. 2 vols., Cambridge, 1890.

Seward, A. C. (ed.). *Darwin and Modern Science*. Cambridge, 1909.

Seward, Anna. *Memoirs of the Life of Dr. [Erasmus] Darwin*. London, 1804.

Shaler, Nathaniel Southgate. *Autobiography*. Boston, 1909.

Sherrington, Charles. *Man on His Nature*. 2nd ed., New York, 1953.

Simpson, James Y. *Landmarks in the Struggle between Science and Religion*. New York, 1926.

Smiles, Samuel. *Josiah Wedgwood*. London, 1894.

Spectator.

Spencer, Herbert. *An Autobiography*. 2 vols., New York, 1904.

———. *Data of Ethics*. New York, 1879.

———. *Essays Scientific, Political, and Speculative*. 3 vols., London, 1891.

———. *Principles of Ethics*. 2 vols., London, 1893.

———. *Social Statics*. London, 1851.

———. "Theory of Population." Pamphlet, London, 1852.

Spengler, Oswald. *The Deadline of the West*. New York, 1939.

Stephen, Leslie. "Evolution and Religious Conceptions," in *Nineteenth Century: A Review of Progress*. London, 1901.

———. *Life and Letters*. Ed. F. W. Maitland. London, 1906.

———. *Some Early Impressions*. London, 1924.

Stimson, Dorothy. *Scientists and Amateurs: A History of the Royal Society*. London, 1949.

Strauss, David Friedrich. *The Old Faith and the New*. Tr. M. Blind. London, 1873.

Swinburne, Algernon Charles. *William Blake*. London, 1868.

Temple, Frederick, "The Present Relations of Science to Religion." Pamphlet, Oxford, 1860.

Tennyson, Hallam. *Alfred Lord Tennyson: A Memoir*. London, 1906.

Tuckwell, W. *Reminiscences of Oxford*. London, 1900.

Wain, John. "The Strategy of Victorian Poetry," *Twentieth Century*, May 1953.

Wallace, Alfred Russel. *Contributions to the Theory of Natural Selection*. London, 1870.

———. *Letters and Reminiscences*. Ed. James Marchant. 2 vols., London, 1916.

———. *My Life*. 2 vols., London, 1905.

———. *Studies Scientific and Social*. 2 vols., London, 1900.

Weismann, August. "The All-Sufficiency of Natural Selection," *Contemporary Review*, LXIV (1893).

———. *The Evolution Theory*. 2 vols., London, 1904.

Wells, William Charles. *Two Essays*. London, 1818.

West, Geoffrey. *Charles Darwin*. London, 1937.

Whewell, William. *Astronomy and General Physics*. London, 1864.

———. *History of the Inductive Sciences*. 1st ed., 3 vols., London, 1837; 2nd ed., 3 vols., London, 1847.

Wilberforce, Samuel. *Essays Contributed to the 'Quarterly Review.'* 2 vols., London, 1874.

———. *Life*. Ed. R. B. Wilberforce. 3 vols., London, 1881.

Wilson, David Alec, and MacArthur, David Wilson. *Carlyle in Old Age*. London, 1934.

Wilson, David Alec. *Carlyle to Threescore and Ten*. London, 1929.

Winstanley, D. A. *Early Victorian Cambridge*. Cambridge, 1940.

Wright, Chauncey. *Darwinism*. London, 1871.

Zahm, J. A. *Evolution and Dogma*. Chicago, 1896.

Zirkle, Conway. "Natural Selection before the 'Origin of Species,'" *Proceedings of the American Philosophical Society*, LXXXIV (1941).

INDEX

EUROPEAN HISTORY TITLES IN
NORTON PAPERBOUND EDITIONS